Roots of Empire

Brill's Series in the History of the Environment

Series Editor

Aleks Pluskowski

VOLUME 4

The titles published in this series are listed at *brill.com/bshe*

Roots of Empire

Forests and State Power in Early Modern Spain, c.1500–1750

By

John T. Wing

BRILL

LEIDEN | BOSTON

Cover illustration: Inset from a chorographic chart of forests and other resources within the naval provinces of the Department of Cádiz by José Antonio Espelius in 1765. A large coat of arms belonging to King Carlos III (r. 1759-1788) sits on a beach at the forest's edge where one can see a worker chopping wood, four more workers carting logs, a shipyard on the shore with a vessel under construction, a fisherman, a vessel at sea on the horizon, and a small inlet for preserving the wood. (Source: España. Ministerio de Cultura. Biblioteca Nacional de España. MR/45 FACS. 24. Cádiz (Departamento marítimo) Bosques 1765 por José Espelius.)

Library of Congress Cataloging-in-Publication Data

Wing, John T.
Roots of empire : forests and state power in early modern Spain, c. 1500-1750 / by John T. Wing.
pages cm
Includes bibliographical references and index.
ISBN 978-90-04-26136-5 (hardcover : acid-free paper) -- ISBN 978-90-04-26137-2 (e-book)
1. Forest policy--Spain--History. 2. Forest management--Political aspects--Spain--History. 3. Forests and forestry--Spain--History. 4. Forests and forestry--Spain--Colonies--History. 5. Shipbuilding--Spain--History. 6. Spain--Colonies--History. 7. Imperialism--History. 8. Spain--Politics and government--1516-1700. 9. Spain--Politics and government--1700-1746 10. Spain--Colonies--Environmental conditions. I. Title.
SD631.W46 2015
634.9'20946--dc23

2014041315

ISSN 1876-6595
ISBN 978-90-04-26136-5 (hardback)
ISBN 978-90-04-26137-2 (e-book)

This book is printed on acid-free paper.

Contents

Acknowledgements IX
List of Maps, Figures, and Tables XIII

Introducing Spanish State Forestry 1
The Widow's Oak and the Spanish State in the Valley of Carriedo 1
The Politics of Wood Shortage Fears in the Early Modern World 19
Spanish Forest Landscapes from Prehistory to the Reconquest 29
Plan of the Work 38

1 A New State Forestry for the First Global Age 44
Crown and Forests, Redefined 44
Spanish Forest Landscapes from the Reconquest to the Habsburgs 47
Medieval Forest Regulations and Naval Power in the Reign of Carlos I (1516–1556) 54
A New Era of Naval Strategy in the Reign of Felipe II (1556–1598) 65
Forests and the Collection of Geographic Knowledge 70
The Relaciones 71
Mapping Projects 73
Forests and the State Bureaucracy 75
The Enterprise of England: Recovery and Response 76

2 Forests of the Ultramar 85
Colonial Forest Territoriality in the Sixteenth Century 85
Geographic Knowledge Channels Across the Atlantic 87
Spanish Shipbuilding in the Old and New Worlds 91
Regulating Colonial Forest Use in the Sixteenth Century 96
Realejo, Guayaquil, and the Pacific Coastal Region 99
Havana and the Caribbean Region 106
Cavite and the Philippines 109
Naval Forestry and Imperial Rivalries 116

3 The Struggle to Stay Afloat in the Seventeenth Century 120
Transition and Persistence in Spanish Forests under the Later Habsburgs 120
State Forestry in Times of Peace and Global Conflict in the Early 1600s 122

To a New Generation of Forest Inspectors: Barros's Letter to Riva Herrera 127
The Formation of Forest Superintendent Dynasties 130
The Waning of Spanish Imperial Might 138
Toribio Pérez de Bustamante's *Instrucción* of 1650 141
Colonial Forest Territoriality in the Seventeenth Century 145
Havana and the Caribbean Region 145
Guayaquil and the Pacific Coastal Region 149
Cavite and the Philippines 153
Habsburg State Forestry in an Era of Naval Decline 156

4 Bottoming Out and Revival under the First Bourbon, 1700–1746 165
Expansion of State Forestry in the Reign of Felipe V (1700–1746) 165
Voices for Reform from the North Coast 168
Patiño's Tenure and the Reestablishment of the *Almirantazgo* 174
Forest Reconnaissance in the Naval Departments of Ferrol, Cartagena, and Cádiz 178
Ferrol 178
Cartagena 180
Cádiz 182
The Forest Inspections of Juan Valdés y Castro in Segura and Catalonia 183
Segura 183
Catalonia 187
Colonial Forest Territoriality in the Eighteenth Century 193
Accomplishments of Felipe V's Reign 201

5 The Triumph of State Forestry: 1748–1754 203
Ensenada's Push for Peace and Naval Revival, 1743–1748 203
Spain's Naval Forest Conservation Ordinance of January 31, 1748 206
Inspections 207
Planting 209
Fines and Licenses 210
Benefits 211
Territoriality 213
Other Forest Legislation, 1748–1751 215
Resistance to the New Legislation 219
Forest Inspection Reports after 1748 222
Interior Forest Reports 223
Naval Forest Reports 225

Other Reforms and Ensenada's Fall from Power, 1749–1754 232
Accomplishments of Fernando VI's Reign (1746–1759) 236

General Conclusion 240

Bibliography 247
Index 263

Acknowledgements

I would like to thank the people who made the writing of this book a profoundly enjoyable and fulfilling experience. I owe an immense debt of gratitude to Carla Rahn Phillips and William D. Phillips, Jr., for agreeing to advise a geography master's graduate who lacked a clear direction or a research agenda at the time of his reentry into the study of history. My appreciation for the Phillips's patience, guidance, wisdom, good will, warmth, and hospitality continues to grow. Their example remains the gold standard for aspiring scholars, teachers, and mentors. As my interests evolved more definitively into the early modern era, Carla Phillips became my principal advisor. I wish to thank her specifically for preparing me for archival research in Spain, including training in paleography, and for continuing to act as a mentor long after the dissertation defense. Being able to stay at the Phillips's lovely apartment during much of my time researching in Madrid was a real windfall. Wim Phillips continued to provide feedback on my writing and served on my dissertation committee, and I was fortunate to be able to work at the University of Minnesota's Center for Early Modern History (CEMH) under his direction. Some of the fondest memories I have from my time at Minnesota were the departmental gatherings at the Phillips's house. They seemed completely at ease with graduate students sipping red wine over their pristine white carpet.

Due to an inexplicable confluence of good fortune, I entered the PhD program the same year the department hired Giancarlo Casale as their new Ottoman historian. I sought his input and advice whenever I could, and our conversations always infected me with renewed energy. I am grateful that he became my dissertation co-advisor and more grateful still to consider him a colleague and friend. I would also like to thank the other members of my dissertation committee, including the eminent early modern historian James D. Tracy and the distinguished professor of forest ecology, Lee E. Frelich for their valuable insights and helpful guidance.

I'm fortunate to have experienced the salad days of early graduate studies with a fantastic cohort at Minnesota, including Sinem Arcak, Tovah Bender, Luis Morera, Jack Norton, Mike Ryan, and Tim Smit. I'd like to thank Ellen Arnold for setting the example for how to approach writing and teaching preindustrial environmental history. My time working with Jamie Bluestone (née Stephenson) at the CEMH remains a highpoint. Her ability to adapt and reinvent, as well as prepare top-notch speaker series events, was inspiring. I'm grateful to Michael Gold for his thoughtful conversations that blended equal parts sanity and hilarity. I thank all these friends for their jovial spirits, which buoyed mine even through the icy winters.

My research took me to Spanish archives large and small, so I'd like to express my appreciation for the diligent attention and service provided by the personnel at the Biblioteca Nacional and Museo Naval in Madrid, the Archivo General de Simancas near Valladolid, the Archivo General de Indias in Seville, the Archivo Histórico Provincial de Cantabria in Santander, and the municipal archives of Santander, Lekeitio, Castro Urdiales, and Pasajes. I am particularly thankful for the time and patience of Isabel Aguirre Landa at Simancas. During my stay in Valladolid, María Isabel Vicente Maroto and Hilario Casado Alonso were friendly and generous hosts. In Madrid, Elizabeth Powers, Jorge Navarro and Francisco Navarro elevated my rustic ways with their sophistication and fine taste. I am very grateful to Luis Gil Sánchez at the Escuela Técnica Superior de Ingenieros de Montes at the Universidad Politécnica de Madrid for his wisdom, guidance, enthusiasm, and generosity.

I would have been unable to complete the book without financial support from the University of Minnesota's Graduate School, the Center for Early Modern History, the Union Pacific Foundation, the Program for Cultural Cooperation between Spain's Ministry of Culture and United States' Universities, the Professional Staff Congress of the City University of New York, the CUNY Academy, and the Frederick M. Binder Fund.

As the manuscript took shape, I benefitted from the comments and advice at different stages from knowledgeable readers. Thank you especially to Karl Appuhn, Paul Warde, J.R. McNeill, Sam White, and Verena Winiwarter. I'd also like to express my appreciation for comments and discussions from panelists and participants at the conferences of the Association for Spanish and Portuguese Historical Studies, the American Society for Environmental History, the World Congress of Environmental History, "Common Ground, Converging Gazes: Integrating the Social and the Environmental in History" at L'École des Hautes Études en Sciences Sociales in Paris, and "Before Environmentalism" at the Early Modern Center of the University of California, Santa Barbara.

I am fortunate to work in New York City where one can find numerous outlets for exchanging research and tremendous resources for supporting scholarship. I would like to thank the following individuals and groups for providing me opportunities to share and discuss works in progress: Nicole Eustace and Hayley Negrin at New York University's Atlantic World Workshop, Liz Hill and the Renaissance Seminar group at Columbia University, Michael Rawson and the New York Metro Seminar in Environmental History, and the Faculty Fellowship Publication Program (FFPP), sponsored by the Office of Recruitment and Diversity at CUNY. My FFPP colleagues deserve individual mention, so thank you to Jorge Alves, Vandana Chaudhry, Veronica Michel, Yadira Perez, Ahmed Reid, Kathleen Wentrack, and our mentor Mojúbàolú Olúfúnké Okome.

My home institution and department at the College of Staten Island (CUNY) has been collegial and supportive from day one. I am indebted to Jonathan Sassi, Calvin Holder, Richard Powers, and Eric Ivison for the wisdom and guidance they provide with unstinting conviviality and kindness. Donna Scimeca provided refreshing, enlightening, and entertaining conversation during our timely walks to the campus coffee shop. Serendipitously, I found myself entering this department along with an exceptional unit of junior scholars. We formed a reading group and actually followed through to the benefit of all our research agendas. Thank you to Zara Anishanslin, John Dixon, Marcela Echeverri (now at Yale University), Mark Lewis, and Ben Mercer (now at Australian National University).

Part of Chapter 1 appeared previously as "Keeping Spain Afloat: State Forestry and Imperial Defense in the Sixteenth Century," *Environmental History* 17 (January 2012): 116–145, and is reprinted here by permission of Oxford University Press. Part of Chapter 4 appeared as "Spanish Forest Reconnaissance and the Search for Shipbuilding Timber in an Era of Naval Resurgence, 1737–1739," *Journal of Early Modern History* 18, 4 (2014): 357–382, and is reprinted here by permission of Koninklijke Brill NV. I extend my sincerest gratitude to Aleks Pluskowski, the General Editor of Brill's Series in the History of the Environment. It has been a pleasure working with Sabine Steenbeek, Michiel Thijssen, and their colleagues at the History and Biology Department of Brill Publishers. I owe deep and sincere thanks as well to the anonymous reviewers who read and responded to earlier drafts of the manuscript. The final product is much better as a result of the input from all reviewers and readers along the way, but any errors that remain are my own.

I am immeasurably grateful to José de la Cruz and Ana Sanmartín for their friendship, good humor, and hospitality. My research trips to Spain are made so much better when we see each other. Bradley Wilson and Cynthia Gorman have been dear friends and role models in the private as well as professional worlds. My "East coast family" has been an unwavering support system and I extend to all of them my heartfelt thanks.

I have only ever wanted to be like my big brother, so any success I encounter in life also partly belongs to him. I have been truly blessed to have a Mom and Dad who continue to give unconditional love and support, except when the Portland Timbers play the New York Red Bulls.

I will never force my son Jonah to read this book, but if he ever gets as far as the acknowledgements, I want him to know how special he is and how important our time together has been while I finished writing it. My wife

Sharon has already read far more about Spanish forests than she ever could have imagined, but the debt of gratitude I owe her is due to so much more than her patient and brilliant editorial assistance. I thank her for being who she is, for making our life extraordinary, and for helping me keep things in perspective.

List of Maps, Figures and Tables

Maps

1 Spain (Political) 11
2 Spain (Physical) 12
3 The Americas 101
4 The Philippines 111
5 1748 forest ordinances 218

Figures

1 Guernica oak tree 15
2 Spanish galleon 17
3 Forests of Valivana and Valcanera 82
4 Tree trunks and branches for shipbuilding 93
5 Map of Western Cuba including Jagua 150
6 Juan Valdés's reconnaissance of the Ebro River system 194
7 Coatzacoalcos, México 196
8 Manila, Philippines 199
9 Cagayán, Philippines 200
10 Illustration of cutting and pruning methods 212
11 Forest inspection of Tarifa, 1749–52 233

Tables

1 Interior forest reports, 1749–55 224
2 Middle zone forest reports, 1749–53 225
3 Department of Cartagena forest reports, 1749–51 229
4 Department of Cartagena forest report, 1749 229
5 Department of Ferrol forest reports, 1749–51 230
6 Department of Cádiz forest reports, 1749–53 234
7 Department of Cádiz forest report, 1753 234

Introducing Spanish State Forestry

The Widow's Oak and the Spanish State in the Valley of Carriedo

In the hours before dawn on October 7, 1740, in the northern Spanish town of Selaya (Cantabria), Juan de Cuero y Arze started cutting down his neighbor's oak tree. The tree had grown large enough to block sunlight to de Cuero's garden, and he required assistance from several men to remove it, including Vicente de la Concha, the acting mayor of the town. The disturbance awoke the widow Catalina de Pando, the owner of the tree. When she and her children looked outside, the men threw stones at them, knocking down and seriously injuring the "poor helpless widow of the highest valor."[1] The men finished removing the tree, and Catalina de Pando soon took the men to court for the damages they caused that night. In the past, or in another part of Castile, this kind of dispute would have remained a matter to be settled by municipal authorities. However, the case of the widow's oak wound up at the royal *junta* of the navy in Madrid, which was overseen by Zenón de Somodevilla, the Marqués de la Ensenada, who would soon become the most powerful minister in the Spanish empire.[2] Why did the removal of a single oak tree in this town capture the attention of Ensenada and the Spanish navy's top decision-makers? The full answer to this begins with the origins of naval forest regulation in Spain, when a shift in the crown's interest in trees occurred in the sixteenth century initiating numerous efforts to protect and enhance shipbuilding timber supplies. Such efforts over the course of the ensuing era of imperial expansion and defense led to increased bureaucratic oversight, forest reconnaissance missions to locate new supplies, more intensive mapping of the monarchy's resources, and new laws and regulations, all of which contributed to the enhancement of state and naval power. As this case illustrates, state officials had grown concerned with the condition of certain forests down to individual trees by the mid-eighteenth century, so that the lost potential worth of Catalina

1 Archivo General de Simancas (Hereafter, AGS), Secretería del Despacho de la Marina (Hereafter, Marina), Legajo (Hereafter, leg.) 553, Guarnizo, June 15, 1741, Jacinto Navarrete. Other documents related to this case are in the same legajo.

2 John Lynch, *Bourbon Spain, 1700–1808* (Oxford: Basil Blackwell, 1989), 97–98, 158–178; José Luis Gómez Urdáñez, *El Proyecto Reformista de Ensenada* (Lleida: Milenio, 1996); Fernando Bordeje y Morencos, "El poder marítimo en la concepción política del Marqués de la Ensenada," in *La Marina de la Ilustración: Ciclo de Conferencias*, Nov. 1988 (Madrid: Instituto de Historia y Cultura Naval, 1989), 5–22.

 | DOI 10.1163/9789004261372_002

de Pando's oak to the nearby naval shipyards and iron foundries, regardless of the injuries sustained by the widow, made the case a matter of concern for Ensenada. In the end, Ensenada ordered the case to be handled by Manuel García Gómez, the top naval minister of the region, who ordered Juan de Cuero y Arze and his accomplices to plant another oak tree of the same quality in the very same location at their own cost.[3] The culprits completed this task the following June of 1741 in the presence of a notary. This decision did little to help Catalina de Pando in the short term, but it served to reinforce a long-term investment for naval interests, which rested on the hopes of benefitting from sturdy timber in the coming generations.

What follows traces the development of naval forest regulation policies underpinning one of the world's first truly global empires. It examines how the problem of a perceived crisis related to shipbuilding timber scarcity was identified, interpreted, and addressed within the Spanish monarchy by pursuing a history of one of the earliest and largest bureaucracies dedicated to forest use regulation in the early modern world. The storied galleys, caravels, galleons, and frigates of Spain's early modern empire defended sea lanes and transported settlers, slaves, plants, animals, and treasure around the world. The importance of these voyages for Spain and world history is well known, but how the Spanish were able to meet the material demands of supporting its maritime fleets is less well known. Access to shipbuilding timber became one of the foundations of Spain's maritime presence, and from the sixteenth century the crown began taking a more active role in prioritizing its interests in strategically located forests. The crown's forest policies did not arise solely from motives of imperial defense, but the protection of its maritime connections remained a high priority for Spain during the early modern era. European rivals, economic contraction, and inadequate resources tested the integrity and security of the global monarchy, and forest conservation continued to evolve over time in its legal and administrative formulations to adjust to such challenges.

The significance of the Spanish case rests not only on its sizable naval material demands, but also on its relatively early development of long-term conservation efforts – a topic that has not received the dedicated attention it deserves, particularly in English.[4] In 1748, Spain issued its first national forestry codes,

3 Manuel García Gómez was a *Comisario de la Marina* in Cantabria.

4 I use the term "conservation," because it appears in the primary sources, but it should not be equated with the modern term as defined by the founder of modern scientific forestry in the United States, Gifford Pinchot, in his work *The Fight for Conservation* (New York: Doubleday, Page & Co., 1910). Pinchot laid out the principles of conservation, which included the

including the *Ordenanza para el aumento y conservación de montes y plantíos* (Ordinance for the increase and conservation of forests and plantations) which would, with some modifications, remain in place until the early nineteenth century. There had been many forest laws in the past, but these ordinances were the first attempt to place all of Spain's forests under the watchful eye of a government that prioritized naval interests. This book argues that the regulations were based on strategies, methods, and aims that took shape over the previous two centuries guided by the experiences, insights, and lessons communicated principally by royal forest superintendents. It begins, then, with the crown's assertion of control over strategic forests for naval shipbuilding in the mid-sixteenth century and ends with the immediate implementation of the 1748 ordinances under the guidance of its principle author, the Marqués de la Ensenada.

The goal of state forestry by the middle of the eighteenth century was to gain and maintain primary access to all suitable forests within economical range of Spanish and colonial shipyards by extending and refining state forestry methods that were developed in the previous century and a half.[5] The assertion of control over natural resources has been acknowledged by historians as a key component of colonialism, and this was true of the Spanish empire, but a similar process was also occurring within Spain, during which the crown asserted its priorities over the traditional forest harvesting practices at the local level. Even by the end of this study in the eighteenth century, there still remained a dizzying array of overlapping legal jurisdictions. As a result of the tension between an expanding state power and its political and economic

---

fullest use of natural resources, the prevention of waste, and the protection of public interests. The goal of the twentieth century's conservation movement was to achieve national efficiency and the greatest good for the greatest number of people for the longest time. By contrast, "conservation" as used in the early modern centuries by the Spanish navy aimed to prevent shortages for naval provisioning by limiting damages to trees and by replacing trees that were destroyed.

5 State forestry and naval forestry are defined here as the combined goals and practices of the crown that aimed to conserve and access trees for naval use in this period. Like "conservation," the term "forestry" as used in this work differs from the modern use of the term. Luis Gil Sánchez writes that forestry in the contemporary sense of the word is "when rational use is exercised, which is achieved through the application of dasonomic principles to regulate its use and subsequent exploitation, what has today become 'sustainable use'." From Luis Gil Sánchez, *Pinares y rodenales: La diversidad que no se ve* (Madrid: Real Academia de Ingeniería, 2008), 12. By contrast, naval forestry of early modern Spain did not apply the principles of scientific forestry, but rather asserted the political primacy of the navy in forests to obtain material resources.

limitations, early modern state forestry had its own peculiar territoriality dependent on and constituted by local societal and geographic conditions. Gaining primacy over the trees and over what people could do to them in the forests resulted in a process of internal territorialization beginning in the mid-sixteenth century that slowly enhanced state power and altered the functionality of coastal woodlands from local, mostly domestic supply uses to mostly naval shipbuilding needs, which included the forest spaces and tree species best suited for naval access and use. Internal territorialization is understood here as a political process of establishing control of resources and behavior within states through the articulation and enforcement of access and usage rights.[6] What this meant in practice was a progressive deepening and extension of knowledge of forests through surveys and mapping. The crown gained access to forests and control over them at increasingly local levels while at the same time enahancing centralized management. Studying this process informs us of state and society relations regarding natural resource accessibility and connects political actions and the rise of state power.

Through the lens of internal territorialization, this book analyzes the intensification of forest resource use for Spanish naval shipbuilding and its relation to early modern state formation from the sixteenth century to the middle of the eighteenth century. Physical mapping is the most literal way of visualizing internal territorialization, but to truly understand the process as a whole, it must be understood as something encompassing much more than just drawing lines on a map. Critical to the formulation and implementation of state forestry policies was the historical process of gaining geographic knowledge, but the generation of this knowledge during this period, as the king and his ministers increasingly saw a need for locating additional sources of shipbuilding timber, came from numerous crown-sponsored reconnaissance missions using various methods to suit the needs of naval interests. These were not all daring voyages of global exploration, but frequently involved crown

6 Human territoriality is defined by geographer Robert David Sack as "the attempt by an individual group to affect, influence, or control people, phenomena, and relationships, by delimiting and asserting control over a geographic area." Sack argues this occurs through three interdependent relationships. First, it must involve a form of classification by area. Second, its conditions, such as its territorial boundaries and its restrictions on activities, must be communicated somehow. Third, it must involve an attempt at enforcing control over access. Robert David Sack, *Human Territoriality: Its Theory and History* (Cambridge, 1986), 19. The political process of establishing control over forests within Spain by crown authorities is what I call internal territorialization. This process has been studied recently in a different context by Chandra Mukerji in *Impossible Engineering: Technology and Territoriality on the Canal du Midi* (Princeton, 2009).

representatives trekking up river valleys into the mountains of Spain itself and mapping or recording physical and socio-economic conditions. Still, such activities added to the crown's knowledge of potentially critical resources. With closer observation and regular inspection of forests by royal bureaucrats, the search for and identification of valuable sources of timber enabled an extension of state power and centralization of geographic knowledge of Spain and its empire.

Even where individual trees were not immediately useful for the navy the process of territorialization would assert crown control over the management of trees and forests. Consider, then, a scenario in which the widow's oak survived but turned out to be unsuitable for the navy. The documents related to de Pando's case do not specify the size of the tree that was lost nor its shape. We know the oak tree grew near de Pando's house and it cast shade in an otherwise sunny area, so it probably grew somewhat isolated from the forest. It likely grew, therefore, more branches, producing numerous knots in the trunk, which caused difficulties for naval officials who wanted a long and uniform grain for ship planks and other pieces. However, rather than being of material value to the navy, the widow's oak still proved to possess value as a symbol of royal power. Ensenada's commissary in Cantabria, García Gómez, gave de Cuero a minor sentence, but it was more important for the navy that he was able to assert the crown's authority in Selaya by getting de Cuero to replant the tree. Both the old tree and the newly planted tree had become emblems of territorial control. The achievement of naval forest territoriality as examined in this book, then, was not limited to the numbers of trees that wound up floating downriver and into ships, but also included other forms of authority exercised by the state in forests and over trees that may not have even been useful for the navy. Such trees served a dual purpose for the crown. At the local level, a tree like the one planted to replace de Pando's oak is a reminder of what the state could do. At the state level, the crown received additional information about the kingdom while the tree became a site for the imposition of royal authority regardless of the final destination of the tree's timber.

A 'mapping' of Spain developed in the early modern era, as agents of the crown located and described forests and other geographic features that had been previously unfamiliar to state authorities.[7] Much like the explorations and mapping of the New World, the forest reconnaissance missions within Spain cannot be understood outside of a broader process of imperial

---

7 I use the word mapping to include cartographic representations as well as texts that described geographic features and locations, because the crown utilized various forms of geographic information to learn about its territory.

expansion.[8] The timber they were looking for, after all, was to be used to build the galleons and other vessels that maintained imperial communication, trans-oceanic trade, and naval defense. Similar internal explorations occurred in France, the Veneto, and elsewhere, but the Spanish covered more ground and counted more trees, whether they would actually be utilized by the navy or not, into the hundreds of millions by the middle of the eighteenth century.[9] If Spain's maritime and imperial strength depended on wooden sailing vessels built with planks, beams, and masts sawn from suitable trees and sealed with the resin of pines then the roots of Spanish imperial power were literally growing in its forested mountains.

Looking at this 200-year history, there are certain themes that emerge as a result of state power infiltrating forests to help determine which forests were targeted, which people were affected by Spain's naval timber needs, and how these needs were met. Each theme intersects with broader phenomena characteristic of the early modern era. First, the expansion of the crown's bureaucracy, which was so characteristic of state power generally, extended into the realm of forest management where superintendents with experience in naval service or shipbuilding, or both, worked to connect the needs of the shipyard with the material resources of the forest.[10] This forestry bureaucracy in Spain first emerged in the reign of Felipe II (1556–1598) out of the broader naval bureaucracy, and clearly, by 1740, had an inclination and capacity to respond to seemingly minor affairs such as the removal of de Pando's oak. As purveyors of royal policy, such bureaucrats worked at the important interface of imperial defense strategy and municipal-level forest regulation. The forest superintendents of Spain reflected a broader contemporary trend of the emergence of actors in the extension of centralized power to monitor and manage valuable

8 Ricardo Padrón, *The Spacious Word: Cartography, Literature, and Empire in Early Modern Spain* (Chicago: University of Chicago Press, 2004). Padrón wrote about how mapping and empire were inseparable by focusing mainly on examples from Mexico, but a similar argument could be made about geographic information-gathering in Spain itself.

9 For French naval forest inspections, see Hamish Graham, "For the Needs of the Royal Navy: State Interventions in the Communal Woodlands of the Landes during the Eighteenth Century," *Proceedings of the Western Society for French History* v. 35 (2007): 135–148. For Venice, see Karl Appuhn, *A Forest on the Sea: Environmental Expertise in Renaissance Venice* (Baltimore: The Johns Hopkins University Press, 2009).

10 Karl Appuhn, "Inventing Nature: Forests, Forestry, and State Power in Renaissance Venice," *The Journal of Modern History*, 72, no. 4 (Dec., 2000): 863; Jan Glete, *War and the State in Early Modern Europe: Spain, the Dutch Republic, and Sweden as Fiscal-Military States, 1500–1660* (London: Routledge, 2002); Paul Warde, *Ecology, Economy, and State Formation in Early Modern Germany* (Cambridge: Cambridge University Press, 2006).

natural resources. The Spanish formed the largest of such forest bureaucracies in the early modern world, which warrants greater attention to its origins, growth, and function.

Next, the dramatic escalation of material needs for the navy in the early modern centuries, part of a broader phenomenon often referred to as the Military Revolution, placed tremendous pressure on forests within economical range of shipyards.[11] Even as Spain's naval power waned in the later seventeenth century, access to shipbuilding timber remained an acute governmental concern. Selaya's proximity to the coast had sustained the crown's interest in conserving its surrounding forests in the Valley of Carriedo even during the bleak shipbuilding years of Carlos II's reign (1665–1700), indicating that fears of timber shortages did not always correspond with high demand or even actual depletion. Not long afterwards, greater control over forest resources in Spain and increased exploitation of colonial forests in the early eighteenth century contributed to a Spanish naval resurgence, with ramifications for warfare in the Atlantic world and imperial governance in the Americas. By the 1780's, Spain was operating the second largest navy behind Great Britain, in part due to an effective state forestry.[12]

The increased sophistication of record-keeping and collection of quantifiable data by Spain's forest bureaucracy meant that individual trees of the right species, like de Pando's oak, mattered to state authorities.[13] Such records were used by superintendents to devise and monitor planting quotas or collect fines in thousands of forest communities. More reliable records meant more reliable estimations of timber supplies for Spanish arsenals. The monarchy's ability to have a predictable supply of timber in the major ports in any given year enhanced the state's ability to plan and organize naval activity. The selection of wood was the primary preoccupation of the shipwright.[14] Generally, hard and sturdy woods were required for the keel and hull while strong but lightweight

11 Geoffrey Parker, *The Military Revolution: Military innovation and the rise of the West, 1500–1800* (Cambridge: Cambridge University Press, 1988 and 1996); J.R. McNeill, "Woods and Warfare in World History," *Environmental History* 9 (July 2004), 393.

12 It was effective in the sense of shipbuilding, not in the sense of any contemporary notion of environmental management. Lynch, *Bourbon Spain, 1700–1808*, 315; John D. Harbron, *Trafalgar and the Spanish Navy* (Annapolis: Naval Institute Press, 1988); Álvaro de la Piñera y Rivas and Jacqueline Thial Boisière, "La construcción naval en España durante el siglo XVIII," *Revista de Historia Naval* 79 (2002): 17–33.

13 Alfred W. Crosby, *The Measure of Reality: Quantification and Western Society, 1250–1600* (Cambridge: Cambridge University Press, 1997).

14 José Luis Casado Soto, *Barcos y Astilleros: La Construcción Naval en Cantabria* (Santander: Puerto de Santander, 1993), 63.

wood was required for the superstructure of the vessel.[15] The masts, for example, needed to be lightweight, strong, and somewhat flexible. Spain exploited the pines of its Mediterranean highlands, but Baltic pines were on the whole better-suited and highly sought after by all European maritime powers. Spain imported some of its pine for masts from the Baltic, Italy, and Albania and, during this period, increasingly relied on the supply zones of colonial shipyards.[16] Pines also provided the pitch that helped waterproof the sides of wooden vessels. Oak was strong and durable enough for ships' hulls. In Spain, suitable oak trees grew principally in the wetter north, particularly the Basque provinces of Vizcaya and Guipúzcoa. They also grew in a great variety of shapes and forms, allowing for greater correspondence between the natural structure of the wood and the shape of the piece necessary for construction.[17] Well-seasoned oak was resistant to rot and other diseases, due to its high content of tannin. The downside, however, was that the tannin facilitated the oxidation of iron, causing nails and hinges to rust. Europe's maritime empires, including Spain, prioritized the tree species with the desired qualities for different parts of sailing vessels in their regulation of domestic and colonial forests. In addition to oak and pine in Europe, the highly esteemed timbers typically included beech for oars; fir for masts, spars, and planking; larch (in northern Europe) for decks and frames; elm for spars; and walnut for rudders.[18] Teak in South and Southeast Asia and mahogany in the Caribbean became preferable to oak in those regions. The notion that a monarch could order the clear-cutting of a forest to build a navy fails to consider, then, that the state aimed to gain access and conserve specific types of forests or species within forests rather than all trees.

While the themes of bureaucratization, militarization, and quantification underscore the forces that contributed to the expansion of state power, there were also limiting factors in the crown's ability to control its forests and the people who lived there, such as geographic, technological, and social conditions of the time. The crown may have wished to manage each and every suitable tree, but an overall balance had to be struck between meeting the needs of the

15 Carla Rahn Phillips, *Six Galleons for the King of Spain: Imperial Defense in the Early Seventeenth Century* (Baltimore: Johns Hopkins University Press, 1986), 79.

16 Fernand Braudel. *The Mediterranean and the Mediterranean World in the Age of Felipe II*, trans. Sîan Reynolds, 2 vols. (New York: Harper Colophon Books, 1976), 142–143; Y. Eyüp Özveren, "Shipbuilding, 1590–1790," *Review (Fernand Braudel Center)* v. 23, n. 1, Commodity Chains in the World-Economy, 1590–1790 (2000), 23 and 30.

17 Casado Soto, *Barcos y Astilleros*, 65.

18 Phillips, *Six Galleons*, 49; Özveren, 52; Braudel. *The Mediterranean*, 142; Appuhn, *A Forest on the Sea*, 54.

navy and preventing unrest in a society dependent on forests for daily material needs. State authorities ultimately relied on the forced or cooperative labor of people in forest communities to carry out the arduous and long-term care of trees, which required planting the seeds, cultivating the saplings, guarding against hungry cattle and other livestock, managing their growth, cutting down the grown trees desired by the navy or extracting the resin from pines, sawing and preparing the timber, and transporting the timber to the shipyard. There was no professional class of foresters in Spain yet, just representatives of the crown who visited forests to enforce royal policies. The difficulty of entirely fulfilling the goals of forest conservation legislation is discernible from the numbers of times such legislation reappeared. Upon demand from the royal naval *junta*, Juan de Cuero y Arze replanted a tree for Catalina de Pando, but the tree could have eventually grown to be judged worthy for the arsenal, at which point the tree's owner and the people of Selaya would have been called upon to procure it for the navy. Not surprisingly, state authorities often had trouble enforcing royal policies such as the creation of new plantations, claiming priority over timber for the king, or collecting fines for not meeting planting quotas. Superintendents faced strong resistance or defiance as a result wherever they operated.[19] On multiple occasions, resistance forced superintendents to alter their more heavy-handed methods, and even caused the crown to revise its regulations. In the colonies, where slave labor or a draft labor system was used, resistance and uprisings also occurred, such as in Samar, in the Philippines, in 1649, when a levy of workers for the Manila shipyards at Cavite caused the local population to resist.[20]

Extremes of physical geography and the challenges of transportation limited the state's ability to exploit forest resources (Map 1 and 2). The greatest impediments to forest access were mountainous terrains and great distances to navigable rivers. Spain's extreme topography contrasts strongly with the wide plains of northwestern Europe and posed barriers to royal forest inspectors interested in extracting large timbers. Spain has the second highest mean elevation on the continent (2,200 feet) behind only Switzerland. At the same time, more than a third of this land is composed of flat plateau called meseta. The meseta is divided in two by the Cordillera Central, or Central Mountains, just north of Madrid, made up of several ranges including the Somosierra, Guadarrama, and Gredos, with peaks reaching 7,200 feet. Together, mesetas

19 Ricardo Gómez Rivero, "La superintendencia de construcción naval y fomento forestal en Guipúzcoa," *Anuario de historia del derecho español* v. 56 (1986), 617.

20 John A. Larkin, "Philippine History Reconsidered: A Socioeconomic Perspective," *American Historical Review* v. 87, n. 3 (June, 1982): 605.

form about 36% of the Iberian Peninsula, and support some of Spain's best agricultural land, but they are almost entirely surrounded by mountain ranges. The northern meseta, making up most of Old Castile, is divided to the east by the Iberian mountains, which reach heights of 7,500 feet. Spain's north is the most highly forested region, and is also quite mountainous. Iberia is divided from France by the extreme topography of the Pyrenees mountains, which extend for nearly 150 miles and reach altitudes of over 11,000 feet. To the west of the Pyrenees are the lower mountains of the Basque Country, followed by the more formidable Cordillera Cantábrica, or Cantabrian Mountains, with altitudes reaching 8,500 feet. This chain divides a narrow coastal plain from the interior and, like the Pyrenees, poses a daunting impediment for travelers. Continuing westward, much of the Asturian coast is characterized by rocky cliffs, while Galicia's coastline is made up of numerous *rías*, deep inlets that resemble Scottish firths or less dramatic Scandinavian fjords. The southern meseta, which includes much of New Castile, is also bounded to the east by the Iberian mountains and to the south by the Sierra Morena, which reaches altitudes of 4,600 feet. Extremes along the Mediterranean coast range from desert between Murcia and Alicante, to the snowcapped mountains of the Sierra Nevada, part of the Baetic mountains (11,411 feet at its highest point), each only a few miles from the sea. In between these extremes, one can find ideal conditions for human settlement, such as the large fertile plain, or huerta of Valencia. The average elevation of the meseta range from 2,300 to 2,600 feet in the north and from 2,000 to 2,300 feet in the south. As a result, travelers from many parts of the meseta can actually descend to the surrounding mountains.[21] In general, Spain's mountainous topography made many of its forests inaccessible until long-distance transportation and communication improved with the arrival of the railroad.

Erratic rainfall in this topography forced most of Spain's rivers to descend rapidly to the coast and carry widely-fluctuating amounts of water, making dependable river transportation virtually nonexistent. Such difficulties hindered economic interactions between the interior and the coasts. The navigable parts of major rivers, however, were used heavily. The Ebro River, which originates in the Cantabrian Mountains, flows for 565 miles between the Pyrenees and the Iberian mountains through Aragón and Catalonia into the Mediterranean. The Júcar flows 316 miles out of the Montes Universales range in the Iberian

21 Carla Rahn Phillips and William D. Phillips, Jr., Spain's *Golden Fleece: Wool Production and the Wool Trade from the Middle Ages to the Nineteenth Century* (Baltimore: Johns Hopkins University Press, 1997): 11; and William D. Phillips, Jr., and Carla Rahn Phillips, *A Concise History of Spain* (Cambridge: Cambridge University Press, 2010), 5.

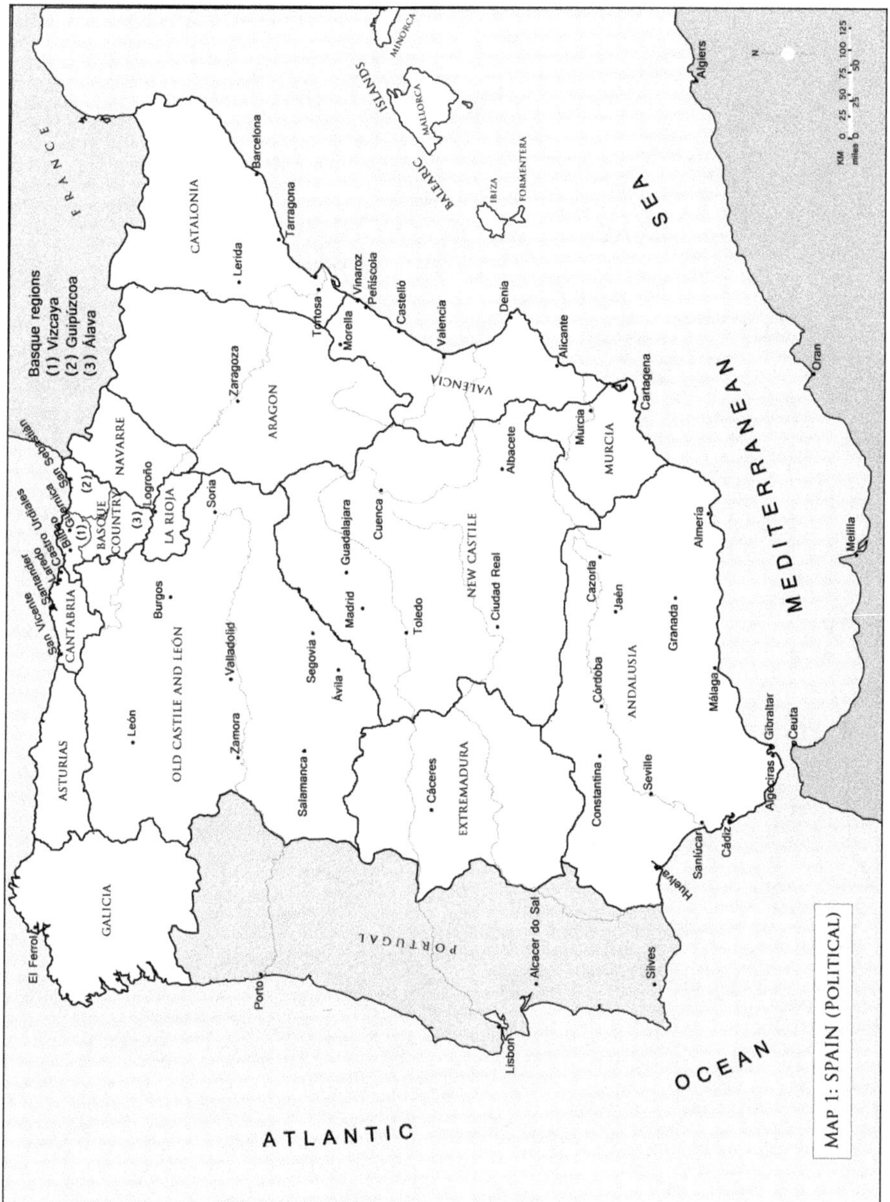

MAP 1 *Spain (Political)*

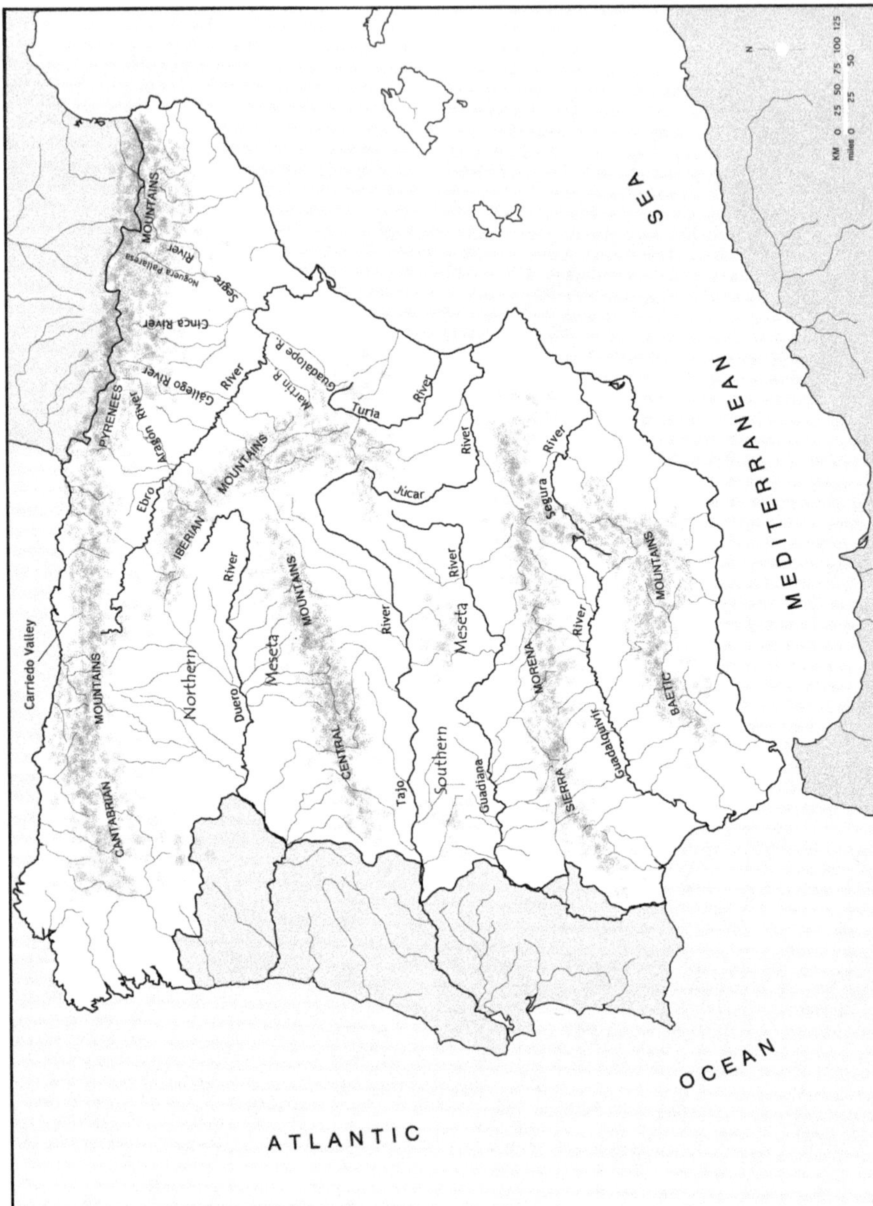

MAP 2 *Spain (Physical)*

mountains past the scenic city of Cuenca and empties into the Mediterranean south of Valencia. The Segura flows 202 miles out of the Baetic mountain system in the province of Jaén, past Murcia and into the Mediterranean north of Cartagena. Spain's other main rivers all empty into the Atlantic. The Duero cuts through the grain and wine-producing region of the northern meseta, past Valladolid and Zamora, emptying at Porto, Portugal, 557 miles from its source in the Sierra de Urbión near Soria. Farther south, the Tajo originates near the source of the Júcar, but flows towards the Atlantic 645 miles past Toledo and into Portugal, ending at Lisbon. The Guadiana flows 508 miles through the middle of the southern meseta and forms part of the southern border between Spain and Portugal. The only major Atlantic river that starts and ends in Spain is the Guadalquivir, which rises near the source of the Segura, but flows mostly southwest through Andalusia between the Sierra Morena (with elevations reaching over 4,300 feet) and the Baetic mountains, past the cities of Cordoba and Seville. These major rivers were the main entry ways into the forested sierras of the interior, but the pattern of rainfall in much of Spain made even the larger rivers only seasonably capable of transporting timber. During the hot months of summer, the riverbeds would become too dry, and seasonal rains had the potential to cause flash flooding.

Accessibility reflected a cost-benefit analysis that on the one hand put many adequate forests out of reach for the navy, such as in the inland Sierra de Guadarrama northwest of Madrid, but simultaneously drew focused attention from the crown on other forests, like near Catalina de Pando's Carriedo valley where the right combination of suitable timber, river transport, and close proximity to the sea made it ideal for the navy. The historical geography of her hometown of Selaya helped ensure that its sturdiest trees were of interest to the navy, a fact that de Pando knew and hoped to work to her benefit. The valley of Carriedo is located in modern day Cantabria about twenty miles from the sea.[22] According to the *Catastro de Ensenada*, a survey of Spanish population and resources conducted in the early 1750s, the Valley of Carriedo was mostly *realengo* land, or part of the royal domain, and made up of a mix of

22 The Valley of Carriedo was also the home of the parents of the great Spanish poet, Lope de Vega. They left for Madrid shortly before he was born, but Lope referenced his ancestral homeland in some of his work, on one occasion calling its people "pure and faithful." See Lope de Vega, *Virtud, pobreza y mujer*, in *Comedias Escogidas de Frey Lope Félix de Vega y Carpio*, vol. 4, edited by Juan Eugenio Hartzenbusch, Juan Pérez de Montalván, Antonio Gil y Zárate, and Adolfo de Castro, 213–232 (Madrid: Hernando y Compañía, 1901), 219: Act 2, scene 1, line 18, "Montaña limpia y leal." La Montaña was also used to refer to the old region of Cantabria.

pasture land, gardens, orchards, and forests of oak and beech trees, which were planted in obedience to the recent naval ordinances and still used at regional shipyards.[23] Similar conditions existed in well-forested coastal towns within economical range of water transport. In colonial contexts, most successful imperial coastal settlements and active ports thrived at least in part from their proximity to accessible stands of timber. Due to the economical restrictions of overland transport, the crown typically limited its claim to forest resources that were at a distance of twenty to thirty leagues from navigable waterways, a practice shared by other kingdoms such as England and France.[24]

Well-entrenched socio-economic and legal conditions imposed further limitations on the state's ability to do what it wished in the forests. People at all levels of society relied on forest resources, and centuries-old protections against preventing access to such resources meant the crown could not simply assume exclusive control everywhere.[25] Municipal laws, or *fueros*, existed in various forms all over Spain and as many of them guaranteed access to resources of the commons, they were strongly defended by municipalities. The most famous symbol of this local autonomy may be, in fact, a tree. In the city of Guernica in the Basque province of Vizcaya local elders gathered under an oak tree, made laws and insisted that as soon as a new monarch ascended the throne of Castile he or she or a representative had to swear to uphold Vizcaya's *fueros*. The practice of gathering there to reinforce local autonomy is what gave the Guernica oak its symbolic power.[26] In *La prudencia en la mujer* (1633),

---

23 Respuestas Generales del Catastro del Marqués de la Ensenada, AGS, Catastro de Ensenada, Respuestas Generales, leg. 40, folio 304. For the history of Guarnizo's shipyard in this era, see Ramón Maruri Villanueva, "Ensenada y el Real Astillero de Guarnizo," *Brocar: Cuadernos de investigación histórica* 25 (2001): 123–136.

24 A league was a unit of distance that a person or a horse could walk in about 1 hour, approximately 3.5 miles or 5.5 kilometers.

25 On local land use and jurisdictions, see David Vassberg, *Land and Society in Golden Age Castile* (Cambridge: Cambridge University Press, 1984); Helen Nader, *Liberty in Absolutist Spain: The Habsburg Sale of Towns, 1516–1700* (Baltimore: The Johns Hopkins University Press, 1990).

26 The oak tree remains a symbol of Basque identity to the present day; it is the subject of its unofficial anthem, it is depicted at the center of several Basque coats of arms, and it is an emblem of the Basque nationalist party. In 1810, William Wordsworth wrote a poem about the tree where "those lofty lawgivers meet, Peasant and lord in their appointed seat, Guardian's of Biscay's ancient liberty." There have been four "Guernica oak" trees since the fourteenth century. After one tree dies, another one, grown from the old one's acorns, is transplanted to the location where the old one once stood. The first one, "el Padre" lived until the middle of the eighteenth century. Then, "el Viejo" stood from 1742 until 1892.

the playwright Tirso de Molina (1571–1648) wrote of a Basque man's defiant response to a Castilian's attack on his heritage with a statement that the Guernica oak itself protects his people from the tyranny of royal authority (Figure 1).[27]

FIGURE 1 *Guernica oak tree – From the Middle Ages, leaders of the Basque province of Vizcaya gathered under the tree to make laws and to witness future kings of Castile promise to uphold the local* fueros *of the region. The current tree is the "great-grandchild" of the one planted in the fourteenth century.*
PHOTO BY AUTHOR

The third tree survived the bombing of Guernica in 1937, but died after a 2004 heat wave. Currently, a tree that had been sown in 1986 and planted at the traditional assembly site in 2005 now stands.

27 Tirso de Molina, *La Prudencia en la Mujer* (1633), in *Obras dramáticas completas de Tirso de Molina* (*En tres tomos*), vol. 3, ed. Blanca de Los Ríos (Madrid: Aguilar, 1958), 905–906. "Don Enrique: Vos, caballero pobre, cuyo estado / cuatro silvestres son, toscos y rudos, / montes de hierro, para el vil arado, / hidalgos por Adán, como él desnudos, / adonde en vez de Baco sazonado, / manzanos llenos de groseros ñudos / dan mosto insulso, siendo silla rica, / en vez de trono, el árbol de Garnica...don Diego: ...El árbol de Garnica ha conservado / la antigüedad que ilustra a sus señores, / sin que tiranos le hayan deshojado, / ni haga sombra a confesos ni a traidores. / En su tronco, no en silla real sentado, / nobles,

While privately held forests required consent from the owner and an agreed price before timber could be taken for the king, in *realengo* lands, the king theoretically had the right to use the resources of the forest. However, even in the royal domain, when the king claimed trees for the navy, residents often complained that access to the commons had been a right of theirs practiced since time immemorial. The crown's reassertion of its rights to municipal forests related directly to its growing shipbuilding interests, symbolized most recognizably in the form of the Spanish galleon. In the poetry and artwork of the time, the vessels of Spain's maritime Golden Age, particularly the galleons, symbolized the wealth and power of the monarchy. The playwright and poet, Lope de Vega (1562–1635), called the *Armada Invencible* a "forest of the sea" in a sonnet that glorified the "enterprise of England" of 1588.[28] Spanish galleons were floating fortresses and symbols of defense against hostile and heretical enemies. They served as Spain's commercial and military lifeline in this period, and even as the galleon gave way to other types of vessels in the eighteenth century, naval power still depended on resources of the forest. As galleons' hulls were made of Spanish oak trees, forests, by extension, became important and valuable sites for the preservation of Spanish imperial power (Figure 2).

The Guernica oak and the galleon hull were two political symbols made of the same material, but they represent two very different forms of territoriality. Guernica's oak tree stood as a symbolic contrast to crown authority and represented the interests of the local scale. As the Basque lawmakers' assembly point, it symbolized the importance of the local decision-making process in determining regulations of resource use.[29] On the other hand, the galleon

puesto que pobres electores / tan sólo un señor juran, cuyas leyes / libres conservan de tiranos reyes." In referencing the "montes de hierro," or forests of iron, the passage also helps to illustrate that due to the difficulties in transporting construction wood, forests were mainly used for making charcoal and for fueling the iron industries.

28 Lope de Vega, "Soneto XLVI: A la Jornada de Inglaterra a borde del 'San Juan'," *Rimas* (1602), Biblioteca Virtual Miguel de Cervantes, edited by Ramón García González, 2003, last retrieved on July 2, 2014, from http://www.cervantesvirtual.com/servlet/SirveObras/13527274323929275754491/p0000001.htm?marca=A%20la%20jornada%20de%20inglaterra#238.

29 The crown did, however, regulate the use of certain resources. For example, from the Middle Ages the crown granted privileges to families involved in the iron industry, including the right to exploit the forests for timber. King Alfonso X granted the first privilege in 1262, allowing the Mondragón iron industry to exploit the village mineral seams and to use the necessary timber for charcoal making: "E Yo por faservos merçed tengo por bien e mando que ansy estas como otras algunas, sy en vuestro termino las oviere [venas], que se labren en vuestra villa e non en ningund otro lugar. E otrosy tengo por bien que el

FIGURE 2 *Spanish galleon – Built for both trade and war, galleons helped connect Spain and its empire primarily in the sixteenth and seventeenth centuries. This is a model of one of the last galleons made in Spain, the* Nuestra Señora de la Concepción y de las Ánimas, *with 90 cannons (1687–1705) held at the Naval Museum of Madrid. This galleon sailed as the royal flagship of the Ocean Sea. The reconstruction was based on a study of Antonio Gaztañeta's* Arte de fabricar Reales *(1688). Galleons would eventually be replaced by faster and more maneuverable vessels in the eighteenth century.*
PHOTO BY AUTHOR. MODEL BY JESÚS MARÍA PERONA LERTXUNDI (2000).
ACTUAL DIMENSIONS: LENGTH: 48M, WIDTH: 12 M. WOOD AND HEMP.
MUSEO NAVAL, MADRID.

carbón et la lenna que oviere menester para las ferrerías, que quanto más syn dapno lo pudiesen fazer de los montes que lo fagan. Et defiendo que ninguno sea osado de cortar nin de quemar las bustalisas antiguas; ca qualquier que lo fisiese a el e a todo quanto que oviese me tornaría por ello." Miguel Angel Crespo Rico, José Ramón Cruz Mundet, José Manuel Gomez Lago, and José Angel Lema Pueyo, *Colección Documental del Archivo Municipal de Mondragón*. V.I (1260–1400) (San Sebastián: Eusko Ikaskuntza, 1992), 4.

symbolized interests on a national or international scale. As the vessel that protected kingdom and empire for much of this period, it represented the might of the Spanish monarchy, a force that could be increasingly felt in the forests of the Basque Country during the early modern centuries. The contrast between local and imperial interests in forest resources at times caused direct conflict. In certain regions, like Cantabria, the effects of crown involvement had soured relations with the local population to the point that it caused serious damage to its own interests by the late eighteenth century.[30] However, rather than representing a simple dichotomy, these two symbols are part of an interdependent and complicated relationship between the crown and the forest communities of Spain. Despite clashes between crown and local authorities, the expansion of state forestry unfolded within a context of diverse and overlapping interests. Understanding the importance of forests as sites of multiple uses for the operation of early modern societies and economies is critical to defining the approaches taken by the crown to meet its needs realistically. Guernica is in the Basque province of Vizcaya, which is near the coast and is well-forested. Many of its people made their living by fishing and through maritime trade, but also by serving the crown in the navy. Likewise, the crown continued to reconfirm the *fueros* of Vizcaya under the Guernica oak each reign, in part because it needed *vizcaínos* to serve in its navy and to build its vessels.[31] The crown also knew it could not completely restrict forest communities from using forest resources; to prevent a population from gaining access to its primary fuel and building supplies would have generated calamitous social upheaval. The Basque region was not the only part of Spain with climate suitable for forest growth and well-represented local interests. Catalonia, for example, contained many of the forests used to build the galleys of the Mediterranean. The crown and Catalonia struck a similar balance of interests as in the Basque country. Municipalities in other provinces in the forested regions of northern Spain, such as Galicia, Asturias, and Cantabria, traditionally managed their forests without much involvement from the crown. Forest management goals on the local scale were not uniform, where farmers and shepherds competed for land, while church and mill construction projects competed for timber. Pig farmers, tanners, coopers, and all sorts of craftsmen

30 C. Diego Liaño and J.C. García Cordón, "La corona y los pueblos en la explotación de los montes de Cantabria: Deforestación y gestión del bosque en la segunda mitad del siglo XVIII," *Cuadernos de la Sociedad Española de Ciencias Forestales* n. 16 (2003), 215–220.

31 In addition to Vizcaya, the Basque region consisted of two other provinces, Guipúzcoa and Álava. All three were part of the Kingdom of Castile. Vizcaya and Guipúzcoa were strongly maritime, whereas Álava is landlocked.

sought a wide range of forest products to earn their livelihood. At the same time on the state level, officials in the central government disagreed often about allocating resources for the navy. They sought to balance the expenses and interests of the army, the court, and the bureaucracy at once. So, rather than serving as an indication of two sides that faced off in direct competition over control of forests, the Guernica oak and galleon hull, and indeed the widow's oak as well, introduce the political and social structures in which the struggle over limited resources occurred in early modern Spain.

As the Spanish crown faced limitations to its ability to know its forests down to the last tree, and to enforce its forest legislation, it never fully achieved the ideal goal of becoming entirely self-sufficient in naval timber. Fortunately for Spain, none of its rivals did either. However, for over two centuries, by identifying specific areas that contained suitable and accessible timber both within Spain and in the colonies, by communicating the rules and regulations of planting and cutting, and by developing techniques to enforce control over access – all the elements of internal territorialization – the crown was able to address the material needs of the navy without upending the social balance it sought to maintain.

The Politics of Wood Shortage Fears in the Early Modern World

In the late sixteenth century, King Felipe II of Spain (r.1556–1598) wrote, "One thing I'd like to see done is about the conservation of the forests…I fear that those who come after us will have much to complain of if we leave them depleted, and please God we do not see it in our time."[32] Heads of state like Felipe II welcomed forest clearings to a certain degree, as they promised more tax revenues in the form of new agricultural land,[33] but statesmen also expressed fear of a future without adequate forest resources. Recent research points to similar contemporary examples, such as Karl Appuhn's work on Renaissance Venice and Paul Warde's work on southwestern Germany.[34] While

32 Henry Kamen, *Philip of Spain* (New Haven: Yale University Press, 1997), 182–183 for the English translation. Guillermo Muñoz Goyanes, *Crónica sobre bosques y montes de la Península Hispánica* (Madrid: Fundación Conde del Valle de Salazar, 1983), 103. Kamen and Muñoz Goyanes provide different dates for this statement, 1582 and 1572 respectively.

33 John F. Richards, *The Unending Frontier: An Environmental History of the Early Modern World* (Berkeley: University of California Press, 2003), 25–38 on Mughal India, for example.

34 Appuhn, *Forest on the Sea*; Warde, *Ecology*.

vastly different geographically and economically, in both regions the fear of wood scarcity spurred the involvement of state authorities to make sure adequate supplies of forest resources existed. These and other examples allow us to begin speaking in general terms about how the state became involved in the regulation of forests in early modern Europe. As in Spain, regulations elsewhere on the continent emerged in the twelfth and thirteenth centuries motivated by fears of wood shortages. The archival record in Spain and in growing numbers of Spanish forest histories indicate that the state's role in forestry evolved through a familiarity with local conditions that developed over generations to prevent wood shortages rather than having emerged as a sudden imposition of centrally designed, imported, borrowed, or purely rational techniques from the distant court in Madrid to meet exclusive demands of the state. Some historians are studying concepts of sustainability with greater historical perspective, particularly of woodlands, looking back to the sixteenth and seventeenth centuries, before the actual term was in use. From the late Middle Ages, state authorities began to encourage behavior that would leave enough forest resources to future generations.[35] Also as in Spain, but some time afterwards, European maritime states began expanding their role in forest management to assert their own naval and military interests.

The intensification of forest land use has received a lot of attention from historians of the early modern world. Forests disappeared in this era along new settlement frontiers as populations grew or migrated great distances. Consumption of domestic and industrial fuelwood, along with construction timber increased worldwide. In colonial settings especially, forests were transformed into plantations for cash crops such as sugar and tobacco, used up in the process of mining precious metals, or were themselves exploited by colonists looking for "green gold" and easy profits.[36] With access to seemingly

35 Paul Warde, "The Invention of Sustainability," *Modern Intellectual History* 8, no. 1 (2011), 153–170; J. Soriano Martí, "La documentación Medieval y la sostenibilidad de los aprovechamientos forestales mediterráneos," *Cuadernos de la Sociedad Española de Ciencias Forestales* n. 16 (2003), 73–78; J.M. Rodríguez Villa, "Historia forestal y desarollo forestal sostenible," *Cuadernos de la Sociedad Española de Ciencias Forestales* n. 16 (2003), 297–301.

36 Jennifer L. Anderson, *Mahogany: The Costs of Luxury in Early America* (Cambridge, MA: Harvard University Press, 2012); Londa Schiebinger, *Plants and Empire: Colonial Bioprospecting in the Atlantic World* (Cambridge, MA: Harvard University Press, 2007); Sidney W. Mintz, *Sweetness and Power: The Place of Sugar in Modern History* (New York: Viking Press, 1985); Richard Grove, *Green Imperialism: Colonial Expansion, Tropical Island Edens, and the Origins of Environmentalism, 1600–1860* (Cambridge: Cambridge University Press, 1995); John Perlin, *A Forest Journey: The Story of Wood and Civilization*

endless colonial supplies and little incentive to conserve, the Spanish have been portrayed as among the most reckless destroyers of forests in this period.[37] As with other Atlantic maritime empires, Spain exploited colonial forests and took part in the global market to meet its naval demands. This did not mean, though, that the crown could afford to disregard the conservation of domestic forests.

The many people in the pre-modern world who relied on wood resources long identified the potential loss of woodlands as a serious problem, necessitating the implementation of socio-legal safeguards against scarcity. From the thirteenth century increasing numbers of regulations were being written down to quell disputes over wood distribution spurred by competition between social groups.[38] While agriculture expanded in late medieval Europe, clearing forest cover at a rapid pace, people still depended on forest resources for shelter, tools, food, water, and fuel.[39] Trees were cut to build homes, churches,

---

(Woodstock, VT: The Countryman Press, 1989 and 2005), 249–262; Michael Williams, *Deforesting the Earth: From Prehistory to Global Crisis, An Abridgment* (Chicago: University of Chicago Press, 2006), 191–225; Shawn William Miller, *Fruitless Trees: Portuguese Conservation and Brazil's Colonial Timber* (Stanford: Stanford University Press, 2000); Warren Dean, *With Broadax and Firebrand: the Destruction of the Brazilian Atlantic Forest* (Berkeley: University of California Press, 1995); Daviken Studniki-Gizbert and David Schecter, "The Environmental Dynamics of a Colonial Fuel-rush: Silver Mining and Deforestation in New Spain, 1522–1810," *Environmental History* 15, no. 1 (2010), 94–119; Jack P. Greene and Philip D. Morgan, eds., *Atlantic History: A Critical Appraisal* (Oxford: Oxford University Press, 2009).

37 The environmental historian Joachim Radkau highlights this line of thinking in Joachim Radkau, *Nature and Power: A Global History of the Environment*, tr. Thomas Dunlap (Cambridge: Cambridge University Press, 2008), 190.

38 Paul Warde, "Fear of Wood Shortage and the Reality of the Woodland in Europe, c.1450–1850," *History Workshop Journal*, Issue 62 (Autumn 2006), 41; María del Carmen Carlé, "El bosque en la Edad Media (Asturias-Leon-Castilla)," *Cuadernos de historia de España* (Buenos Aires) 59–60 (1976), 308–319.

39 The literature on medieval and early modern European forest history is vast. Some major works in the field include W.G. Hoskins, *Making of the English Landscape* (London: Hodder and Stoughton, 1955); Oliver Rackham, *Trees and Woodland in the British Landscape* (London: J.M. Dent, 1976); Charles R. Young, *The Royal Forests of Medieval England* (Philadelphia: University of Pennsylvania Press, 1979); Jean R. Birrell, "The Medieval English Forest," *Journal of Forest History* 24, no. 2 (Apr., 1980), 78–85; N.D.G James, *A History of English Forestry* (Oxford: Blackwell, 1981); Oliver Rackham, *The History of the Countryside* (London: J.M. Dent, 1986); Grenville Astill and Annie Grant, eds., *The Countryside of Medieval England* (Oxford: Basil Blackwell, 1988); John Langton and Graham Jones, *Forests and Chases of England and Wales c.1500–c.1850: Towards a*

mills, and other public structures. Woodlands provided much more than wood, of course, as people extracted medicinal herbs, fruits, nuts, cork, meat, and fish. Beekeepers maintained their bees in the forest. People made soap from wood ashes, and made cordage, mats and paper from grasses. People extracted timber for industrial needs as well, including the tanners who used bark to tan leather, and the iron founders, miners, potters, brick makers, glass blowers, and other artisans who needed charcoal for heating fuel. Acorns from oak trees fed pigs, while leaves and branches from beech and elm trees fed sheep and cattle. Herders used the forest to shelter their animals in the winter or during times of harsh weather. Boars, bears, deer and other game were hunted in forests, including royal and noble reserves. Some forest communities depended on profits from selling their timber to large urban centers or regions with little forest cover.[40]

Almost any of these activities could cause irreparable damage to forests if not done in moderation. Grazing, for example, had great potential to damage saplings and prevent the natural regeneration of forests.[41] Because all of these potential uses of woodlands could not be exploited equally by everyone at once in the same forest, the various interests involved meant that a complicated set of relationships developed among people over the use of and access

survey and analysis (Oxford: St. John's College, 2005); John M. Gilbert, *Hunting and Hunting Reserves in Medieval Scotland* (Edinburgh: J. Donald, 1979); T.C. Smout, *People and Woods in Scotland: A History* (Edinburgh: Edinburgh University Press, 2003); Andrée Corvol, Pierre Chaunu, and Roland Mousnier, *L'homme et l'arbe sous l'Ancien Régime* (Paris: Economics, 1984); Andrée Corvol, *L'homme aux bois: histoire des relations de l'homme et de la forêt, XVIIe–XXe siècle* (Paris: Fayard, 1987); Roland Bechmann, *Trees and Man: the forest in the Middle Ages* (New York: Paragon House, 1990); Ellen F. Arnold, *Negotiating the Landscape: Environment and Monastic Identity in the Medieval Ardennes* (Philadelphia: University of Pennsylvania Press, 2013); Jaap Buis *Historia Forestis: Nederlandse bosgeschiedenis,* 2 vols. (Utrecht: Landbouwhogeschool-Wageningen, 1985); María del Carmen Carlé, "El bosque en la Edad Media (Asturias-Leon-Castilla)," *Cuadernos de historia de España* (Buenos Aires) 59–60 (1976), 297–374; Julián Clemente Ramos, coord., *El medio natural en la España medieval: Actas del I congreso sobre ecohistoria e historia medieval,* [Celebrado en Cáceres entre el 29 noviembre y el 1 diciembre, 2000] (Cáceres: Universidad de Extremadura, Servicio de Publicaciones, 2001); Erich Bauer Manderscheid, *Los Montes de España en la Historia* (Madrid: Fundación Conde de Valle de Salazar, 2003; First ed., 1980).

40 Jesús Bravo Lozano, *Montes para Madrid. El abastecimiento de carbón vegetal a la villa y corte entre los siglos XVII y XVIII* (Madrid, 1993).

41 Warde, *Fear,* 35; Carlé, 360. Such damage meant the forest could not regenerate the same way, because it would adopt the "low-forest" structure as a consequence, which is less valuable ecologically, mainly because of regeneration restrictions.

to forest resources. Gerhard Jaritz and Verena Winiwarter have argued how early modern perceptions of nature were not uniform, and how a single landscape like a forest was not viewed as a single, homogenous entity.[42] How one perceived a forest or understood its value depended on one's daily confrontation with nature. Depending on one's place in society, nature posed different problems, and the patterns of problem-solving reflected the economic necessities within a given community.[43] People required specific forms of wood to meet specific needs. Smaller diameter wood, for example, was better for cutting firewood. As a result, the cutting back of the trunk or branches to allow for new growth and the harvesting of new rods, known as coppicing, was widely practiced since the Middle Ages and probably much earlier in certain regions. In the seventh century, Isidore of Seville called this "'Second growth' (*recidivum*)," which "consists of what sprouts again after other parts have been cut."[44] Such methods allowed people to extract resources from the forests without causing serious damage, usually performed and managed in a communitarian system.

Naval ministers, by contrast, were interested in larger, more mature trees, which could serve as the larger pieces of their vessels, such as the keel or a mast. In England, such trees could sometimes be found in a system of forestry known as coppice with standards, where a scattering of mature trees, or standards, provided the local populace with seeds, shade, mast, and their own construction needs.[45] Trees left to grow might form odd shapes over time, which were also highly valued by shipbuilders who needed curved or bent pieces known as compass timber. However, on the whole, there was little incentive for the populace to preserve timber trees suitable for shipbuilding. As Warde has pointed out, the types of wood needed most by the general population were young poles and rods used directly or for making charcoal. Peasants and artisans had a strong interest in making sure that timber supplies for their needs, as well as stall litter for their animals, remained accessible.[46]

---

42 Gerhard Jaritz and Verena Winiwarter, "On the Perception of Nature in a Renaissance Society," in *Nature and Society in Historical Context*, ed. Mikuláš Teich, Roy Porter and Bo Gustafsson (Cambridge: Cambridge University Press, 1997), 91.

43 Jaritz and Winiwarter, 93–94.

44 Stephen A. Barney, W.J. Lewis, J.A. Beach, Oliver Berghoff, Muriel Hill, *The Etymologies of Isidore of Seville* (Cambridge: Cambridge University Press, 2006), 341. "Some say that recidivum is from 'falling' (cadere), because they make new growth after being felled. Others have said that it is from 'cutting back' (recidere) and sprouting again. Therefore, something may be recidivus (i.e. 'restored') where there has been death or falling."

45 Warde, "Fear," 36.

46 Warde, "Fear," 41.

The development of more state oversight in European forests emerged in the fifteenth and sixteenth centuries from a combination of widespread fears of shortages, increased activities and responsibilities by centralized governments, and the pressures of population growth and industrial production on local resources. This phenomenon should be placed within the broader context of governments at the time, which were increasingly regulating socio-economic matters for the polity as a whole. Wood was a central economic resource and something increasingly within the purview of state authority. Warde has also pointed out that the prevention of wood shortages became a task for which the state was uniquely suited, and was something that people generally thought ought to be done. No other institution was in a better position to facilitate the flow of goods and resources over the entire land and serve as arbiter over intra-municipal disputes. For example, since the Middle Ages, Castile's representative assembly, the Cortes, described to the crown the abuses that occurred in the forests and also asked the government to step in to stop the damages. In 1576 the Cortes met in Madrid and presented the following statement to the king:

> The conservation of forests, in all times, was very necessary; but much more in this [time] for they've been poorly guarded, they are in notable decline, and, although the Council anticipated remedies that have seemed possible, the malice of the poachers have found ways to defraud what is intended. To remedy this, the cities and towns of these kingdoms have made many ordinances, a motion has been brought to the Council to be confirmed, and many have spoken against and hindered the malice. We beseech your Majesty to order all ordinances that are in Council touching upon the conservation of forests be sent very soon, in preference to the other things that will be affected, due to how much this matters.[47]

In addition to matters of the forest, the state simultaneously became more deeply involved in the regulation of all kinds of other resources, including foodstuffs, livestock, money markets, and much else.[48]

While there was very little fundamental resistance to the idea that the state could address the problem of timber scarcity, people at all levels of government expressed very different ideas about how to achieve results.[49] Early

47 Bauer, *Montes*, 62–63.

48 Warde, *Fear*, 28–57, especially 42; Warde, *Ecology*, 23, 166–167.

49 Warde, *Ecology*, 198–199, 353.

methods were rather indirect, including legislation aimed to curb destructive behavior in the forest, such as starting fires and letting immoderate numbers of livestock pass through.[50] Even if such behavior was customary, the state could enact punishments or legislative changes by arguing that it better served the common good. This appeal to the common good frequently reappeared. In a typical sentiment, from an important forest instruction written in 1650 by a Spanish forest superintendent, Toribio Pérez de Bustamante, it was important to conserve the forests because, "if those who came before us did not conserve them, they could not be enjoyed by us today. And if we do not conserve them, they will not be available to future generations."[51]

A diverse collection of scholarship indicates that persistent wood shortages in the sixteenth and seventeenth centuries increased the activities of personnel in major states throughout the early modern world, although it is difficult to generalize about the relationship between state power and forestry on such a scale. However, there was a trend beginning in the the sixteenth century toward greater centralization of state forestry practices, which meant that policies were set from the center to achieve primarily the goals of state politics. This phenomenon is defined here for Spain as state forestry, but could be seen as well in many other centralizing state governments where the aim was to secure timber for the navy or for other matters of state defense. The Ottoman Empire, for example, set aside the best tracts of timber, called *miri koru*, for provisioning the Imperial Arsenal in Istanbul and installed guards (*kurucus*), who were sometimes janissaries, to enforce state mandated restrictions against unauthorized cuttings, grazing or charcoal-making.[52] The diversity of early modern states carried over into a diversity of approaches to accessing timber and regulating forest use. In Sa'did Morocco, Sultan Muhammad al-Sheikh sought expansion of wartime and merchant fleets for defense against Europeans, Ottomans, and Maghrebi rivals, as well as protection of internal economic interests, requiring spruce, cedar, and pine trees from forests around the Mediterranean outpost of Badis to help with naval construction after 1547. Within thirty years, the Moroccan fleet matched the corsair fleet of Algiers, but timber supplies were low and the Moroccans engaged in long distance trade

---

50 Appuhn, *Forest on the Sea*, 115; Warde, *Ecology*, 175; Carlé, 374; Vassberg, 68; Radkau, 136–142.; María Valbuena-Carabaña, Unai López de Heredia, Pablo Fuentes-Utrilla, Inés González-Doncel, and Luis Gil, "Historical and recent changes in the Spanish forests: A socio-economic process," *Review of Paleobotany and Palynology* 162 (2010), 492–506.

51 AGS, Section Guerra Antigua (Hereafter, GA), leg. 3309, Santander, February 15, 1650.

52 Sam White, *The Climate of Rebellion in the Early Modern Ottoman Empire* (New York: Cambridge University Press, 2011), 29.

with Spain's northern European rivals. In 1581, for instance, Sultan Ahmad al-Mansur refused to export any saltpeter for English weaponry until Queen Elizabeth agreed to send them shipbuilding timber.[53] In precolonial India, sultans declared their exclusive rights to the best hardwood, including teak for its use in shipbuilding, and even established teak plantations.[54] In China, while the Qing dynasty (1644–1911) exploited forests heavily in almost all other frontier zones within its spacious territory, they instituted protectionist policies in their homeland of Manchuria.[55] Leaders in Tokugawa Japan (1603–1868) began responding to widespread deforestation by 1670, instituting measures that would lead to remarkably successful conservation practices. Forest magistrates (*hayashi bugyō*) performed their duties through the finance office, rather than through the navy or military, and their staff carried out surveys, record-keeping, and supervision of logging projects.[56] Based on such information, the magistrates would implement policy either through district administrators or directly through its staff members. French administrators under Louis XIV (r.1643–1715) and Louis XV (r.1715–1774) made woodland management a

53 Dahiru Yahya, *Morocco in the Sixteenth Century: Problems and Patterns in African Foreign Policy* (Atlantic Highlands, NJ: Humanities Press, 1981): 131, 186; For more on Morocco, see Vincent Cornell, "Socioeconomic dimensions of *reconquista* and *jihad* in Morocco: Portuguese Dukkala and the Sa'id Sus, 1450–1557," *International Journal of Middle East Studies* 22 (1990): 379–418; J.R. McNeill, *The Mountains of the Mediterranean World: An Environmental History* (Cambridge: Cambridge University Press, 1992); Weston F. Cook, *The Hundred Years War for Morocco: Gunpowder and the Military Revolution in the Early Modern Muslim World* (Boulder, CO: Westview Press, 1994); Richard L. Smith, *Ahmad al-Mansur: Islamic Visionary* (New York: Pearson Longman, 2006).

54 Williams, *Deforesting the Earth*, 323–324. On a wider range of issues connecting South, Southeast and East Asia, see Richard Grove, Vinita Damodaran, and Satpal Sangwan, eds. *Nature and the Orient: The Environmental History of South and Southeast Asia* (Delhi: Oxford University Press, 1998); Greg Bankoff and Peter Boomgaard, eds. *A History of Natural Resources in Asia: The Wealth of Nature* (Basingstoke, UK: Palgrave Macmillan, 2007).

55 Richards, *Unending Frontier*, 134–143; Also on China, see Mark Elvin, *The Retreat of the Elephants: An Environmental History of China* (New Haven: Yale University Press, 2004); Mark Elvin and Liu Tsui-jung, eds. *Sediments of Time: Environment and Society in Chinese History* (Cambridge: Cambridge University Press, 1998); Patrick Caffrey, "The Forests of Northeast China, 1600–1953: Environment, Politics, and Society" (PhD diss., Georgetown University, 2002); Robert Marks, *Tigers, Rice, Silk, and Silt: Environment and Economy in Late Imperial South China* (Cambridge: Cambridge University Press, 1997); Robert Marks, *China: Its Environment and History* (Lanham, MD: Rowman and Littlefield, 2012).

56 Conrad Totman, *The Green Archipelago: Forestry in Pre-Industrial Japan* (Berkeley: University of California Press, 1989), 91.

priority. In 1669, France's finance minister Jean-Baptiste Colbert issued the *Ordonnance des eaux et fôrets*, which set the standard for French forestry for decades and greatly influenced European forestry in general. Royal officers of Waterways and Forests were charged with monitoring the exploitation of private, church, communal, and royal woodlands. In the eighteenth century, the officers' duties were expanded, and they set out to establish management systems to increase future productivity. Agents of the navy carried out their own inspections of forests for timber suitable for shipbuilding, beginning nearly a century after the first Spanish forest superintendent. The French entered the Baltic naval stores trade as a short term solution in the 1660s while naval inspectors carried out their visitations. For the next century, though, the French rarely faced depletion of sources.[57] Only after 1766 did French Atlantic shipyards increasingly turn to external markets in Italy, Corsica, Catalonia, Albania, and the Baltic and Black Sea regions for timber again.[58] English forests had been greatly reduced by the mid-seventeenth century due to enclosures, charcoal burning, the growth of the merchant fleet, and industrial construction.[59] Naval timber provisioning operated mostly through the open market, and there was no equivalent to France's royal ordinances to regulate the oak trade. Parliament did at times respond to fears of timber shortages, including acts from 1543 and 1570, which strengthened provisions to conserve timber.[60] The establishment of the Navy Commission in 1618 was meant to provide greater defense of the merchant marine, as did the Navigation Ordinances of 1650 and 1651. Naval power expanded rapidly between 1649 and 1651, aided by prizes seized from the Dutch during wartime. The merchant marine also expanded from 115,000 tons in 1629 to 340,000 tons in 1686. Dramatic expansion of shipbuilding in Restoration England contributed to the Royal Society's search for advice from John Evelyn on how to relieve the timber shortage in England. In the years after Evelyn delivered *Sylva, or A Discourse of*

57 Graham, "Royal Navy," 139; Özveren, "Shipbuilding," 61; Gaspar de Aranda y Antón, "La construcción naval en Francia durante los siglos XVII y XVIII," *Revista de Historia Naval* no. 49 (1995): 66, 68–71.

58 Özveren, "Shipbuilding," 62. Mediterranean ports were importing oaks earlier, around the start of the eighteenth century. See Per Eliasson and Sven G. Nilsson, "'You Should Hate Young Oaks and Young Noblemen': The Environmental History of Oaks in Eighteenth-and Nineteenth-Century Sweden," *Environmental History* 7, no. 4 (Oct. 2002), 671, and Paul W. Bamford, *Forests and French Sea Power, 1660–1789* (Toronto: University of Toronto Press, 1956), 89–92, 95–112.

59 Eliasson and Nilsson, 671.

60 Robert Albion, *Forests and Sea Power: The Timber Problem of the Royal Navy, 1652–1862* (Cambridge, MA: Harvard University Press, 1926), 123.

Forest Trees and the Propagation of Timber in His Majesty's Dominions in 1662 reforestation efforts noticeably increased. During this time, though, the share of foreign-built ships also increased from one-third to one-half between 1654 and 1675.[61]

Many preventive measures initiated for the most part by highly militarized states suggest the physical destruction of forests in the medieval and early modern eras was remarkable. At least in Western Europe, though, there is good reason to remain suspicious of claims for a general scarcity of wood before the late eighteenth century.[62] Local shortages did occur, but since the timing and geographic scope of so much forest legislation across Europe occurred when there apparently was no general scarcity, historians must look to factors other than responses to local ecological pressures for causing and motivating their promulgation.[63] European forest historians assert, then, that forestry legislation was more a product of political rather than ecological concerns. In Western Europe, the politics of state were driven increasingly by the funding, supplying, and fighting of war, so the legislation of state forestry overwhelmingly aimed to meet these demands.[64] Perhaps it could be argued that even on a global scale in the sixteenth century, growing involvement of state authorities began altering forest management techniques, redefining the problem of resource allocation by prioritizing state interests. In Europe's maritime empires, including Spain, naval interests were behind the growing amount of forest legislation.

Due to the attention given to shipbuilding timber in forest legislation, the amount of wood consumed by Europe's naval powers has been assumed to be comparatively greater than other sectors of the economy. However, compared to the amount of wood consumed for domestic fuel, the amount of timber used in shipbuilding was quite small, about one percent of European domestic fuel demand in the late eighteenth century.[65] The gravity of the success in getting and maintaining that one per cent is what led to greater state intervention, not the quantity itself. Shipbuilding demand caused considerably less forest

61 Özveren, "Shipbuilding," 37.

62 Warde, *Fear*, 39; also see Williams, *Deforesting the Earth*, 151–154; and David Bruce Young, "Forests, Mines, and Fuel: The Question of Wood and Coal in Eighteenth-Century France," *Proceedings of the Western Society for French History* v. 3 (Oct., 1975), 328–336.

63 Warde, *Fear*, 42.

64 Andrée Corvol, Pierre Chaunu, and Roland Mousnier, *L'homme et l'arbe sous l'Ancien Régime* (Paris: Economica, 1984); Bamford, *Forests and French Sea Power*; Bauer, *Montes*; Albion, *Forests and Sea Power*.

65 Warde, *Fear*, 40–41.

cover loss than domestic fuel and other industrial needs, but was the principal motivation behind the most elaborate forest legislation in maritime-oriented states, including Sweden in 1558, France in 1669, and Spain in 1748.[66] These and most other forest laws claimed their purpose was to prevent imminent scarcity using language similar to Pérez de Bustamante's 1650 Instruction. Jason W. Moore notes that the timber requirements for shipbuilding were "quantitatively modest but qualitatively demanding."[67] The qualitative demands imposed by the crown were far-ranging in early modern society, requiring alterations in legal and administrative structures, property regimes, and forest economies. The study of state forestry in Spain, then, leads us towards understanding broader changes in the operation of the global monarchy.

Spanish Forest Landscapes from Prehistory to the Reconquest

Before the demands of crown-directed naval forestry were exacted in the sixteenth century, Iberian peoples had already carried out extensive deforestation, devised ways to regulate disputes over the use of forest lands, and even implemented attempts to conserve forest resources. The forests of Habsburg Spain were products of a long history of numerous evolving property systems, successive political regimes, and variations in climate, rainfall and terrain. Spain's geography did not determine the course of its people's history, but it did condition their interactions and regional differences.[68] Landscapes and cultures have coevolved throughout the Holocene epoch, or about the past 11,500 years.[69] Humans have been the primary cause of change to the vegetation in Spain for at least 4,500 years.[70] The evolution of Spain's forest landscapes is not a familiar subject to an English-speaking audience nor to an early modernist audience, yet this background conditioned the crown's strategies for accessing and managing forests suitable for naval shipbuilding. The composition and distribution of Iberian plant life by the sixteenth century was the

66 Warde, *Fear*, 44; Bauer, *Los Montes de España*, 57–62; Eliasson and Nilsson, 661; James, *English Forestry*, 126, 166.

67 Jason W. Moore, "'Amsterdam is Standing on Norway' Part I: The Alchemy of Capital, Empire, and Nature in the Diaspora of Silver, 1545–1648," *Journal of Agrarian Change* 10, no. 1 (Jan. 2010), 55.

68 Phillips and Phillips, *Concise History of Spain*, 1–20.

69 William R. Dickinson, "Changing Times: The Holocene Legacy," *Environmental History* 5, no. 4 (2000), 483–502.

70 Valbuena-Carabaña, et al., 492.

result of both the physical environment and stochastic changes brought about by plant and animal (including human) interactions.[71]

The Iberian Peninsula juts out to the European continent's southwestern limits situated between the Atlantic Ocean and the Mediterranean Sea.[72] The climate is affected by interacting conditions from these bodies of water and from the Sahara Desert. The total number of days of measurable precipitation in a typical year can range dramatically from 140–180 days in the northern regions to fewer than 20 days in the desert of the southeast. While Spain's climate and topography are characterized by extreme variations in temperature, rainfall, and elevation, it can be divided into two climatic regions: the Eurosiberian and the Mediterranean.[73] In the Eurosiberian climate regions of north and northwest Spain, pressure fronts moving from west to east bring high humidity and abundant rainfall.[74] Winters are mild and summers are cool, each moderated by the ocean currents. Some of this precipitation falls as snow, making the higher altitudes inaccessible in the winter. The northern mountain ranges pose a barrier to the Atlantic weather systems, which means the south-facing slopes and land east of Portugal get far less rain, and experience a more semi-arid climate. These Mediterranean climate territories, covering the rest of Iberia, receive abundant sunshine, unpredictable rain in the fall,

71 Valbuena-Carabaña, et al., 492.

72 Much of this section relies on the chapter "The Ecological Context of Spanish Transhumance," in Phillips and Phillips, *Spain's Golden Fleece*, 7–23, as well as works of economy, geography and climatology, including Jaime Vicens Vives, with Jorge Nadal, *An Economic History of Spain*, tr. Frances M. López-Morillas (Princeton: Princeton University Press, 1969); Catherine Delano Smith, *Western Mediterranean Europe: A Historical Geography of Italy, Spain, and Southern France since the Neolithic* (London: Academic Press, 1979); Manuel de Terán, et al., *Geografía de España y Portugal*, v. 1 and 2 (Barcelona: Montaner y Simón, 1952); Joan Vilà Valentí, *La Península ibérica* (Barcelona: Ariel, 1968); Inocencio Font Tullot, *Climatología de España y Portugal* (Madrid: Instituto Nacional de Meteorología, 1983).

73 On recent research regarding Spain's paleoecology, see work by José S. Carrión, including José S. Carrión, Santiago Fernández, Penélope González-Sampériz, Graciela Gil-Romera, Ernestina Badal, Yolanda Carrión-Marco, Lourdes López-Merino, José A. López-Sáez, Elena Fierro, Francesc Burjachs, "Expected trends and surprises in the Lateglacial and Holocene vegetation history of the Iberian Peninsula and Balearic Islands," *Review of Paleobotany and Palynology* 162 (2010), 458–475; and José S. Carrión, ed., *Paleoflora y paleovegetación de la peninsula Ibérica e Islas Baleares* (Murcia: Ministerio de Economía y Competitividad-Universidad de Murcia, 2012).

74 José Miguel González Hernández, "Nature Conservation in Spain," *Thirteenth Annual Cabrillo Festival Historical Seminar* (San Diego, CA: Cabrillo Historical Association, 1984), 1.

winter, and spring, and experience hot dry summers. The mild winters of the coasts become more severe as one moves inland.

There is no doubt that climate influenced Holocene vegetation sequences of the Iberian Peninsula, but the diversity of the landscapes, ecological history, fire activity, and human activity have often generated unexpected vegetation responses, especially in the Mediterranean climate region. The history of Spanish forests begins prior to the written record. Much of what we know about them, then, comes from paleobotanical techniques, such as palynology, anthracology, and dendrochronology.[75] A the time of the Last Glacial Maximum, when temperatures and precipitation were at their lowest about 22,000 years before the present (BP), northern European landscapes were dominated by treeless tundra and prairie steppe lands. The last glacial retreat occurred about 16,000 BP and forests recovered under generally warmer climatic conditions about 10,000 BP, which also raised sea levels 10 km into the northern coastal belt changing much of the coast from shingle beaches to the rocky cliffs that are so characteristic today along the coasts of Asturias. During the Lateglacial (14,700–11,500 BP) and Early Holocene (11,500–8,500 BP), before humans were the dominant agents of landscape change, most studies indicate that the vegetation changed as a direct result of rapid shifts in the physical properties of the ecosystem. Under generally warming conditions, pine (*Pinus*), juniper (*Juniperus*), birch (*Betula*), and then oak (*Quercus*) spread northward from southern European localities and then to higher altitudes. Additionally, pollen sequences indicate that, since the Lateglacial, oaks increased at the expense of steppe lands and pine forests. The expansion of deciduous trees and woody scrub continued to expand in the Early Holocene, and witnessed the expansion of angiosperm trees including hazel (*Corylus*), alder (*Alnus*), ash (*Fraxinus*), elm (*Ulmus*), maple (*Acer*), fir (*Abies*), beech (*Fagus*), and oak (*Quercus*).[76]

---

75 A. Pèlachs Mañosa and J.M. Soriano López, "Las fuentes paleobotánicas y la historia forestal: el ejemplo de los valles de la Coma de Burg y Vallferrera (Pallars Sobirà-Lleida)," *Cuadernos de la Sociedad Española de Ciencias Forestales* no. 16 (2003), 155.

76 Karl W. Butzer, "Paleolithic Adaptations and Settlement in Cantabrian Spain," *Advances in World Archaeology* 5 (1986), 206, 209–212, 223, 242; María J. Iriarte Chiapusso and A. Arrizabalaga Valbuena, "El bosque en el País Vasco prehistórico," *Cuadernos de la Sociedad Española de Ciencias Forestales* no. 16 (2003), 85–90; M. Cristina Penalba, "The History of the Holocene Vegetation in Northern Spain from Pollen Analysis," *Journal of Ecology* 82, no. 4 (Dec., 1994), 815–832; Pèlachs Mañosa and Soriano López, 155–160; Hellmuth Hopfner, "La evolución de los bosques de Castilla la Vieja en tiempos históricos: contribución a la investigación del primitivo paisaje de la España central," tr. M. de Terán, *Estudios Geogáficos* v. 15 (Aug., 1954), 415–416; José Luis Casado Soto, "Historia del bosque," in

The human population before the last glacial retreat (16,000 BP) was primarily dependent on animal foods, like those depicted in the caves of Altamira (Cantabria), as well as other products of the forest, including the nuts of the stone pine (*Pinus pinea*), which were likewise depicted in the caves of Nerja (Málaga). Resources were probably reasonably dependable and concentrated.[77] From the Mid to Late Holocene (after 8500 BP), forest declines and expansions can not be explained solely by climatic change or differences in physical settings, but require the consideration of the role of humans in disturbing the landscape. As ice sheets reached their modern limits and most European forests expanded to their maximum extent around 6000 BP, humans opened up landscapes through fires, overgrazing, and agriculture; spread grasslands; expanded heaths in the Eurosiberian Region and garrigas in the Mediterranean Region; and decreased arboreal diversity.

Neolithic peoples developed early techniques of agriculture and animal husbandry about 6–8,000 years ago. The conversion of forests into farmland occurred first along Spain's central and southern Mediterranean coasts about 6850–5700 BP. Over time, no other activity would be as destructive to forests as the expansion of agriculture. Stockbreeding of goats, sheep, and pigs, who fed on acorns, developed farther west in the central meseta and in the north, while horses and cattle were bred in the valleys of the south and west. Where humans brought their animals and started their fires, changes in vegetation composition also occurred. As a result of local fire disturbances by the Neolithic populations, evergreen oak scrub replaced pine forests. For example, maritime pine (*Pinus pinaster*) woodlands expanded in the eastern mountains during the early Holocene, but were replaced by evergreen oak species and Mediterranean maquis (shrubland) after an increase in fire rates.[78] In general, the Mediterranean climate region of Iberia exhibits a greater extension of pine woodlands than in the Eurosiberian region. The definitive

---

El bosque en Cantabria, ed. Carlos Aedo (Santander: Universidad de Cantabria, Asamblea Regional de Cantabria, D.L., 1990), 35–37; Valbuena-Carabaña, et al., 492–493.

77 Butzer, 212; E. Badal, "La recolección de piñas durante la prehistoria en la Cueva de Nerja (Málaga)," in V. Villaverde, ed., *De neandertales a cromañones. El inicio del poblamientohumano en las tierras valencianas* (Valencia: Universidad de Valencia, 2001), 101–104. Similar issues are found in Gibraltar: J.S. Carrión, C. Finlayson, S. Fernández, E. Allué, J.A. López-Sáez, P. López-García, G. Gil-Romera, and P. González-Samperiz, "A coastal reservoir of biodiversity for Upper Pleistocene human populations: palaeoecological investigations in Gorham's Cave (Gibraltar) in the context of the Iberian Peninsula," *Quaternary Science Reviews* 27, no. 23–24 (2008), 2118–2135.

78 Valbuena-Carabaña, et al., 493–494.

distribution of tree species in Spain was well established by two or three thousand years ago.[79]

Bronze Age peoples arrived in Iberia about 2200 BP, with the Iberians migrating from North Africa in the south, settling near the Mediterranean, and the Celts from over the western Pyrenees in the north, settling near the Atlantic. By the time the Greeks and Romans arrived, a composite Celt-Iberian culture blended with other peoples of the peninsula, but no unified polity existed. They practiced herding and agriculture, which continued to reshape the vegetation in the northeastern part of the peninsula. Holm oak and Aleppo pine advanced and various deciduous species retreated as grazing and fire clearance rates increased.[80] Celt-Iberians also mined metals, which put pressure on forests to produce charcoal to fuel the smelting process. There were 154 mines operating in pre-Roman times, mostly in the southern provinces of Huelva and Córdoba. During the Roman period, 87 more mines opened in the Sierra Morena.[81] Rio Tinto mines produced 20 million tons of slag in the ancient period.[82] Metals such as gold, copper, and silver entered Mediterranean and northwest European trade networks and attracted the major commercial powers of the ancient Mediterranean to Iberia, beginning with the Phoenicians around 800 BCE. Eastern merchants coming from more arid climates were also drawn to Iberia for its timber.[83]

Ancient writers added new kinds of sources to our understanding of Spanish forests, particularly the Greek writer Theophrastus (c.371–c.287 BC), the Greek geographer Strabo (63 BCE–24 CE), the Roman-Spanish agriculturalist Columella (4–70 CE), and the Roman writer Pliny the Elder (23–79 CE). The Phoenicians and Greeks settled new colonies along the coasts, where they cultivated grains, olive trees, and grape vines, creating landscapes to support

79 Casado Soto, "Historia del bosque," 36–37.

80 J. Gordi Serrat and J. Vilaginés Segura, "La evolución del paisaje forestal de la Llanura del Vallés (Barcelona) hasta el Siglo XII," *Cuadernos de la Sociedad Española de Ciencias Forestales* no. 16 (2003), 169.

81 C. Domergue, *Les mines de la Péninsule Ibérique dans l'antiquité romaine* (Rome: Ecole Française de Rome, 1990), carte 2.

82 J.M. Luzón, "Instrumentos mineros de la España Antigua," in *La minería Hispana e Iberoamericana. Contribución a su investigación histórica*. Vol. 1 (León: Cátedra de San Isidoro, 1970), 232.

83 Phillips and Phillips, *Concise History of Spain*, 13–16. For the prehistoric era, also see María Cruz Fernández Castro, *Iberia in Prehistory* (Oxford: Blackwell, 1995); Ann Neville, *Mountains of Silver and Rivers of Gold: The Phoenicians in Iberia* (Oxford: Oxbow Books, 2007); Richard Harrison, *Spain at the Dawn of History: Iberians, Phoenicians, and Greeks* (New York: Thames and Hudson, 1988).

the classic Mediterranean triad. The Romans expanded cultivation, and the olives and grapes were well-suited for the frost-free climate of southern Spain.[84] Atinian elm (*Ulmus procera*), tolerant of summer drought, and brought from Italy, was used to support and train grape vines, a practice recommended by Columella in the first century.[85] Elsewhere, much of Spain was well suited to grazing, perhaps better than to any other economic use due to the extremes in topography and climate. Domesticated sheep in Spain already produced high quality wool by the first century. Strabo, Pliny the Elder, Martialis (d.260), and Columella all praised the high quality of Iberian wool. The two mesetas, Spain's central plateaus surrounded by high mountain ranges, promoted a mobile pastoralism, or transhumance, which brought herds to exploit pastures with the changing seasons. However, it is probable that herding migrations in ancient times did not involve the long distances characteristic of the medieval and early modern eras.[86]

Beyond the agricultural and pastoral landscapes, classical writers commented on the abundance of forests in Iberia. Strabo remarked that much of Spain was so forested and mountainous that it was inhospitable to settlers. He also said that people of the north spent three quarters of the year making an acorn-based bread.[87] Along the Mediterranean coast, according to the poet Mosco in the second century BC, pinewoods dominated the landscape.[88] Strabo and Pliny the Elder mentioned the abundance of pines on the island of Ibiza, a name derived from the tree itself.[89] Roman silvicultural works by Cato, Columella, Marcus Terentius Varro, Pliny the Elder, and others spread throughout the empire, and intended to rationalize the use of forest resources.[90] Avieno, an Italian writer and translator from the fourth century, remarked on great forests along the Guadiana and Guadalquivir rivers, in the sierras between Málaga and Granada, and in the Pyrenees mountains.[91]

84 Mount Testaccio in Rome was a dumping site for more than 26 million oil amphoras, of which 80% had Iberian origin. Amphora production demanded large amounts of firewood and contributed to deforestation.

85 Luis Gil, Pablo Fuentes-Utrilla, Álvaro Soto, M. Teresa Cervera, Carmen Collada, "English Elm is a 2,000-year-old Roman Clone," *Nature* 431 (Oct. 2004), 1053. Valbuena-Carabaña, et al., 494–495.

86 Phillips and Phillips, *Golden Fleece*, 9, 19, 24.

87 Casado Soto, "Historia del Bosque," 37–38; José María Rábade Blanco, ed., *Atlas Forestal de España* (Madrid: Grupo Tragsa, 2002), 93; Valbuena-Carabaña, et al., 494.

88 Valbuena-Carabaña, et al., 493.

89 Valbuena-Carabaña, et al., 493–494. The Phoenicians called the island Ebussus and the Greeks called it Pytiusas, both meaning "pine island."

90 Casado Soto, "Historia del Bosque," 39.

91 Rábade Blanco, 92.

Ancient writers also remarked on the loss of forest cover in many places, particularly surrounding major colonies. Pliny the Elder indicated that the consumption of forest resources was generally moving westward from the Apennines, to the Alps, to Corsica, and then to the Pyrenees. Palynological studies close to important Iberian settlements, such as the Sinus Tartessii (Bay of Cádiz), confirm that littoral pinewoods disappeared quickly. As the mining process required the use of charcoal, forests even at great distances from mining centers were affected. Pomponius Mela and Pliny the Elder enumerated and praised the mineral wealth of Spain, but Strabo pointed out a lack of vegetation around mining sites by the first century. Shipbuilding, too, had an effect on coastal forests. Phoenicians, Greeks, and Romans built some of their commercial and naval fleets from Iberian wood, pitch, and esparto grass. Theophrastus wrote that they used conifer timber for planking and masts and oak or beech for oars and keels. Julius Caesar sent a delegation from his army to the forests of Spain to provision his ships with wood, esparto, and pitch.[92] Strabo, Pliny the Elder, Avieno, and later, Plutarch remarked on evidence of the disappearance of coastal pines due to shipbuilding and tar production. Only the forests in the inaccessible mountain ranges and far from the hinterlands of ancient settlements were left unaffected.[93]

Ancient peoples in Spain, therefore, established the agricultural staples of the Mediterranean, expanded the herding economy, and began a process of deforestation near settled areas. The Visigothic era in Spain (418–711 CE) was characterized by much continuity from the Roman era. In the fourth century, Germanic groups facing invasions by nomadic Huns from Central Asia sought to enter the Roman Empire. The Visigoths, who were made up of several Gothic groups as well as some Huns and Alans, would eventually sack Rome in 410, and move west to control Spain and parts of Gaul. The economy remained predominantly pastoral and agrarian, although trade and urbanization apparently declined easing pressure on Iberian forests.[94] Under the Romans, the continued sale of the *ager publicus* and rather rapid privatization in certain places did not eliminate the importance of *compascua* (common pasture) property, or the right to utilize the commons. The Visigoths, though, consolidated the use of the commons and installed Germanic customary law.[95]

The complexity and confusion of customary law induced the Visigoths to generate new law codes in the sixth and seventh centuries. The *Liber*

92 Rábade Blanco, 93.

93 Valbuena-Carabaña, et al., 494.

94 Phillips and Phillips, *History of Spain*, 31–33; Valbuena-Carabaña, et al., 495.

95 Casado Soto, "Historia del Bosque," 39.

iudiciorum (649–672) spelled out various administrative regulations and civil legislation, and is perhaps the earliest legislation on woodlands in Spain.[96] It would have a long-term impact on the laws of succeeding Catholic kingdoms, and has come to be known by its thirteenth-century Romance title, the *Fuero Juzgo*. It specified severe penalties for those who damaged woodlands, particularly stealing wood or starting forest fires, which indicates a strong need to preserve those resources. In one example, drawn from Book Eight, "Acts of Violence and Injuries," Title II, "Concerning Arson and Incendiaries," Law II, "Where forests are set on fire," it states,

> Anyone who burns a grove belonging to another or any pine or fig tree, or any other tree…shall be arrested by order of the judge; shall receive a hundred lashes; and shall render pecuniary satisfaction for the injury done;…If a slave should commit this offence without the knowledge of his master, he shall receive a hundred and fifty lashes. Where the master is unwilling to render satisfaction for the act of his slave, he shall surrender him, in full amends for the same, even if the loss occasioned by said slave should amount to double or triple his value.[97]

The importance of the livestock economy in Visigothic times had an impact on some Spanish forests. For example, people facilitated the spread of the holm oak (*Quercus ilex*) over much of the landscape by selecting the trees with the best and sweetest acorns for themselves as well as for their pigs, a process called "forest fruitalization." Three laws within the *Liber iudiciorum* regulated the procedure of herding pigs to forage under the oak trees at the time the acorns ripened. Other laws showed preferences for certain tree species by fining different amounts: "Where anyone…cuts down a tree belonging to another; if it is a fruit tree, he shall pay three *solidi*; if it is an olive, five *solidi*; if it is an oak of large size, two *solidi*."[98]

The greatest scholar of Visigothic Spain was Archbishop Isidore of Seville (560–636) whose *Etymologies* transmitted ancient knowledge about agriculture, arboriculture, viticulture, and much more. For example, Isidore provided an early description of the practice of coppicing, or the cutting back of the trunk or branches to allow for new growth and the harvesting of younger

---

96 Casado Soto, "Historia del Bosque," 40–41; Valbuena-Carabaña, et al., 495; Rábade Blanco, 94–95.

97 Valbuena-Carabaña, et al., 495, from S.P. Scott, *The Visigothic Code* (*Forum Judicum*) (Boston: Boston Book Company, 1910).

98 Valbuena-Carabaña, et al., 495. This law can be found in Book VIII, Title III, Law I.

branches. This practice would remain commonplace for centuries. For each encyclopedic entry, Isidore provided the etymological background and all that was known about the subject. Much of his knowledge of natural history was based on the work of Pliny the Elder, but he inserted Christian authorities when he could.[99] The *Etymologies* found its way into libraries throughout medieval Europe, and would continue to spread with the advent of print. Between 1470 and 1529, for example, ten editions were published.[100]

Just as Isidore's work combined pagan and Christian knowledge of the natural world, many of Spain's forests remained home to a blend of Christian and pagan beliefs, legends, and practices. Popular stories, carnival celebrations, and other public festivities in the forests of northern Spain venerated numerous trees such as oaks and yews as symbols of life and wisdom. Similar to other regions of Europe, forests were often seen as places of refuge for divinities or spirits of various kinds.[101] There were many stories of miracles that occurred in the forest including one that tells of a repentant bandit named Muño Oñez who received a revelation from an angel in a mountain valley to go see a hidden image within the hollow of the tallest oak tree in the forest of Valvanera, located in the Rioja region. He took his companion Dominico with him through rough terrain and thick forest to reach the tree, and they saw an image of the Virgin, now called Our Lady of Valvanera. They built an altar at the tree and wooden cabins in the surroundings, and she became the patron saint of La Rioja.[102] She is recognizable in works of art by the oak hollow that typically can be seen behind her. Other stories reinforced the image of the forest as an environment conducive to a religious life, one of solitude and proximity to nature. Forests could also be the shelter of the persecuted or society's delinquents, and home to various kinds of fierce beasts.[103]

The Islamic invasion of 711 spelled the end of Visigothic rule and the beginning of centuries of religious warfare on the Iberian Peninsula. The Middle Ages were also a formative time for forest regulation. The condition of forests and forest management preceding the reign of the first Habsburg king, Carlos I, will be treated in the next chapter. As the crown was interested in different types of trees for different shipbuilding purposes, adequate timber could be found in both the Atlantic and Mediterranean climates. Successive eras of human settlements on the Iberian Peninsula each left distinctive marks on the

99 Phillips and Phillips, *History of Spain*, 40.
100 Bauer, *Montes*, 189.
101 Casado Soto, "Historia del Bosque," 40.
102 Carlé, 304–305.
103 Carlé, 304–308.

landscape, which conditioned the political and economic decisions made by the crown in the early modern period.

Plan of the Work

Spain's historical position as one of Europe's first trans-oceanic empires and the largest and most powerful European state in the sixteenth century warrants a closer look at the ways in which the crown gained access to and utilized its timber resources. Spanish naval power waned in the seventeenth century, especially in relation to the rising powers of the Netherlands, France, and England. Due in large part to this transition, scholars continue to overlook the recovery and reassertion of Spanish naval power in the second half of the eighteenth century. The historical background to that resurgence remains an obscure topic, but has its roots in the promotion of conservation techniques that developed in the sixteenth century and persisted in the seventeenth. This book, then, asserts a stronger sense of continuity in royal territorial authority for the early modern era, spanning Habsburg and Bourbon dynastic rule. It also emphasizes the crucial significance of the forested northern provinces of Spain in the eyes of the crown during an era in which much historical attention towards imperial authority is placed on the central and southern regions surrounding Madrid and Seville, thereby providing a more complete peninsular view than is usual in Spanish history.

Previous studies on the history of early modern Spanish forests have tended to fall into three different categories. In the first category, scholars emphasize changes in forest cover, including causes and rates of deforestation and success or failure of reforestation efforts.[104] Such studies tend to utilize a long time-frame. In the second category, scholars maintain a local scale analysis, in both time and space.[105] These studies contribute detailed evidence about local disputes and management practices, but tend to overlook broader trends. In the

104 Carlos Aedo Pérez, *El Bosque en Cantabria* (Santander: Universidad de Cantabria, 1990); Carlos Tarazona Grasa, *La Guardería Forestal en España: Conservación de nuestros bosques a través de los tiempos* (Barcelona: Lunwerg, 2002).

105 Baudilio Barreiro Mallón, "Masa arbórea y su producto en Asturias durante la Edad Moderna," en J. Cavero y otros, *El medio rural español: Cultura, paisaje y naturaleza* (Salamanca, 1992); J.M. Garayo, "Deforestación del territorio: el hayedo de los montes de la Parzonería General de Encía (siglos XVIII–XX)," *Agricultura y Sociedad*, no. 62 (1992); A. Herrera García, "Labradores, ganadores, y aprovechamientos comunales: Algunos de su conflictividad en las tierras sevillanas durante el Antiguo Régimen," *Agricultura y Sociedad* no. 17, (1980); V. González Muñoz, "Vigo y su comarca en los siglos XVI y XVII," *Vigo en su Historia* (Vigo, 1980), 151–276; C.M. Manuel Valdes and A. Rojo y Alboreca, "Valsaín

third category, scholars connect forest history to periods that were particularly active for the navy. These have included the period from the Invincible Armada (1588) to the end of the reign of Felipe IV (1665) and the period from the reign of King Carlos III (r.1759–1788) through an era of naval revival until the Napoleonic invasion (1808).[106]

As a result of such approaches, Spanish forest history tends to suffer, respectively, from (1) an overemphasis on strictly material change that overlooks the evolving social and political structures shaping forest territoriality; (2) an overemphasis on the local scale, missing the development of state-wide forestry policies; or (3) an overemphasis on the drama of human events that overlooks the long-term continuity of forest functionality. These shortcomings have helped to shape the organization of this book, which connects social and political conditions to the process of material changes in forests. It also connects the formulation of centralized forest policies with their implementation, successful or not, in various local contexts. In order to treat fully the establishment of state forestry in early modern Spain, the book is not overly focused on specific events, such as naval battles, nor on the history of ecosystems on an epochal scale. It focuses on a two-hundred year period beginning with the formation of a new state forestry under the early Habsburgs and ends with the creation and initial implementation of Spain's first national forestry codes of 1748, under Fernando VI (r.1746–1759). Context for this crucial period is provided by discussing developments related to state forestry both before and after the period of primary focus. There is much more that can be said about forests in the Spanish monarchy that falls outside the scope of this work, much of which has already been treated by other scholars.[107] In particular, there are

Forest in the XVIII Century: An Example of Forest Management in the Pre-industrial Era," *Investigación Agraria*, Sistemas y recursos forestales, fuera de serie no. 3, (1994); Álvaro Aragón Ruano, *El Bosque guipuzcoano en la Edad Moderna: Aprovachamiento, ordenamiento legal y conflictividad* (Donostia: Sociedad de Ciencias Aranzadi, 2001); Miguel-Juan Pereda Hernández, "Conservación y repoblación de arbolado en Almansa a mediados del siglo XVI," en *Conflictos sociales y evolución económica en la Edad Moderna* (*Actas del I Congreso de Historia de Castilla-La Mancha*) 1 (Ciudad Real, 1985).

106 David Goodman, *Spanish Naval Power, 1589–1665: Reconstruction and Defeat* (Cambridge: Cambridge University Press, 1997); Luís Urteaga, *La tierra esquilmada: Las ideas sobre la conservación de la naturaleza en la cultura española del siglo XVIII* (Madrid: CSIC, 1987).

107 Some of the best regional studies of historical transformations in Spanish forests emerged from the Third National Forest Inventory, 1997–2007, overseen by Spain's Ministerio de Agricultura, Alimentación y Medio Ambiente. For information on the project and its publications, visit http://www.magrama.gob.es/es/biodiversidad/servicios/banco-datos-naturaleza/informacion-disponible/ifn3.aspx (last accessed July 3, 2014).

studies on struggles over fuel sources for Spain's iron foundries, microhistories of municipalities' reactions to rising state power and crown authority, case studies of crown policy implementation in the colonies, rates of deforestation over time, and Spanish forest history since the late eighteenth century.[108] Material from these sources has been incorporated into the present work in as much as they relate to the history of crown involvement in early modern forest management.

This book does not focus exclusively on Spanish Iberia. To better understand the process of internal territorialization as it unfolded in early modern Spain, it is important to connect it to what was occurring in the colonies. Often, the crown's forest policies in Spain were influenced directly by experiences elsewhere in the monarchy. For example, methods of gathering new geographic information, including questionnaires distributed to towns and villages, were implemented in the Americas first and then reworked for use in Spain. Also, Felipe II, while still a prince, brought ideas for how to map Spain back with him from his travels to the Habsburg Low Countries. Forest policies devised in Spain were frequently meant to be applied in the colonies as well, necessitating a comparative approach considering each context. The rise of certain colonial shipyards, such as Havana, brought about a closer interrelationship with peninsular yards, and regular exchange of materials and personnel occurred between the arsenals of the old and new worlds especially by the eighteenth century. Treatment of colonial forests in this book is not exhaustive, but focuses on the contexts in which the crown sought to act in its own interests, ostensibly by preventing timber shortages.

Initially, state forestry operated in a small fraction of Spanish woodlands. Forest management regimes may undergo a series of transitions over time, at different rates in different places, changing the basic function that forests serve in society. An example of a change in forest functionality includes a transition from an integrated agrarian (agro-silvo-pastoral) system that serves a variety of local interests to a commercial management system that transfers surplus to a wider national or international market. As a result, changes in functionality

---

108 José Alcalá-Zamora y Queipo de Llano, *Historia de una empresa siderúrgica española: Los altos hornos de Liérganes y La Cavada, 1622–1834* (Santander: Centro de Estudios Montañeses, 1974); Liaño and Cordón, "La corona y los pueblos en la explotación de los montes de Cantabria;" Reinaldo Funes Monzote, *From Rainforest to Cane Field in Cuba: An Environmental History since 1492*, tr. Alex Martin (Chapel Hill, NC: University of North Carolina Press, 2008); Iñaki Iriarte-Goñi and María Isabel Ayuda, "Wood and Industrialization: Evidence and Hypotheses from the Case of Spain, 1860–1935," *Ecological Economics*, 65, no. 1 (March 2008), 177–186.

are decisive for understanding forests' biological and economic evolution.[109] The political imposition of state forestry in the early modern period altered the functionality of Spain's coastal woodlands. Local and royal forest legislation from the early Middle Ages through the fifteenth centuries sought ways to defend and conserve forest resources within a mixed-use agro-silvo-pastoral landscape. Chapter 1, "A New State Forestry for the First Global Age," analyzes the earliest phases of a new forest functionality for naval shipbuilding along Spain's northern coasts. Royal oversight in the sixteenth century shifted from settling domestic concerns to meeting the needs of naval shipbuilders by regulating the use of forest spaces and tree species best suited for naval access and use. The closer regulation of forests by and for the crown was the result of increased imperial commitments, such as the defense of Castile's Indies trade, and a persistent concern for wood shortages. The role of the crown began to change during the reign of Carlos I (r.1516–1556), challenging the notion that Carlos was uninterested in long-term naval commitments, as overseas conflicts and population growth began to place a greater strain on timber supplies. During the reign of Felipe II (r.1556–1598), forest legislation alone could not ensure crown access to timber for shipbuilding, so the king took steps to enhance naval power for the long term, and appointed Cristóbal de Barros as Spain's first superintendent of forests and plantations in 1574. Through its bureaucracy, as well as mapping projects and questionnaires, the Habsburg government gathered all the geographic information it could, especially on the forests of primary shipbuilding regions. Such methods were instrumental in Spain's recovery from the disastrous Armada sent to England in 1588. The chapter closes with discussion of new pressures on forests that emerged in the sixteenth century, arguing that the crown had little reason to alter forest management practices at the municipal level before it gave priority to maritime foreign policy concerns.

Chapter 2 examines forest territoriality in the Spanish colonies of the Americas and the Philippines in the sixteenth century. Colonial environments posed challenges to Spanish efforts to recreate familiar landscapes, but Spanish settlers brought with them many of the same forest management practices as those that developed in medieval Europe. The methods and technology of shipbuilding remained the same, and colonial shipbuilders learned quickly

---

109 Antonio Ortega Santos, "Commons and Rural Communities: An Environmental History of Conflicts in [a] Mediterranean Ecosystem." A paper presented at *Common Ground, Converging Gazes: Integrating the Social and the Environmental in History*, International Conference held at l'École des Hautes Études en Sciences Sociales, Paris, September 11–13, 2008, 4.

how to adapt local resources to traditional ways. The key difference between Spanish and colonial forestry remained the use of forced labor to procure materials and the disregard for indigenous forest management practices. As in Spain, the crown supported a range of efforts to gather intelligence about strategic resources. Colonial legislation appeared later than in Spain due to the abundance of New World vegetation, but emerged where localized shortages occurred, typically near important shipyards. Nothing like Spain's network of forest superintendents and their regular inspection of plantations ever developed in the colonies, but naval forest legislation issued in the Indies was restrictive and selective, prioritizing the needs of the navy. Colonial developments after the sixteenth century are discussed in sections within subsequent chapters.

Chapter 3 analyzes the intensification and expansion of operations of the Habsburg state forestry bureaucracy in a period of high naval activity during the first half of the seventeenth century. In 1598, the forestry bureaucracy expanded from one superintendent covering the entire north coast, to one each in Galicia, Asturias and Cuatro Villas (Cantabria),[110] Vizcaya, and Guipúzcoa. Catalonia's forest administration operated slightly differently, but the principality also came under closer observation by the royal bureaucracy in the early seventeenth century. Then, it argues that during a period of naval decline in the second half of the seventeenth century, despite a decrease in demand for shipbuilding timber, the crown remained concerned with shortages and continued to enhance its authority over coastal forests. Spain's forest superintendent offices continued to be occupied by the descendents of the first appointees who were mostly accomplished naval officers. Even though many superintendents were from the regions under their jurisdiction, they were still often viewed as outsiders and as representatives of the crown. As a result, superintendents faced resistance to their attempts to establish new plantations and collect fines. Forest communities and superintendents had to balance their interests to limit social unrest.

Within the context of naval renewal under a new Bourbon dynasty, Chapter 4 treats an unprecedented effort on the part of the crown to gather information about Spanish timber supplies in the years 1737–1739 and extend state forestry beyond northern coastal forests. After the Habsburg king Carlos II (r.1665–1700) died without an heir, Felipe V (r.1700–1746), grandson of Louis XIV of France, ascended to the throne and defended his place as Spain's first Bourbon king during the War of Spanish Succession (1701–1714). Numerous

110 Cuatro Villas referred to the Cantabrian cities of Santander, Castro Urdiales, Laredo, and San Vicente de la Barquera.

naval ministers voiced renewed concerns for timber shortages and the breakdown of forest oversight after the war. By the 1710's, the Bourbon government began taking steps to reform naval administration and recover naval power by improving forest management practices. Naval secretary José Patiño established three territorial naval departments in Spain, in part to organize coastal forest management, and laid plans to create a new arsenal in each department. After Patiño's death, the king established the *Almirantazgo*, or admiralty, and appointed the Marqués de la Ensenada as Secretary. The *Almirantazgo* oversaw the reconnaissance of most of Spain's coastal forests beginning in 1737, which proved to be influential in Ensenada's 1748 forest legislation.

Chapter 5 analyzes Spain's first national forestry codes from 1748 and their initial implementation. This ordinance was part of a series of reforms under the direction of the Marqués de la Ensenada between 1746 and 1754 that aimed to enhance state power. The first piece of forest legislation from 1748 concerned coastal forests in the three naval departments. At the end of the year, the crown issued further legislation that extended naval forestry practices to the interior so that all of Spain's forests came under the rules and regulations of the crown, although the navy itself continued to limit its concern to coastal forests and navigable rivers. In a sense, then, this can be seen as the triumph of Spanish state forestry, but a triumph that was only limited to the immediate interests of naval supply. The influence of the previous two centuries on the principles of the 1748 ordinance is clear in the methods of inspections, planting guidelines, licensing requirements, fining practices, and the motivation to prevent timber shortages in the future. The chapter then presents the first six years of implementing the ordinances, including the resistance to the new laws, ending when Ensenada was removed from power and the laws were suspended temporarily. The ordinances caused some heavy resistance in places, particularly in interior regions completely unfamiliar with naval regulations, but the ordinances led to the planting of over one million trees and helped lay the groundwork for naval expansion in the second half of the eighteenth century under Carlos III.

CHAPTER 1

A New State Forestry for the First Global Age

Crown and Forests, Redefined

In 1496, King Fernando of Aragón (r.1479–1516) and Queen Isabel of Castile (r.1474–1504) issued a decree (*pragmática*) in Burgos on the "Conservation of forests and plantations for the common good of the people."[1] This decree exemplifies key elements of the crown's role in forest regulation for most of the medieval period. Like monarchs before them, Fernando and Isabel had the difficult task of balancing the interests of local municipalities, artisans, and herders who all sought to use the limited resources of the forest, as well as cultivators who extended arable land at the expense of woodlands and pastures. The pragmatic ordered the protection of large trees used by citizens for firewood, tools, and construction. People could cut off branches, but not to the extent that would prevent new growth. The document remarked on the particular value of oak trees, which produced acorns for foraging pigs, and the monarchs aimed to protect forests used by herders who sought shelter and forage for their animals in the winter. Anyone wishing to cut down such valuable and useful trees required special license. As the title of the document suggests, the general purpose was to protect the common interests of the people. The monarchs decreed, "And if some of these cities, towns, and locales think that something else would be more agreeable, send us a report of it to our Council." The monarchs then went on to state that they aimed to understand conditions fully in order to be of service to the needs of the common good. "It is not our intention to impede" what was best for users of the forests. The *Reyes Católicos* (Catholic Monarchs), as Isabel and Fernando were styled, relied on their representatives in major towns, the *corregidores*, and local judges to ensure that such conservation occurred. In a reign concerned primarily with internal order and security after years of civil war, their chief aim in forest matters was to allow different social groups reliable access to valuable forest resources.[2] People needed fuel,

1 *Novissima recopilación de las leyes de España*, tomo III, libro VII, titulo XXIV, ley I. "Conservación de los montes y plantíos para el bien común de los pueblos."

2 On the reigns of Isabel and Fernando, see John Edwards, *The Spain of the Catholic Monarchs, 1474–1520* (Oxford: Blackwell, 2000) and Felipe Fernández-Armesto, *Ferdinand and Isabella* (New York: Dorset Press, 1975).

 | DOI 10.1163/9789004261372_003

building supplies, food, and shelter, all of which existed, generally for little cost, in the forest. If the towns had the initiative to suggest better ways of meeting the needs of their citizens, the monarchs at least expressed their willingness to listen.

A century later, the traditional role of the crown in forest affairs had changed dramatically, the result of new realities that were unimaginable in 1496. A *real cédula* (royal dispatch)[3] from 1597 was drafted by a new figure in the administration, a Superintendent of Forests and Plantations, named Hernando de la Riva Herrera. The king appointed the superintendent, who worked specifically for the crown's interests. The superintendents were the first state officials in Spain responsible for forest control, and their job was to enforce forest legislation and to procure timber for the navy. Riva Herrera and other superintendents initially worked only in Spain's well-forested northern regions. Once there, the superintendents either replaced local municipal officials in overseeing forest stewardship practices, or closely observed their behavior. The *cédula* of 1597 said nothing about the defense of the common good, nor did the king offer an audience with representatives of the town to discuss best practices. Mostly, the document discussed the need to punish those who destroyed trees without license. Rather than the *corregidores* or local judges being the authorities resolving forest conservation conflicts, the *cédula* stated that the ultimate authority lay with the Council of War in Madrid.

In the time between these two royal regulations, the crown had redefined its relationship with forest resources, shifting from its traditional role of defending against municipal common law abuses to a position of increasingly centralized control over timber access. During the first half of the sixteenth century, the crown began treating forests as manageable sites of extraction for the shipbuilding industry to supply naval arsenals, which helped to ensure Spanish strength in transoceanic defense, colonialism, trade, and migration. Changes in forest regulation led to new understandings of the monarchy's geography, the extension of royal authority to more local levels, and the reshaping of crown and local relations. Shipbuilding had long been important to the coastal regions of Spain, but the demand for more and larger vessels, shaped by conditions both internal and external to the Iberian Peninsula, affected the crown's resource management

3 Guillermo Muñoz Goyanes, *Crónica sobre bosques y montes de la Península Hispánica* (Madrid: Fundación Conde del Valle de Salazar, 1983), 107. A *real cédula* was a royal dispatch that, from the reign of Juan II of Castile (r.1406–1454), monarchs issued to apply to public and private authorities regarding matters of justice and government. See Enrique Martínez Ruiz, dir., *Diccionario de historia moderna de España, vol. II: La administración* (Madrid: Ediciones Istmo, 2007), 83.

strategies for its global empire in new ways. The changes in forestry brought about by the crown in this period, however, did not alter the kinds of environmental problems people faced throughout the Middle Ages. Rather, the scale of these problems changed, brought on by Spain's global commitments and the ability of the state to confront them in a more systematic and centralized way.

Other factors besides the discovery of the Indies contributed to the closer control of coastal forests. First, the population growth of the period led to more clearings of forests for arable land, placing a greater generalized pressure on woodlands. A similar pressure on forest resources occurred in the thirteenth century, but the political and economic context then differed greatly from the sixteenth. Population decline in the fourteenth century meant a temporary respite from demographic pressures. Second, war escalated in the Mediterranean during the century as the Ottoman Empire expanded its influence, requiring the need for nearly constant vigilance along Spanish coastlines.[4] Third, in the sixteenth century, wealth from the New World attracted other European powers to harass Spanish shipping, leading the state to devise new ways to defend its trans-oceanic empire. Spain did so by building more heavily-armed vessels to defend its coasts and Atlantic trade routes. The crown's involvement in coastal forest regulation was an outgrowth of efforts to improve and regulate shipbuilding after the discovery of the New World. Transatlantic crossings to and from Spanish America, called the *Carrera de Indias*, required vessels that were larger than the caravels typical of early Portuguese explorations of African coastlines. These smaller reconnaissance vessels were lightweight, narrow, and could maneuver their lateen sails to allowing for travel with or against the wind, but venturing out into the vast expanse of the ocean required larger cargo holds for supplies and, later, for trade goods, as well as larger square-rigged sails. Royal policy of this kind of naval construction became very important in the reign of Isabel and Fernando in the late fifteenth century, and royal patronage increased under Carlos I and Felipe II, almost exclusively in the ports of the Cantabrian coast. The trend for most of the century was to build increasingly larger vessels. Full-rigged caravels sailing off Andalusia could be as large as 150–160 *toneles*, but by the end of the century it was more common to see *naos* and galleons of 300–600 *toneles*, with some as large as 1,200 or even 1,500 *toneles* making the voyage across the Atlantic.[5] The increase in size and number of vessels being built placed a lot of

---

4 For an overview of the period, see John Lynch, *Spain 1516–1598: From Nation State to World Empire* (Oxford: Blackwell, 1991).

5 In Spanish usage, the *tonel* was a measure of a ship's carrying capacity used in the Basque province of Vizcaya. It was often called the *tonel macho* or the *tonel de Vizcaya*, and was about

importance on the condition of forests in northern Spain. The increased activity of the shipyards for the *Carrera* and naval defense in the sixteenth century led to significant changes to royal policy in the coastal forests.

Finally, as Spain's imperial wealth became central to Spanish policy in Europe, and as shipbuilding became central to the defense of imperial wealth, the crown sought more information about the geography of the monarchy. Geographic information regarding natural resources would be used to carry out ever more effective methods of forest conservation, setting in motion a process of territorialization to gain and maintain access to valuable stands of timber. Studying early modern state formation through the lens of this territorialization reveals a process of legal reform and geographic intelligence-gathering meant to enhance state control over forest use. The territorialization of forest resources therefore occurred in a close and complementary relationship with the development of state power in early modern Spain.

Spanish Forest Landscapes from the Reconquest to the Habsburgs

The kinds of forest uses and regulations expressed in the Catholic Monarchs' 1496 decree took form over centuries, emerging from a period of religious warfare that affected the politics, economics, and physical landscapes in nearly every region of the peninsula. In 711, Islamic forces crossed the Strait of Gibraltar from North Africa, pushing the Visigoths out of power and a large number of Christians farther north into Asturias, Cantabria, the Basque Country and the Pyrenees. Islamic states existed in Spain for the next eight centuries, but over time, their territory was reduced by what has became known as the Reconquest, or the process of Christian expansion southward, a "reconquering" of Catholic Iberia.[6] In this time, then, Spain consisted of three varying-sized regions, depending on the stages of the Reconquest, including the Christian north, the frontier zone, and the Islamic south. The forests were affected greatly by the moving frontier, the shifting populations, and the nearly constant warfare of medieval Spain. Forests were altered in some way by the

---

1.2 times larger than the *tonelada*, which was used in Andalucía until 1590 when the crown adopted the *tonel macho* as the standard *tonelada*. See Carla Rahn Phillips, *Six Galleons for the King of Spain: Imperial Defense in the Early Seventeenth Century* (Baltimore: Johns Hopkins University Press, 1986), 228.

6 Catholicism was widespread in late Roman Spain, but the ruling class of the Visigoths remained Arian until King Reccared converted to Catholicism in 586. Phillips and Phillips, *History of Spain*, 36–37. Also see Roger Collins, *Visigothic Spain, 409–711* (Oxford: Blackwell, 2004).

moving frontier of the Reconquest. In the ninth century, the frontier zone had moved south to the Duero River and by the tenth century had moved to the Central Mountains. Some scholars have called the Duero valley at this time a veritable no-man's-land. After fighting Muslims, according to the Chronicle of Albelda (883 CE), Christian soldiers under Alfonso I turned the land along the river east of Toro into a wasteland.[7] Soldiers in the frontier zones would often set fire to forests in order to prevent an enemy ambush. The Spanish word for ambush, *emboscada*, literally means "from the forest." The material demands of war also led to heavy exploitation of the woodlands, as thousands of soldiers needed material to set up camp, build weapons, and provide sustenance. The unsettled nature of the frontier zone did not encourage the establishment of permanent agricultural land, but rather encouraged the continued use of a more mobile form of capital, such as livestock. The herds of pigs, goats, and sheep may have also played a significant role in damaging forests or preventing new forest growth in the frontier. The term transhumance has an uncertain origin, but likely derives from either *trans humus*, "across the ground," or *trasfumo*, "after smoke," which indicates the opening of pasture land by setting fire to the forest.[8] A general expansion of herding accompanied the shifting frontier southward over the ensuing centuries. In the eleventh century, the frontier extended from the Central Mountains to the Tajo River, and in the twelfth and thirteenth centuries, the frontier moved from the Tajo to the Sierra Morena. The great guild of sheepherders, the Mesta, emerged in the thirteenth century as a royal attempt to regulate herding. Settled agriculture would become viable in the wake of the frontier and mobile livestock.

Islamic Spain, known as al-Andalus, was one of the most heavily wooded regions of the *dar al-Islam*, and served as a strategic timber resource base during the period of Islamic dominance in the Mediterranean world. Much of North Africa and the eastern Mediterranean relied on Andalusian timber. Wood was sent down the Ebro, Júcar, Segura, Guadalquivir, and Guadiana rivers.[9] The twelfth-century geographer Muhammad al-Idrisi wrote from Sicily

7 J.M. Sierra Vigil, "Bases de partida para estudios sobre los montes españoles en la Edad Media; problemas a resolver, periodización cronológica," *Cuadernos de la Sociedad Española de Ciencias Forestales* no. 16 (2003), 251; María Valbuena-Carabaña, Unai López de Heredia, Pablo Fuentes-Utrilla, Inés González-Doncel, and Luis Gil, "Historical and recent changes in the Spanish forests: A socio-economic process," *Review of Paleobotany and Palynology* 162 (2010), 496.

8 Valbuena-Carabaña, et al., 496.

9 Emilio de la Cruz Aguilar, "La provincia marítima de Segura de la Sierra," *Boletín del Instituto de Estudios Giennenses* 107 (1981), 52.

that ships were built using timber from the hills around Tortosa near the mouth of the Ebro River and in the mountains of Cuenca at the source of the Júcar River, which would carry the timber down to the Mediterranean Sea.[10] Shipbuilding centers, called atarazanas, existed in numerous places along the Mediterranean coast. The atarazanas of Almería were established by Abd al-Rahman II (821–852), and could be supplied by black or Aleppo pines from interior forests or the forests of the Rif mountains in Morocco.[11] Seville's atarazanas were built shortly after a Norman attack in 849–850, and perhaps benefitted from shipbuilding knowledge from the north. Local oaks and perhaps ash trees from forests near Seville were used in place of the northern Europeans' use of English oak (*Quercus robur*), also called common or pedunculate oak. Pines were delivered from the Sierra de Segura near the source of the Guadalquivir and from North Africa. The pines of the Sierra de Segura ceased to be exploited for shipbuilding timber after 973 and would remain out of the purview of the Spanish navy until 1733.[12] New walls and naval protection in ninth-century Seville enabled the city to defend itself against another Norman attack in 858. The third major atarazanas were built at Tortosa on the Ebro River under Abd al-Rahman III (891–961). Tortosa was a major exporter of ships and timber, using nearby pine forests that would remain crucial for Mediterranean shipbuilding throughout the early modern era. Other important shipbuilding sites in Islamic Spain were located in Denia, Alicante, Málaga, and Algeciras on the mainland, and on the island of Ibiza. The Portuguese cities of Silves and Alcacer do Sal were also significant shipbuilding centers. Pitch, used to caulk the seams on ships' hulls, was produced from pine forests, including those near Huelva and Barcelona.[13]

After the resurgence of the Christian Reconquest in the thirteenth century, timber and ship exports from Islamic Spain decreased, impacting the wider region. Historian Olivia Remie Constable argued that the Christians' control of Iberian timber resources helped tip the balance of naval power in the medieval

---

10 Olivia Remie Constable, *Trade and Traders in Muslim Spain: The Commercial Realignment of the Iberian Peninsula, 900–1500* (Cambridge: Cambridge University Press, 1994), 196–197.

11 Enrique Martínez Ruiz, "Los montes en la cultura islámica. El bosque del Al-Andalus desde el siglo VIII al XIII," in Francisco Marín Pageo, Juan Domingo Santos, and Ana Calzado Carretero, eds., *Los Montes y su Historia: Una Perspectiva Política, Económica y Social* (Huelva: Universidad de Huelva, 1999), 72; Thomas Glick, *Islamic and Christian Spain in the Early Middle Ages*, 2nd ed. (Leiden: Brill, 2005), see Chapter 2.

12 de la Cruz Aguilar, 52, 68.

13 Martínez Ruiz, 72.

Mediterranean world.[14] Northern Europeans did not need to look far for forest resources, and once Spain was in Christian hands authorities were not interested in trading timber to the enemy. Both Islamic and Christian states imposed bans on the sale of goods that their enemies could potentially use against them in war, including timber. The philosopher and statesman, Ibn Rushd ruled in the early twelfth century that Islamic traders leaving Islamic lands should be searched to ensure they were not smuggling arms or other prohibited items. Beginning with the Third Lateran Council in 1179, papal bans had existed on commerce in war materials. Moreover, Jews leaving from Christian ports caught trading in timber in Islamic states would incur major fines, to which several cases of Jews leaving Barcelona for Egypt in the fourteenth century attest.[15] In the Cortes of Valladolid in 1351, King Pedro I (r.1350–1369) outlawed the export of wood, which he felt would only be used to build ships for the monarchy's enemies.[16] Perhaps more crucially, the king added that foreign commerce in timber contributed to deforestation of the kingdom's hills.[17]

Shipbuilding and maritime trade on the Atlantic and Mediterranean coasts contributed to the further exploitation of coastal forests. In the late Middle Ages an Atlantic coast trade involving Spain, France, England, and the Low Countries helped several Spanish towns to prosper, including the Castilian ports of San Vicente de la Barquera, Santander, Laredo, and Castro Urdiales, often referred to collectively as the 'Cuatro Villas', as well as the Basque ports of Bilbao and San Sebastián.[18] The trade consisted of a wide variety of goods. Spanish merchants carried many products of the forest, such as iron, wool, leather, wax, honey, and wood.[19] Mining and iron foundries existed in the region since antiquity, especially in the Basque province of Vizcaya, and fishing had been an important component of the local economy for centuries. The

14 Constable, 197.

15 Constable, 254.

16 Cortes (Corts in Catalonia) were local governing bodies, or parliaments, in Castile, Aragón, Catalonia, and Valencia. Primarily, their function was to consent to the levying of taxes, but they also were able to "present grievances to the king, seek confirmation of their privileges, and to discuss matters relating to the dynastic succession, war and diplomacy, or taxation and legislation." See Simon Barton, *A History of Spain* (New York: Palgrave Macmillan, 2004), 79. Also, see Enrique Martínez Ruiz, dir., *Diccionario de historia moderna de España, vol. II: La administración* (Madrid: Ediciones Istmo, 2007), 127–129.

17 Constable, 237.

18 William D. Phillips, Jr., "Spain's Northern Shipping Industry in the Sixteenth Century," *The Journal of European Economic History*, v. 17 n. 2 (Fall 1988), 269; Phillips and Phillips, *History of Spain*, 85–89.

19 Casado Soto, "Historia del Bosque," 43.

region specialized in shipbuilding, designing vessels suited to merchant interests and the wide range of conditions faced by Spanish merchants in the Atlantic and Mediterranean. Commercial activity on the Cantabrian coast also facilitated the development of a naval force capable of assisting Fernando III (r.1217–1252) during the reconquest of Seville and other cities in southern Spain.[20] The king urgently ordered "a fleet of armed *naos* and galleys as numerous and as quickly as possible."[21] The commission was carried out for much of the thirteenth century in the best equipped shipbuilding centers on the Cantabrian coast, including Pasajes and Guetaria (where Magellan's flagship would later be built) in Guipúzcoa, Portugalete and Pobeña in Vizcaya, and Santander and San Vicente de la Barquera of the Cuatro Villas. Fernando III's successor, Alfonso X (r.1252–1284), known as "el Sabio" (the Wise), rebuilt the arsenal in Seville because he understood its strategic location as a protected port for trade and defense in both the Atlantic and the Mediterranean. Initially, Alfonso placed under the jurisdiction of the arsenal "all the forests of this region [around Seville] that grow trees reserved for shipbuilding material."[22] However, the forests near the arsenal were no longer suitable for shipbuilding, and it became common for the shipbuilders of Seville to import timber from Galicia, in northwestern Spain. By the late fifteenth century, though, shipbuilders in Seville were exploiting the forests of Constantina, in the Sierra Morena.[23] In the fifteenth and sixteenth centuries, the Cantabrian ports of Castro Urdiales, Colindres, Santoña, and Guarnizo, as well as Orio (Guipúzcoa), and El Ferrol (Galicia) would build significant numbers of vessels for commerce and war.[24] In the last centuries of the Middle Ages, the shipyards of Vizcaya became the most important in Castile, aided by its iron foundries and forest resources.

Spanish fortunes in the Mediterranean varied in the later Middle Ages. Catalonia's trading empire peaked in the early fourteenth century, but saw a sharp contraction caused by a combination of exclusion from Atlantic markets of Europe and Africa, Genoese competition, financial insecurity, and demographic crisis linked to the Black Death.[25] Most of Spain recovered rather

20 María del Carmen Carlé, "El bosque en la Edad Media (Asturias-Leon-Castilla)," *Cuadernos de historia de Espana* (Buenos Aires) 59–60 (1976), 323–324.

21 Casado Soto, "Historia del Bosque," 43.

22 Carlé, 324.

23 Archivo General de Simancas (Hereafter, AGS), SELLO, Catálogo XIII v.II (1478 y 1480) 265, February 6, 1478, Cantillana, (1) Folio (Hereafter, fol.), 16 and 2357, January 26, 1480, Toledo, fol. 118.

24 Casado Soto, "Historia del Bosque," 44.

25 Angus MacKay, *Spain in the Middle Ages: from frontier to empire, 1000–1500* (New York: St. Martin's Press, 1977), 165–166.

quickly after the plague, but Catalonia entered a century of commercial readjustment. Valencia prospered in the fifteenth century, aided in part by the Genoese use of the port on their Atlantic runs. Shipbuilding on the Mediterranean continued in several places. The atarazanas standing in Barcelona today were built in the thirteenth century, but an earlier version was built on a site nearby in the eleventh century. The timber for the galleys constructed in Barcelona came from the surrounding hills or from the Pyrenees via the Ebro River and coastal transport from Tortosa. Fifteenth-century documents mention the existence of forest reserves near Constantí for the atarazanas.[26] Hemp for rigging came from the outskirts of Tarragona, Lérida, and Balaguer. Iron came from near Barcelona and Sardinia. Other fifteenth and sixteenth-century Mediterranean shipyards existed in the Catalonian ports of Tarragona, Tortosa, Badalona, Arenys, San Feliú de Guixols, Mataró, Masnoy, Palamós, Vallgorguina and Tordera, as well as in Málaga, Cartagena, Alicante, Valencia, Ibiza and Mahón (Menorca).[27]

The landscapes of Christian and Islamic Spain reflected cultural, technological, and agricultural differences. Islamic Spain was based on a productive and diverse agriculture, aided by sophisticated irrigation techniques.[28] Learned texts on agriculture, agronomy, geography, and medicine made their way through the Islamic world. Well-known authors with connections to Spain in their lives and work included the Andalusian physician and botanist al-Gafiqi (d.1166), the geographer al-Idrisi (1100–1166), and the Sevillian agriculturalist Ibn al-'Awwam (fl. late twelfth century).[29] Spanish Muslims continued to cultivate wheat, grapes, and olives, and introduced new crops from the east including citrus fruits, sugar, saffron, and rice.[30] Animal husbandry and mining continued to flourish under Islamic rule, and trade expanded thanks to strong connections to the rest of the Islamic world. Society was more urban than the north, but still

26 AGS, Registro de sello, fol. 118, actas de 1478 y 1480.

27 Erich Bauer Manderscheid, *Los Montes de España en la Historia* (Madrid: Fundación Conde de Valle de Salazar, 2003; First ed., 1980), 165.

28 Roman irrigation systems provided valuable models in Mediterranean Spain, but Islamic innovations brought new technology and cultivars. See Karl W. Butzer, Juan F. Mateu, Elisabeth K. Butzer and Pavel Kraus, "Irrigation Agrosystems in Eastern Spain: Roman or Islamic Origins?" *Annals of the Association of American Geographers* 75, no. 4 (Dec., 1985), 479–509.

29 Bauer, *Montes*, 192–193.

30 Andrew M. Watson, *Agricultural Innovation in the Early Islamic World: The Diffusion of Crops and Farming Techniques, 700–1100* (Cambridge: Cambridge University Press, 1983), throughout; Phillips and Phillips, *History of Spain*, 61–64.

strongly rural.[31] Greater attention to managing forests combined with the greater consumption of wood in artisan industry and urban centers have led to mixed conclusions about the impact on forests during the Islamic period.

Christian Spain had a sparser population, which was supported by cereals and livestock. The famous long-distance seasonal migrations of sheep herders definitively emerged by the late Middle Ages, when the *cañadas*, or sheepwalks, extended for hundreds of miles across both mesetas. In the sparsely-settled landscape, sheepherding continued to thrive, both aiding and encouraged by foreign demand for Spain's fine Merino wool. The wool trade with Flanders and northern Europe thrived by the mid-fifteenth century. As the population grew and agriculture expanded, conflicts arose more frequently between farmers and shepherds over the trafficking of animals through farms and town commons. Livestock owners who were unaffiliated with the Mesta would attempt to enclose their pastures to prevent the passage of transhumant herds. Open wood pasture land in Spain is called a *dehesa*, which has its origins as *defesa*, or defended pasture, related to these early attempts to close off the land. Royal privileges granted herds the right to graze in these landscapes in time, though, removing such defenses. Herds could also graze in the *baldíos*, common lands where plowing was forbidden. Some grazing herds would enter agricultural fields at certain times of year after the harvest to graze on stubble, a practice known as the *derrota de mieses*. The animals were able to eat without having to travel too far, and the fields would benefit from their manure.[32] The *derrota de mieses* blurred the division between privately owned fields and public grazing areas, demonstrating that in some cases several different parties had access to the same piece of land.

To encourage settlement of newly conquered land during the Reconquest, the crown pursued multiple approaches. Kings granted charters (*cartas pueblas* or *fueros*) to new towns, which included control over the surrounding countryside. The commons included woods and pastures, and were part of the royal domain (*realengo*). The concessions granted by the crown established in detail the parameters for acceptable use of the land, and town councils often collected much needed income from the commons. Kings also granted land in *señorío*, sometime in large amounts, to members of the nobility, the church, and especially the Military Orders, who were seen to be effective at securing newly-acquired territory close to the frontier. This group had a tendency to be

31 Martínez Ruiz, 75.

32 David Vassberg, *Land and Society in Golden Age Castile* (Cambridge: Cambridge University Press, 1984), 13–18; Valbuena-Carabaña, et al., 496; L. Gil Sánchez and M. Torre Antón, eds. *Atlas forestal de Castilla y León*, 2 vols. (Valladolid: Junta de Castilla y León. Consejería de Medio Ambiente, 2007), 273 and 276.

less generous in concessions of land for communal use because they were more directly involved in promoting their own interests.[33] At the beginning, though, the availability of large amounts of uncultivated land promoted the creation of different kinds of communal property. In Andalusia a process of land privatization occurred over time due to pressure from the largest land owners, including members of the upper classes, the Military Orders, and urban oligarchies.[34] By contrast, most land on the north coast remained officially in the royal domain (*realengo*), and communal land was predominant. Even by the late sixteenth century, 90.5% of land in Asturias was part of the *realengo*. There were mostly small farms, unlike the great *latifundia* of Andalusia, and the population was denser than in the south. The difficult terrain of northern Spain did not make for much productive farmland, and in sixteenth-century Cantabria, the Basque Country, and Navarre the percentage of uncultivated land stood between 75–85%; its title and regulation corresponded to the town councils of the communities in the valley. Conflicts arose frequently between farmers, herders, forgers, and colliers over the use of the land, including especially forests and water sources.[35]

By the end of the Middle Ages, then, Spain witnessed unprecedented pressures on forest resources brought about by centuries of warfare during the Reconquest, the rise and expansion of naval forces, the growth and success of transhumant sheepherding and its associated maritime trade, the repopulation of conquered territories, and the general expansion of the population as a whole. Forests would continue to be exploited in similar ways after 1500, but demands of global empire and the assertion of crown control of woodlands for naval shipbuilding added a new dimension and ushered in a new era of Spanish forest history.

Medieval Forest Regulations and Naval Power in the Reign of Carlos I (1516–1556)

The Habsburg succession to the Spanish crown was not straightforward, but through largely fortuitous events and untimely deaths by 1516 Fernando and

33 Carlé, 310–311; Sierra Vigil, 252–253; María-Pilar Laso and Erich Bauer, "La propiedad forestal en España," *Revista de Estudios Agrosociales* n. 49 (1964), 9–11; Antonio Miguel Bernal Rodríguez, "La tierra comunal en Andalucía durante la Edad Moderna," *Studia historica. Historia moderna*, n. 16 (1997): 106–111.

34 Bernal Rodríguez, 101–128.

35 Baudilio Barreiro Mallón, "Montes comunales y vida campesina en las regiones cantábricas," *Studia historica. Historia moderna*, n. 16 (1997): 17–56.

Isabel's grandson Carlos I (r.1516–1556) inherited Castile, its American possessions, Aragón, the Low Countries, and Habsburg lands in Central Europe; and in 1519, he became Charles V, Holy Roman Emperor. As a result, Carlos had to face the collective European crises of religious divisions, war with Islamic powers, and challenges by his European rivals in the Americas. Each crisis would intensify during his reign, posing great strains on the finances and military capabilities of his realms.

Since Carlos I did not create a permanent navy, historians have overlooked the evolution of his approach to forest management for shipbuilding.[36] During his reign, however, naval power became crucial for Spain's foreign policy. It should not be surprising, therefore, that Carlos established important precedents regarding the need for closer management of forests for naval interests. He sought the help of local experts in forest matters, and he made the first royal legislation in Spanish history that called for increasing forests specifically for shipbuilding. In comparison to the subsequent reign of his son, Felipe II, such developments seem minor, but they were significant breaks from the crown's traditional approach to forests.

Carlos I proclaimed his first important forest legislation in May of 1518 in front of an Aragonese Cortes suspicious of their new foreign-born king. Earlier in the year in Valladolid, the Castilian Cortes forced Carlos to promise to respect the laws of Castile, remove foreigners from his service, and learn Castilian.[37] Carlos faced even more challenges in Aragón, and the Cortes would not recognize him as their king until the next January. During his visit in 1518, however, Carlos issued the decree, "Formation of new forest plantations and ordinances to conserve old and new forests," in his and his mother Juana's name,[38] and was not a major break from the Spanish tradition nor other European ordinances of the day.[39] In fact, similar decrees were made

36 From Lynch, *Spain 1516–1598*, 323, "The resources of Charles V had been diverted to his interests in central Europe and elsewhere: in the Mediterranean his policy had been bankrupt of ideas; neglecting permanent naval power, he had used improvised forces in occasional and largely fruitless expeditions against the bases of Islam in North Africa. Lacking the continuity which naval power demands, Felipe II had to begin almost from scratch. He thus started at a disadvantage, and the initiative remained for some years in the hands of the enemy."

37 Lynch, *Spain 1516–1598*, 51.

38 *Novissima recopilación de las leyes de España*, tomo III, libro VII, titulo XXIV, ley II. "Formación de nuevos plantíos de montes y arboledas, y de ordenanzas para conservar los viejos y nuevos." Juana was still alive, and ruled jointly with her son. Due to her mental instability, however, she was kept away from court in Tordesillas.

39 Paul Warde, "Fear of Wood Shortage and the Reality of the Woodland in Europe, c.1450–1850," *History Workshop Journal*, Issue 62 (Autumn 2006), 28–57.

elsewhere in his realms, including the Low Countries in 1517.[40] According to historian Erich Bauer Manderscheid, the legislation in Castile and Aragón had been influenced greatly by the work of Gabriel Alonso de Herrera, who had published his *Agricultura General* in 1513 under the patronage of Cardinal Cisneros. Alonso de Herrera's work would become influential in other parts of Europe as well. Book three of this six-book work dealt with different tree species and silviculture. He expressed concern for the need to replant trees in the forests and gave advice on how to establish successful plantations with sensitivity towards proper environmental conditions for different species.[41]

The 1518 decree indicated that Carlos listened to the assessments of the *procuradores*, who worked as royal representatives in the parliaments of the various peninsular kingdoms and reported to the crown about petitions sent by the towns.[42] The pragmatic begins,

> ...We are informed by the *procuradores* of the kingdom...that in the cities, towns, and places of our kingdoms and lordships forests are damaged and destroyed without new plantings, and that there is great disorder as forests disappear. As a result, there is no shelter for livestock...and a great shortage of firewood.

The decree required that justices accompany knowledgeable people familiar with the area to help choose places most in need of replanting and, echoing the work of Alonso de Herrera, be mindful of the types of species that would be best suited to the ecology.[43] The king sought the direct collaboration of local inhabitants in replanting trees, and he called for the permanent observation of conditions by the courts and *regidores*. The ordinance confirmed the necessity of new plantations, and it continued in the tradition of equal concern for all segments of the population. Repeated confirmation of this legislation, four times in Carlos's reign alone,[44] indicates that deforestation continued, due perhaps to ineffective oversight. Like many other royal decrees regarding forest conservation, the 1518 pragmatic ostensibly came as a response to disappearing woodlands.

---

40 "Placcaet General op't Stuk der Houtvesterye," in Jaap Buis, *Historia Forestis: Nederlandse bosgeschiedenis*, 2 vols. (Utrecht: Landbouwhogeschool-Wageningen, 1985), 16.

41 Bauer, *Montes*, 201–203.

42 Martínez Ruiz, 303–304.

43 The pragmatic used the phrase "según la calidad de la tierra," or "according to the quality of the land."

44 In 1523, 1525, 1534, and 1543.

The 1518 decree had much in common with a long tradition of forest regulation in Spain and Europe. In the defense of the forest from the Middle Ages, regulations were mostly negative in that they stated partial or complete prohibitions concerning specific action by particular groups regarding particular forests or species. Much rarer were positive attempts to reverse damages to the forest. One of the earliest examples of a reforestation policy in all of Europe came from the Monastery of Oña, north of Burgos, in 1276.[45] In southern Spain, conflicts over access to forests would sometimes fall along religious lines. In 1480, for example, the towns of Úbeda and Cazorla sought protection against attacking Muslims who were trying to stop the construction of a road meant to facilitate timber extraction from a disputed forest.[46] Although, as in most of premodern Europe, attempts to protect the forest for future use were primarily concerned with human behavior and the mitigation of the effects of one person's actions on the ability of someone else's ability to provide for themselves.[47] Starting in the thirteenth century, and increasing in the fourteenth, the promulgation of local laws and ordinances related to economic life and municipal finances proliferated. Nearly all of them referenced modes of exploiting forest products.[48] The right of communities to extract timber and other resources for firewood and individual needs was protected by the *cartas pueblas* or *fueros*.[49] The *fueros* of Logroño (La Rioja) in 1095 and Hinestrosa (Burgos) in 1287, typical of local *fueros* elsewhere, stated, "wherever they find water to irrigate orchards or vineyards, or for their mills, or for their gardens, or for other things they might need, they may take them; and wherever they find firewood, and forests, and trees to burn or for building houses, or for all that they might need, they may take them without hindrance."[50] The church owned extensive forested land in Spain, and several monasteries had their own forest ordinances, which they defended as carefully as the municipalities defended theirs. In one example from the sixteenth century, the abbot and monks from the monastery of Nuestra Señora de la Espina, northwest of Valladolid, took various

---

45 Carlé, 372.

46 AGS, SELLO, Catálogo XIII v.III (1480–84), 72, September 13, 1480, Medina del Campo, fol. 242.

47 Paul Warde, "The Environmental History of Pre-Industrial Agriculture in Europe," in *Nature's End: History and the Environment*, ed. Sverker Sörlin and Paul Warde (Basingstoke, UK: Palgrave Macmillan, 2009), 72.

48 J. Soriano Martí, "La documentación Medieval y la sostenibilidad de los aprovechamientos forestales mediterráneos," *Cuadernos de la Sociedad Española de Ciencias Forestales* no. 16 (2003), 74.

49 Carlé, 345.

50 Vassberg, 10–11.

citizens of the towns of Villabrágima and Tordehumos to court for entering the monastery's forests, destroying several trees, and resisting with arms.[51]

All territory in Castile was incorporated into municipalities, which had ramifications for forest management. A municipal lord could be the crown, a nobleman, the church, or a military order, but city and town councils owned and managed the land. Councils made decisions about crop rotations, grazing, who could plow new ground, and how to use municipal assets. Typically, a municipality consisted of a town and its hinterland, which included arable land, grazing land, and common land or *monte*. Townspeople could establish colonies in the municipality to improve their economic prospects, and these new settlements could develop into thriving villages in their own right. However, decisions affecting a new village would still be made by the old town council. Over time, the newer villages might seek greater autonomy. This could be done by purchasing a town charter for themselves from the king, who would also benefit from the payment for the charter. Land ownership could be much more complicated than this, however. Towns and their surroundings typically had a hybrid property structure, allowing the coexistence of private and public property.[52] Crown lands, seigniorial property, private property, municipal commons, and intermunicipal property could exist in a single place. Municipalities or the crown had to balance the multiple interests involved in regulating land use in order to maintain order.

Forest regulations, whether applied to the kingdom as a whole, like in the *Liber iudiciorum*, or applied to the municipal level, like in the *cartas pueblas*, became much more widespread by the fifteenth century. In a study of several municipalities surrounding medieval Castelló, in the Kingdom of Valencia, J. Soriano Martí argues that an authentic concern for the sustainability of forest resources for future use is evident in their *cartas pueblas*, privileges, and ordinances. Like elsewhere in Spain, licenses were required in this region to cut down all or part of a tree, which required a municipal authority to visually analyze and approve the site of the proposed cutting. The one seeking the license had to declare the amount of wood he would take, and whether he intended to use it himself or sell it somewhere else. Generally, it was unlawful

51 AGS, Consejo Real de Castilla: 367, 7, 1568, 56 fols., A-147-21, "Abad y monjes del monasterio de Nuestra Señora de la Espina, contra varios vecinos de Villabrágima, Tordehumos y otros lugares comarcanos, por haber entrado en sus montes a talar, y resistirse con armas;" and 1569, 31 fols., A-147-21, "Información hecha a pedimiento del monasterio de la Espina, sobre las ordenanzas que han de hacer para la guarda y conservación de los montes."

52 Vassberg, 83.

to cut holm oaks of a certain height (ten *palmos* high or higher) down to the stump, but exceptions were granted for special cases, like when massive building projects required large beams. Rules were made against shaking the acorns out of oak trees. Like in the *Fuero Juzgo*, municipalities set the timing of retrieving fallen acorns on the calendar. In La Torre d'En Besora, firewood cut from holm oaks needed to be a diameter larger than a man's arm. Private livestock herds were not allowed to pasture in the commons, and if they were to do so, a townsperson had the right to behead one or two of the animals to send the bloody message that such behavior was not tolerated. For people who were accused of breaking these laws, if they could not prove their innocence, they paid the set fine for that infraction. Any citizen could be the accused or the accuser. In general, the *cartas pueblas* of the Castelló region reflect a mixed-use agro-silvo-pastoral landscape and the employment of meticulous legislation to defend and preserve forest resources.[53]

Municipalities appointed guards who swore to protect the commons, uphold the local ordinances, and defend against illegal cuttings and fires. In keeping vigilance over the forests, the guards normally worked in pairs and on horseback. They swore to uphold the local ordinances and defend against illegal cuttings and forest fires. Some oversaw the annual acorn harvests to ensure the towns' pigs got equal shares and did no permanent damage to the oak trees by hastening the fall of the acorns.[54] In Arjona, east of Córdoba in Andalusia, guards inspected all forested properties once a week, and they reported damages promptly, or else they risked being fined themselves for not doing their duty. Guards required many animals to wear bells so they could hear when a herd unlawfully entered certain forests. They regularly counted the herd to prevent excessive numbers from walking through. Another rule, typical of many localities and meant to benefit everyone in the long term, was the protection of new forests or new growth after a fire. An ordinance from Córdoba prevented livestock from entering such forests for three years.[55] Some municipal ordinances specified different fines for different infractions, depending on which animals were involved, the size of the herd, and the time in the forest's growing season. According to such ordinances from the city of Soria, the city would impose the highest fine, in theory, if more than 150 head of livestock entered a forest at night and in its vulnerable early growing period.[56] Being a guard could be quite demanding and dangerous; larger forests required

53 Soriano Martí, 75–77.

54 Vassberg, 37.

55 AGS, SELLO, Catálogo XII: 4472, December 20, 1494, Madrid, fol. 59.

56 Vassberg, 68.

extensive traveling, and any forest could be a hideout for outlaws, thieves, or murderers. However, it was an important job, it came with a salary, and it required the appointment of respectable and trustworthy people of the community.[57]

Royal laws promulgated for the entire kingdom were meant to rationalize the local laws, but the *cartas pueblas* and municipal *fueros* would remain important. In the ninth century, Catalonia was part of Charlemagne's Empire, and was subject to an important piece of legislation, the *Capitulare de Villis* (802). Among many other things, this legislation regulated the use of the forests, established penalties for those who damaged the forest, protected game, and even called for new plantings of juniper, apple, pear, plum, rowan (*Oleaceae*), loquat (*Rosaceae*), chestnut, hazelnut, almond, mulberry, laurel, pine, fig, walnut, and cherry trees.[58] A resurgence of Roman law in the thirteenth century gave new strength to royal power, leading to broadly proclaimed laws for the defense of the king's resources in all his lands. Alfonso X sponsored the creation of a new law code for his realms, based mostly on a revival of Roman law, to help centralize and enhance royal power. The *Siete Partidas*, completed in 1265, reinforced the Roman legal idea that public lands, including most forests, belonged ultimately to the state, an idea that kings defended throughout the early modern era.[59] Alfonso X ordered that people could not cut down, burn, uproot, or damage trees, vines, or other useful plants in any way.[60] By this time, the crown took forest destruction very seriously and set penalties, even including the death penalty in some cases, for those who destroyed trees belonging to town councils.[61] In a famous and often-quoted decree, Alfonso X ordered anybody who started a destructive and unlawful fire to be thrown into the flames.[62]

Alfonso XI (1312–1350) produced two important works related to forests. The *Ordenamiento de Alcalá* (1348) prohibited the sale of communal lands and specified punishments for other damages to the forests, including death for starting forest fires.[63] The *Libro de Montería* was a different kind of document entirely, which aimed to educate a noble audience in the proper ways of hunt-

---

57 AGS, SELLO, Catálogo XIII v.I (1454–1477), 1000, March 25, 1476, Zamora, fol. 114.

58 Bauer, *Montes*, 195–196; Rábade Blanco, 97.

59 Vassberg, 121; Carlé, 369; Alfonso X, King of Castile and León, *Las Siete Partidas*, eds. S.P. Scott and Robert Ignatius Burns (Philadelphia, University of Pennsylvania Press, 2001), v. 3, "Medieval Law."

60 Carlé, 369.

61 Phillips and Phillips, *Golden Fleece*, 337, footnote 29.

62 Bauer, *Montes*, 49.

63 Casado Soto, "Historia del bosque," 42.

ing, including the equipment, care of hunting dogs, and ways to recognize various game animals. The third part of the *Libro* named forests from many parts of Castile, León, and Andalusia, the types of species most prevalent, and the game available, especially wild boars and bears. Some modern scholars have used these detailed descriptions of fourteenth-century forests to study deforestation over the centuries.[64] The fifteenth century also saw the 1405 regulations of Enrique III (r.1390–1406) and the Montalvo Codex of 1480, both of which stipulated methods of preventing the destruction of forests and fauna.[65]

Sometimes the crown would act as intermediary between parties disputing forest use or access. In the Cortes of Toledo in 1480, the municipalities requested crown assistance to combat the usurpation of common lands by nobles and other outsiders. The Law of Toledo ordered *corregidores* (crown representatives in major towns) and other royal justices to hear complaints from the municipalities about usurped property.[66] The law allowed municipalities to recover properties lost to powerful local interests through the help of the crown, which began establishing royal magistrates in the municipalities. In 1492, the crown returned unlawfully privatized land back to the commons.[67] However, delays in many cases meant that the law did not always work effectively. Sometimes parties reached temporary solutions while settling legal disputes. In September of 1499, the city of Toro (Zamora) was in the midst of a dispute with the *regidor* (town councilor) Diego de Ulloa over access to the forest for cutting firewood. Ulloa had prohibited any felling in the forest of San Miguel de Gros, much to the dismay of the people of Toro. A royal council stepped in and declared that the people of the city could cut thirty cartloads of firewood before Christmas, and if the case had not been resolved by then, they could take one load of firewood each week until the case settled.[68] Some municipal councils challenged certain ordinances proclaimed by the crown. In 1503, the council of Seville petitioned against a new royal ordinance prohibiting the entrance of livestock into the forests of Carmona for two years after a fire clearing.[69] Such laws could vary from place to place, no matter how close

---

64 Bauer, *Montes*, 197, 199; Casado Soto, "Historia del bosque," 42–43; Rábade Blanco, 97; Luis Gíl Sánchez, "La voz 'montes' y la transformación histórica del espacio arbolado," *Cuadernos de la Sociedad Española de Ciencias Forestales* 16 (2003), 20–21, 24.

65 González Hernández, 3.

66 Vassberg, 77.

67 Vassberg, 67.

68 AGS, SELLO, Catálogo XIII v.XVI (1499) 1843, September 14, 1499, Valladolid, fol. 403; and 2008, September 25, 1499, fol. 47.

69 AGS, Consejo Real de Castilla, 21, 14 (1,503 fols., A.15-6).

they were to each other. As mentioned above, Córdoba had a law against entering new forests for three years, but in Seville, two years must have seemed too long.

Royal forests were lucrative for the government, which received receipts for sales of timber, coppice, charcoal, ashes, and fallen trees, as well as fines and dues paid by users.[70] Some people from rural areas could make money by selling firewood in cities.[71] However, the chief beneficiaries of the forest communitarian system appeared to be the landless poor and the small landowners.[72] The range of forest resources for people, especially the poorer classes, could provide just enough for a family to eke out a living. It was a vital part of the fabric of society and was a system that even the upper classes within municipal governments defended.[73] The system balanced the interests of stock raisers and cultivators, and provided a base for the social and economic structure of rural life. Not surprisingly, people vigorously defended their rights to utilize the resources of the forest.

A strong legal tradition developed over centuries to ensure that nearly everyone could benefit from the forests of Spain. While the system was imperfect, and suffered from lengthy delays, favoritism, and corruption, the primary interest of the crown for centuries remained the common good of forest use at the local level.[74] The crown, therefore, lacked strong motivation to remove the diverse local ordinances and regulations that governed forest use and access. The most heavily forested regions of Spain, including the Basque region, Cantabria, Asturias, Galicia, Navarre, and Catalonia also had some of the strongest *fueros* of anywhere in Spain. By 1518, Carlos I continued to support forest regulation to protect livestock and conserve firewood, keeping with a well-established role for the crown.

The importance of naval power for Carlos I became apparent early on in his reign, but would not impact his approach to forests until closer to the end. Like other monarchs of the day, he assembled naval forces as he needed them. Usually, the emperor made contracts with private ship owners and combined forces with allies. Carlos's galley fleet included four permanent squadrons from

70 Bechmann, 201. The author indicates that the revenue from Saint Louis' (r.1226–1270) forests were worth a quarter of the budget.

71 Carlé, 346. Also see Jesús Bravo Lozano, *Montes para Madrid. El abastecimiento de carbón vegetal a la villa y corte entre los siglos XVII y XVIII* (Madrid, 1993).

72 Vassberg, 33.

73 Vassberg, 33 and 83.

74 On similar challenges implementing forest administrations, see Warde, *Ecology*, 187–192, for examples from Early Modern Germany.

Spain, Naples, Sicily, and Genoa. Together, the fleets had about sixty vessels with more than half belonging to private owners and Italian rulers. The squadrons were not administered directly by the crown, but through a series of contracts with the owners or the captains general of individual squadrons.[75] The need to contract out all types of provisioning was due at least in part to the inability of Carlos's bureaucracy to keep pace with the demands of empire.[76] Felipe II would assert greater royal control over the administration of war, including shipbuilding, but the demands on the resources and finances of the monarchy would force the crown to return to a reliance on the contract, or *asiento* by the early seventeenth century.[77]

The 1540s proved to be a turning point for both ship construction and forest legislation. Between 1500 and 1520, seven vessels were built in the Basque provinces of Guipúzcoa and Vizcaya. Fourteen were built between 1520 and 1540. During the 1540s, forty-one vessels were built, a dramatic five-fold increase from the previous decade. By the late 1540s, the crown could count on annual shipments of silver from New World mines, which became essential to finance Carlos's wars in Europe, and they helped make naval expansion possible. Most of the ships on the Indies run were built in the Basque region and in Cantabria, and memoranda on coastal defense from the 1560's already indicated problems of timber shortages there.[78] The growing importance of the Basque region for Spanish shipping and its impact on timber supplies there soon drew the attention of the king. In Valladolid in 1547, Carlos issued a

---

75 I.A.A. Thompson, *War and Government in Habsburg Spain, 1560–1620* (London: Athlone Press, 1976), 163–165. Also see Esteban Mira Caballos, "El Sistema Naval Español en el Siglo XVI: Las Armadas del Imperio," *Revista de Historia Naval*, n. 74 (2001), 39–54.

76 Phillips, *Six Galleons*, 93.

77 The debates over direct or indirect administrative policy are at the heart of Thompson's *War and Government*. "The one was a system of direct administration by ministers and officials of the crown acting largely in a public capacity, financed at all levels by the royal treasury and accountable in detail for all stages of procurement and distribution. This is what in sixteenth-century Spain was called *administración* (cf. the French *régie*; there seems to be no equivalent English usage of 'administration'). The other was a less cohesive system of indirect administration conducted for the crown by non-royal agents, either by private contractors on a commercial basis (*asiento*), or by local powers acting on the strength of an effective authority that derived not from a royal commission but from an antecedent social, economic or institutional position in their own particular localities," 3.

78 Phillips, *Six Galleons*, 80. For shipbuilding in Cantabria see José Luis Casado Soto, *Barcos y astilleros: La construcción naval en Cantabria* (Santander: Puerto de Santander, 1993).

decree titled "Forest Plantations in the Province of Guipúzcoa and the *Señorío* of Vizcaya."[79] It stated,

> Due to the great number of *naos* (vessels) built in the Province of Guipúzcoa and the *Señorío* de Vizcaya that serve us, and due to massive cuttings that destroy the forests, a *relación* was made for us regarding a remedy. No one in the said provinces is allowed to cut down a tree without planting two, and those who have cut down a tree of ten years [or more] are required to plant on all the land around where they have cut. We order...the *Corregidores* of Guipúzcoa and Vizcaya to take special care of the remedy and report to the *Consejo* of any damages that occur.

This is the first case of royal forest legislation specifically meant to help shipbuilding and a turning point in the crown's forest territoriality. The 1547 decree placed new regulations on the people of Guipúzcoa and Vizcaya to increase the amount of trees in the area for shipbuilding. The concern for cutting older trees was especially important for naval interests, because shipbuilders required properly aged trees. Administratively, *corregidores* (local crown representatives) applied this law and reported directly to a royal council, drawing a closer connection between state and local power. The importance of shipbuilding in the Basque region, therefore, attracted special attention and special remedies from the crown to curb deforestation. Since Carlos frequently leased commercial vessels for naval service, healthy commerce in the Basque region meant available resources for naval defense.[80]

However, it soon became clear to Carlos that the 1547 decree was not effective. From Valladolid in 1548, he issued a new one. In it, he said the *corregidores* had been careless and negligent in their execution of the earlier decree, which aimed to plant new forests and conserve old ones. He charged the judges in residence to inspect the actions of the *corregidores* and, if their negligence continued, to send a special person to help with the execution of the order.[81] It is not entirely clear who this person would be, but it is possible that this is the first indication of a royal official charged specifically with

79 *Novíssima recopilación de las leyes de España* (Madrid, 1804–1829), tomo III, libro VII, titulo XXIV, ley VIII. Both Gupúzcoa and Vizcaya are part of the Basque region.

80 Lynch, *Spain 1516–1598*, 170. The Catholic Monarchs prohibited shipment of Spanish merchandise in non-Spanish ships when native vessels were available and introduced subsidies for the construction of larger vessels. King Carlos I also recognized the contribution of the Basque region in supplying the ships that defended the Indies route.

81 *Novíssima recopilación de las leyes de España*, tomo III, libro VII, titulo XXIV, ley VI.

conserving forests – an early version of the forest superintendent used in Felipe II's reign.

Under Carlos I, then, naval power had become essential to crown policies, and these acts of forest legislation indicate a growing appreciation for the well-being of forests for shipbuilding. There were signs that the *corregidores* continued to disappoint, leading the emperor to issue a warning of a greater royal presence in the northern forests. However, the traditional communitarian system for the most part remained unchanged. Pressure on the old system would continue after Carlos's reign, with increasing emphasis on planting new trees and a greater involvement of state officials at the local level. Strictly legislative approaches were proving ineffective, and the crown began to look for new ways to manage its natural resources.

A New Era of Naval Strategy in the Reign of Felipe II (1556–1598)

Felipe II inherited many of the same problems as his father, but with the added element of near financial collapse. While he did not inherit his father's Austrian Habsburg lands, he would continue to fight wars in northern Europe, the Mediterranean and the New World. Initially he shared his father's approach to naval power, but a disastrous expedition to Djerba in North Africa in 1560 ushered in a new era of naval strategy and shipbuilding, which had repercussions for forest conservation.[82] Felipe learned that he could not contain the Ottomans using improvised assemblages of fleets and adventurous crusading campaigns like Carlos I. He responded to the losses at Djerba by initiating strategies for long-term enhancement of naval forces and personnel. He secured funding from a special subsidy from the Pope in addition to the *cruzada* and an extraordinary subsidy from the Cortes of Castile. He initiated shipbuilding projects in Barcelona, but the industry along the Mediterranean coast of Spain had suffered from depressed commerce in Catalonia and deforestation, forcing Spanish shipbuilders to purchase timber in Tuscany and Naples and masts from the Baltic.[83] The decades of previous decline there forced Felipe II to transport experts from the north coast and Genoa and to plant new forests in the area. In 1562, he asked the Basque shipbuilder Martín de Arana to move to

82 Fernand Braudel. *The Mediterranean and the Mediterranean World in the Age of Felipe II*, trans. Sîan Reynolds, 2 vols. (New York: Harper Colophon Books, 1976), 1007; John F. Guilmartin, *Gunpowder and Galleys: Changing Technology and Mediterranean Warfare at Sea in the Sixteenth Century* (Cambridge: Cambridge University Press, 1974), 123–134.

83 Braudel, 141–143.

Catalonia to direct ship construction there.[84] In 1567, he asked for similar help from the *veedor* (shipbuilding supervisor) Juan de Gurruchaga, who brought with him several experts in galley construction to Barcelona.[85]

Felipe was aware of problems in the shipping industry on the north coast as well, and he expressed his belief that they were due at least in part to deforestation or, rather, to negligence in replanting.[86] He might also have believed that lack of administrative control contributed to worsening conditions, because in 1558 he received a report on the "difficulty of governing Guipúzcoa due to the character of its inhabitants."[87] Felipe's commitment to ensuring long-term naval power and his concern with the state of resources in the northern shipbuilding regions led him to appoint Cristóbal de Barros in May of 1563 to investigate forest conditions on the north coast. Felipe ordered Barros to oversee compliance with orders to plant oak trees in the jurisdiction of each *corregidor*, two leagues from the coast in the regions of Guipúzcoa, Vizcaya, Cuatro Villas, Asturias, and Galicia. He was to record in a book the exact quantity of oaks planted annually in each location, in both private and communal forests. Jurisdictions that did not plant new trees would incur fines.[88]

In addition to improving the forests, the king wanted Barros to reward shipbuilders who built larger vessels by providing them with subsidies from the crown. The types of ships the crown wanted did not always correspond to the types of ships investors, shipbuilders, and merchants wanted. While shippers sought the best possibilities for profit, the crown prioritized naval defense over maximizing economic opportunities.[89] Naval defense required large, heavily armed vessels, and the crown wanted to be able to lease such vessels from the commercial sector during wartime. As an incentive to shipbuilders and merchants, the crown gave larger ships in port priority in loading goods for export.[90] A set of navigation acts proclaimed by the Catholic Monarchs had prohibited the use of non-Spanish vessels for commerce when

84 AGS, GA, leg. 70, fol. 296–297.

85 AGS, GA, leg. 71, fol. 179.

86 AGS, GA, leg. 71, fol. 111; see William D. Phillips, Jr., "Spain's Northern Shipping Industry in the Sixteenth Century," *The Journal of European Economic History*, v. 17 n. 2 (Fall 1988), 297.

87 AGS, GA, leg. 67, fol. 7. For more on relations between the crown and Guipúzcoa, see Carla Rahn Phillips, "Naval Recruitment and Basque Resistance in Early Modern Times," *ITSAS Memoria: Revista de Estudios Marítimos del País Vasco* no. 5 (2006), 173–185.

88 AGS, GA, leg. 71, fol. 111; and AGS, GA, leg. 347, fol. 15–16.

89 Phillips, "Shipping Industry," 289.

90 Phillips, "Shipping Industry," 277.

native vessels were available.[91] However, merchants tended to prefer smaller vessels that were more capable of entering inland ports. Barros continued to defend the crown's case that bigger was better,[92] but he could not settle the debates over the best ship design, which continued well into the next century.[93]

Neither could Barros stop the cuttings and burnings that were destroying the forests. The state tried to show communities in Asturias and Galicia that the adherence to plantation quotas had much to offer the local community, but still Barros observed many instances of noncompliance.[94] Certainly, in some regions, plantings occurred successfully, particularly in Vizcaya and Guipúzcoa, but Barros noted that farther west "there is little care, or none, in planting, and great disorder in cutting, burning, and wasting the forests."[95] The shortage of timber had caused a sharp rise in its price, adding to the troubles of a general decline in shipbuilding in the area. On December 7, 1574, as a response to these increasing problems, the king named Cristóbal de Barros superintendent in charge of planting and conserving trees along the whole north coast from France to Portugal. During Barros's tenure, he ordered that oak trees could only be cut for shipbuilding or house construction, that a commissioner should be appointed under each *corregidor* to oversee planting and stewardship, that city councils should be responsible for the costs, and that some new roads would need to be built to generate easier access to forests.[96]

Felipe II continued to issue legislation to improve forest conditions throughout Spain, not only in shipbuilding regions. He reconfirmed the 1518 pragmatic in 1567, but that year he also acknowledged that his father's legislation was not working. He wrote,

> The old forests are cleared and damaged, cut and felled, and they plant few new ones. Neither trees nor plants are placed along rivers, on common lands or private lands. And the land for the most part is deserted or cleared, without any trees. Firewood and timber have become so scarce

91 Lynch, *Spain 1516–1598*, 170.

92 Phillips, "Shipping Industry," 278; *Colección de documentos y manuscriptos*, compilados por Fernández de Navarrete (reprinted Nendeln, Liechtenstein, 1971), vol. 22, part 1, pp. 192–205; Barros argued that smaller ships did not protect the cargo as well as larger ships, loading times of much larger ships were not much greater than smaller ships, and even the Flemings built large ships, despite having shallower harbors than Spain.

93 See José Luis Rubio Serrano, *Arquitectura de las naos y galeones de las flotas de Indias, vol. I (1492–1590), vol. II (1590–1690)* (Málaga: Ediciones Seyer, 1991).

94 AGS, GA, leg. 347, fol. 17.

95 Phillips, "Shipping Industry," 298.

96 Phillips, "Shipping Industry," 298.

> that already in many parts people are unable to live. Not enforcing this remedy, being that forests are the main sustenance for the raising of livestock and the life of men, the damage and detriment would become intolerable.[97]

Five years later, Felipe issued his own *real orden* (royal order), "On the conservation and augmentation of the forests." The king ordered all citizens to plant twenty trees per head in those locations with *regidores* and to ensure they take root.[98] Other scholars have supported the idea that Felipe's reign witnessed increased deforestation, so the king could have been expressing a genuine concern for the condition of Spain's forests. The king responded to such conditions, though, by granting greater powers to personnel who worked directly for the crown in order to assert the king's policies in new areas.

Felipe took special interest in regional forest ordinances near shipbuilding centers. In Galicia, he requested to see the ordinances proposed by the regent and *alcaldes mayores* there, then mandated ordinances for the region himself.[99] In 1579, he ordered a complete prohibition on cutting timber in Galicia and called upon his officials to "redouble their vigilance in the Kingdom of Galicia, Navarre, and the ports of Spain."[100] Felipe also remained committed to renewing ship construction in Catalonia and called for more plantations there, as well as the conservation of wood in the arsenal of Barcelona.[101]

Overall, the 1560's and 1570's witnessed a new direction for naval policy and a new commitment for long-term naval defense. Reports of deforestation and commercial decline only added to the crown's desire to gain greater control over forests. Also, by the later sixteenth century, the overall agropastoral balance had shifted. The distinction between arable and pasture land could be difficult to determine at times, but the conditions favoring pasturage, that is sparse population and uncertain frontier conditions, began to change in favor of arable land. Spain at the start of the century had already expelled the remaining Islamic power on the peninsula and would experience a dramatic increase in population. Across Castile, the population increased by at least fifty percent by the last decade of the century. To augment the food supply, farmers

---

97 Muñoz Goyanes, 102.

98 Carlé, 373.

99 AGS, GA, leg. 78, fol. 76–81, 107–108, (1579). *Alcaldes mayores* were *corregidores* in places that were *señoríos*, or lordships. Martínez Ruiz, 23.

100 AGS, GA, leg. 92, fol. 76–81. "…de redobla la vigilancia en el Reino de Galicia, Navarra, y los puertos de España."

101 AGS, GA, leg. 81, fol. 315 (1576); and AGS, GA, leg. 88, fol. 21 (1578).

had to expand the amount of land dedicated to agriculture. Much of this new arable land came at the direct expense of pasture lands and the *montes*. The crown did not always favor the Mesta, despite the revenues generated by herding interests, such as taxes and tolls. When space allowed, and when the wool trade remained strong, the crown tended to side with the interests of the sheep owners over cultivators. The population growth of the sixteenth century put great strains on available land, and the crown, the Cortes, and law courts began to fear that grain scarcity and high prices could lead to social unrest. As a result, the state began favoring farming over herding interests.[102] Population increases and the establishment of new towns also led to higher demands for building materials and domestic fuel resulting in greater consumption of timber and firewood so that, in Vassberg's words, the *montes* seemed to have fled the cities.[103]

The defeat of the Nasrid ruling dyansty may have removed the last Islamic state from Spain, but Muslims and, later, Moriscos (Muslims converted to Christianity and living in Christian Spain) remained and were concentrated in Granada and Valencia. The Christian settlers in these regions frequently found themselves in conflict with Moriscos over the use of the land. Moriscos were not used to the larger numbers of livestock, for example, and were unaccustomed to the pasture rights and practices of Christian municipalities. After the Morisco uprising in the Alpujarras from 1568 to 1570, Felipe II displaced Moriscos from Granada and sent them to different regions of Castile to induce assimilation. Ironically, it was the Christian population that had difficulties assimilating to the landscape of the Moriscos. The intensive agriculture of the Moriscos contrasted with the extensive agriculture of the Christians, and the new settlers could not maintain the productivity of the land. As a result of Christian settlers extending agricultural land into higher elevations, further deforestation and erosion occurred.[104]

The sixteenth century also witnessed the building of new royal palaces and the establishment of more protected hunting grounds in surrounding forests. El Pardo was a forested region northwest of Madrid and was a favored hunting ground by the time of Alfonso XI (r.1311–1350). Royal control over the forest of El Pardo expanded under Enrique IV (r.1425–1454), and Carlos I built a palace at El Pardo in 1547. The people of Madrid struggled to maintain access to the forest, especially after Felipe II established the city as the capital in 1561.

102 Phillips and Phillips, *Golden Fleece*, 43 and 61; Lynch, *Spain 1516–1598*, 23.

103 Vassberg, 157; on the establishment of new towns, see Helen Nader. *Liberty in Absolutist Spain: The Habsburg Sale of Towns, 1516–1700* (Baltimore: The Johns Hopkins University Press, 1990).

104 Vassberg, 176–183.

The resulting population boom put tremendous pressure on the surrounding forests. While deforestation occurred around Madrid, El Pardo was saved and protected as a royal hunting ground. Pruning and charcoal making was allowed but had to stop once deer hunting season started in the early autumn. Disputes over the use of the forest between court and town would continue throughout the early modern era.[105] Felipe II commissioned other new palaces, including at Aranjuez, Vaciamadrid, Casa de Campo, and San Lorenzo de El Escorial. At these locations, in addition to running the government, the king hunted and fished in forest preserves, strolled through gardens, and enjoyed the fountains. Felipe II expanded and preserved the forest of Valsaín (Segovia) near a favorite hunting lodge, referred to as El Bosque (The Forest). In 1569, the king issued a *real cédula* to plant pines and oaks in Valsaín.[106] The large gardens at his palaces were inspired by his excursions as a prince through the Netherlands, England, France, and Italy. Members of the *Real Junta de Obras y Bosques* (Ministry of Works) managed the palace grounds and royal properties.[107] Trees were brought in from all over Spain, other parts of Europe, and even America. Timber from as far away as Cuba was used in the construction of the Escorial palace. Pressures on the forests, therefore, increased in the sixteenth century and continued to come from many sources other than the navy.

Forests and the Collection of Geographic Knowledge

During the reigns of Carlos I and Felipe II, the crown initiated new projects to gather historical-geographic information about the monarchy's realms. The collection of geographic information derived from the crown's need to become more familiar with its territories, many of which were completely unknown to any Europeans. More reliable information about their realms, including even Old Castile, would provide insight into how to reform government to increase revenue and exert more effective control over natural resources during this

105 Francisco Javier Hernando Ortego, "La lucha por el Monte de El Pardo. Rey, municipio y uso del espacio en el Madrid del Antiguo Régimen," *Cuadernos de Investigación Histórica*, v. 12 (1989), 169–196.

106 Helen J. Groome, *Historia de la Politica Forestal en el Estado Español*, (Madrid: Agencia de Medio Ambiente, 1990), 28; Guillermo Muñoz Goyanes, *Crónica sobre bosques y montes de la Península Hispánica* (Madrid: Fundación Conde del Valle de Salazar, 1983), 103.

107 Geoffrey Parker, *Philip II*, 3rd ed. (Chicago: Open Court, 1995), 38–43; Henry Kamen, *Philip of Spain* (New Haven: Yale University Press, 1997), 182–184; Francisco Javier Díaz González, *La Real Junta de Obras y Bosques en la época de los Austrias* (Madrid: Dykinson, 2002).

crucial period. Information gathered for the specific purpose of improving forest management gave the crown a kind of 'actionable intelligence' from which it could make policy decisions. The crown relied on channels through its bureaucracy to gain all kinds of geographic knowledge, but these channels were fed in part by the first-hand experiences of a diverse group of private merchants, entrepreneurs, explorers, as well as crown representatives who sent reports of their travels and encounters to the king in return for royal grants to support their endeavors.[108] Sometimes, though, the crown commissioned its own projects to produce questionnaires, maps, books, samples, and drawings of the people, plants, animals, and landscapes within the monarchy. The generation of this knowledge is best understood as interconnected and related to similar goals of enhancing the information available to the monarchy's government.

The Relaciones

As early as 1555, a chronicler (*cronista*) under Carlos I named Juan Páez de Castro proposed a plan to collect historical-geographic questionnaires from Spain and its overseas possessions.[109] The investigation would provide information on the various languages, people, forests, rivers, monuments, churches, agriculture, and customs of Carlos's realms. While Páez did not see the completion of the project in his lifetime, his ideas and questions served as the basis for multiple sets of questionnaires that were eventually sent around Spain and the New World. Juan de Ovando ultimately sent out the questionnaires after 1573. The surveys returned to Spain between 1578 and 1586 to form the *Relaciones geográficas de Indias*.

After becoming president of the Council of Finance, Juan de Ovando initiated similar surveys for Spain, resulting in the *Relaciones topográficas de España*.[110] For this massive inquiry, Ovando sent out questionnaires of fifty-seven articles to hundreds of villages, towns, and cities in 1575, the year of his death. Another set of forty-five articles was sent out in 1578.[111] Today, we have

108 Antonio Barrera, "Empire and Knowledge: Reporting from the New World," *Colonial Latin American Review* 15, no. 1 (Jun. 2006): 39–54.

109 Bauer, *Montes*, 113.

110 On Ovando's career, see Stafford Poole, *Juan de Ovando: Governing the Spanish Empire in the Reign of Felipe II* (Norman, OK: University of Oklahoma Press, 2004).

111 Geoffrey Parker, "Maps and Ministers: The Spanish Habsburgs," in David Buisseret (ed.) *Monarchs, Ministers and Maps: The Emergence of Cartography as a Tool of Government in Early Modern Europe*. (Chicago: University of Chicago Press, 1992), 124–152; and Geoffrey Parker, *The Grand Strategy of Felipe II* (New Haven: Yale University Press, 1998), 61–65.

several volumes of responses preserved from the regions of Madrid, Toledo, Cuenca, Guadalajara, Cáceres, Ciudad Real, Jaén, Albacete, Ávila, Alicante, and Murcia. Several questions related to forests in these areas, including one that asked people "if the land is abundant or lacking in firewood, from where do they get their supply, and if it is mountainous, what forests and groves, and what animals, game, and wildlife grow and live there?" Another asked "if there is abundant water or not, and springs or lakes of note that are located in the town, and if there are no rivers or springs, from where do they drink and to where do they go to mill?" One inquired about "the pastures and meadows of note in the limits of the town with the forests, hunting parks, and fishing sites that they have and whose they are and what they are worth." Finally, one asked about "the sorts of houses and buildings that they use in the town, and from what materials are the buildings made, and if those materials are found in the land nearby or brought in from another part."[112] The responses to these questions provide us today with valuable information on sixteenth-century society, economy, material life, geography, and local history from about six hundred communities. In many responses, people took the liberty of saying that they knew forests once existed in many areas that no longer had any trees.[113] With the help of the *Relaciones*, historians have determined that in most places, the agro-pastoral balance had shifted to agriculture by this point, at the expense of the *montes*.[114]

While these documents have been instructional for modern historians, how useful were these projects for the needs of Felipe II during his reign? At the time, "they were intended both to enlighten the king about his subjects and to form the basis of a historical geography of Spain (alas, never completed)."[115] The amount of information gathered through the questionnaires was impressive, but, ultimately, they only helped with tax collecting and provided information for a census in 1591.[116] They probably produced so much information that it took too long to collect for the government to utilize effectively. Felipe wrote to his *corregidor* in Toledo in October of 1575, lamenting that "if we were to send a person to compile the descriptions required, it could not be done with the speed that we desire."[117] Indeed, the responses came from a limited

112 Bauer, *Montes*, 114.

113 Bauer, *Montes*, 114–115.

114 Vassberg, 157.

115 Parker, *Grand Strategy*, 48.

116 Lynch, *Spain 1516–1598*, 267.

117 AGS, Estado, leg. 157, fol. 103, Felipe to the *corregidor* of Toledo, October 1575, minute, cited in footnote 47, Parker, *Grand Strategy*, 321–322.

territory within Spain, including Castilla la Mancha and Murcia, and so do not inform us about the principal shipbuilding regions of the north coast.

Mapping Projects

The questionnaires also resulted in the production of stunning maps of Spain's overseas possessions, including the first map ever of the western Pacific Ocean region.[118] Maps became increasingly important for monarchs and their governments in the sixteenth century.[119] Recent historians of cartography have done extensive research to show us when rulers used maps and what kinds of maps were available. Both Felipe II and Henri IV of France, for example, had maps that even showed locations of small villages in their lands.[120] Maps containing town locations and other phenomena "permitted the government to locate and visualize geographically places that were in the news. They helped ministers to cope with the unforeseen."[121] Nonetheless, maps of Spain from the 1560s suggest that the interior remained poorly understood, and cartographic knowledge did not necessarily indicate control over the area represented.[122]

Attempts to improve this knowledge occurred sporadically in the sixteenth century, but the efforts that were made were ambitious. Spain's coastlines had been well-mapped for centuries dating back to the late medieval Mallorcan School of portolan chart makers. Seville's *Casa de la Contratación* (House of Trade), which was formed in 1503, employed cartographers to map the New World. In 1517, Fernando Colón was commissioned to lead a project based on visitations to villages throughout Spain to record information about their size, economy, and geography. He worked until the crown told him to stop in 1523, leaving the work unfinished. Before then, members of Colón's team were able to visit nearly seven thousand villages. They recorded their findings in the "Description and Cosmography of Spain," composed of eight volumes of description and three volumes of drawings and maps.[123] In 1548, Pedro de Medina produced a work dedicated to Prince Felipe called *Great and Memorable things of Spain*. It included descriptions of natural resources, but was meant to present the wealth of Spain's cultural patrimony.[124]

118 Parker, *Grand Strategy*, 63.
119 Buisseret, 1–4.
120 Buisseret, 4.
121 Parker, *Grand Strategy*, 125–126.
122 Parker, *Grand Strategy*, 126.
123 Parker, *Grand Strategy*, 128.
124 Casey, 15–16.

In the 1550s, Felipe – then still a prince – was in the Netherlands, where he grew interested in Dutch mapmaking. While there in May of 1559, Felipe commissioned the cartographer Jacob van Deventer, who had experience making maps of provinces in the Habsburg Netherlands,

> to visit, measure and draw all the towns of these provinces, with the rivers and villages adjoining, likewise the frontier crossings and passes. The whole work is to be made into a book containing a panorama of each province, followed by a representation of each individual town.[125]

Probably because of his experience with van Deventer, Felipe commissioned a similar project in Spain with the artist-cartographer Anton van den Wyngaerde, which generated views of over fifty Spanish towns. In addition, the king commissioned a map of the entire Iberian Peninsula and ordered a "General Visitation" of his Italian domains.[126] Like the *Relaciones*, these mapping projects received their impetus from Spain's involvement in territories outside of Spain, leading to their application in the Iberian Peninsula. The most ambitious project perhaps was carried out by Pedro de Esquivel, a professor of mathematics at the University of Alcalá. Beginning in about 1566, he set out to survey the entire peninsula. He died in 1575 before the project was completed, but it is probable that his efforts contributed to the completion of the Escorial Atlas, a collection of twenty-one maps begun after 1580. More common than the ambitious mapping projects of entire countries or continents were maps detailing plans for a specific policy or campaign.[127] Most of these have not survived. The ones that have are typically related to some grand military engagement, like the maps showing plans for the Great Armada in 1588. Rarer in the archives are maps such as a hand drawn sketch from 1589 to indicate locations of bakeries in Cantabria that made biscuits for the Spanish navy.[128]

Like the time taken to administer and absorb information from the questionnaires, the time it took officials to gather information, carry out land surveys, and produce the maps were serious drawbacks for their usefulness to administrators in this period. Also, maps contained highly privileged information that could get into the wrong hands quite easily. Still, Geoffrey Parker has pointed out that by the end of Felipe II's reign, "the Iberian peninsula was

125 Quoted in Parker, "Maps," 129.
126 Parker, "Maps," 59.
127 Parker, "Maps," 135.
128 AGS, GA, leg. 264, fol. 34, "Asiento de los lugares de la montaña donde se fabrica el biscocho."

better represented in maps than any other European area of comparable size."[129] Yet another approach to gathering information, particularly about the forests of Spain, came through the bureaucracy.

Forests and the State Bureaucracy

By the middle of the sixteenth century, Spain had long relied on an extensive bureaucracy to carry out royal policies. The crown installed royal representatives as viceroys, ambassadors, town representatives known as *corregidores*, judges, tax collectors, purveyors of materials for shipbuilding, as well as appointees for many other duties. They were held responsible for keeping the crown informed of conditions in the monarchy as well as carrying out royal policy in their jurisdictions. Felipe set Madrid as the monarchy's capital in 1561, and from there, or from palaces nearby, the king could administer his officials through letters and reports.[130]

Felipe relied on this bureaucracy to help inform him of forest conditions.[131] By this period, the crown's concern for forests linked directly to shipbuilding for naval defenses. Logically, the king looked to the bureaucrats involved in shipbuilding to inform him of forest conditions. These officials already had experience in the industry, so they understood exactly what kinds of timber the navy needed, and they knew where they could find it. Felipe merely adapted an established system to new needs when he instructed his purveyors to inspect forest conditions near shipbuilding sites. The use of this bureaucratic network resulted in quicker and more frequent reports on the resources of the monarchy than the questionnaires or the mapping projects. Future kings would rely primarily on the bureaucracy for gathering information until the middle of the eighteenth century, when improved techniques in government and cartography encouraged the crown to make new questionnaires and new maps.

Carlos I had only shown interest in conserving forests marked for shipbuilding in Guipúzcoa and Vizcaya. By the end of the sixteenth century, however, Felipe II would gather information via bureaucratic forest inspections to extend royal power into forests in Galicia, Catalonia, Valencia, Gibraltar, and elsewhere. The appointment of a superintendent of forests and plantations helped with the state's need for geographic knowledge and enforcement of

---

129 Parker, *Grand Strategy*, 134.

130 For an analysis of bureaucracy in action from the perspective of notaries, see Kathryn Burns, *Into the Archive: Writing and Power in Colonial Peru* (Durham: Duke University Press, 2010). Thanks to Marcela Echeverri for bringing this book to my attention.

131 Goodman, 70.

policy. The forest superintendent embodied the extension of royal authority to the forests of Spain, reshaping local conditions, and providing a growing monopolization of information on forests in the hands of the state. Officially, Felipe II did not create the post until 1574, but Cristóbal de Barros was already looking after forest resources on the north coast by 1563.[132]

The remainder of this chapter focuses on bureaucratic reports created for the purpose of gaining 'actionable intelligence' on forest conditions, including reports from the forest superintendent Barros and a reconnaissance of forests in northern Valencia from 1589 in order to go into some detail regarding the complexity of arranging such expeditions as well as the variety of factors such forest inspections took into consideration when judging the utility of a forest for Spanish shipbuilding needs. Taken together, these efforts contributed to an early modern geographic information system that helped the state extend its territoriality over forest resources.

The Enterprise of England: Recovery and Response

As noted above, Felipe II placed a royal representative directly in charge of forest conservation not just in Guipúzcoa and Vizcaya, but along the entire north coast. For the Mediterranean, Felipe sent experienced personnel from the Basque region to revive shipbuilding in Barcelona. These acts prepared Spain for a tremendous output of vessels once a more favorable financial situation developed. Such a windfall occurred in the early 1580s, allowing Spain to build up its shipping capacity sufficiently to plan an invasion of England.

An upswing in the Indies trade after 1578 led to years of prosperity, especially between 1584 and 1586. The volume of trade increased by fifty percent between 1576 and 1586. Moreover, Felipe's accession to the Portuguese throne in 1580 meant a more secure transportation and communication network in

132 These Intendentes were called *Superintendente de fábricas reales y arqueamiento de navíos de alto bordo, visita y aumento y conservación de montes y plantíos.* As mentioned above, their main attribution was linked to shipbuilding control; they authorized the loans for the ship-owners, approved premiums for vessels over 300 *toneles*, and measured the sizes and capacities of the ships of the royal armada. Therefore, a shipbuilding background was needed. They also controlled the timber supply for the shipyards and had forest assistants to help them. This position was held by relevant seamen which often had tough discussions with the local authorities (*instituciones forales*), regarding harvesting licenses. Mario Michel and Luis Gil, *La transformación histórica del paisaje forestal en la Comunidad Autónoma de Euskadi* (Vitoria: Servicio de Publicaciones del Gobierno Vasco, Colección LUR n° 18, 2013), 172.

the Atlantic, helping to make the Indies convoy system work more efficiently. As an added bonus, international politics remained relatively calm in this period. The surpluses from trade and treasure during these years helped to finance the Armada and its subsequent recovery.[133] The idea of an invasion of England appears to have formed in 1583, after the Spanish secured the Azores by defeating a combined French and English fleet. Tensions had been escalating between England and Spain due to Francis Drake's harassment of Spanish shipping in the Indies between 1577 and 1579 and Queen Elizabeth I's support of Drake's activities. The Marqués de Santa Cruz urged Felipe to seize the moment and prepare for an expedition to invade England. Felipe was noncommittal, but indicated that he was taking steps to prepare. He wrote to Santa Cruz,

> The situation is such that it is not possible to speak with certainty at the moment...but I am ordering provisions of biscuit from Italy, and expediting the construction of galleons and the hire of ships in Vizcaya, and everything else which seems necessary in preparation for a favorable opportunity.[134]

Felipe's concern for forests in Catalonia did not slacken during preparations for the Armada, which took place for the most part in Portugal and the north coast. In 1586, a *veedor* and *contador* (accountant) named Pedro de Isunza inspected forests along Catalonia's coast and near Barcelona to learn what wood was available for galley construction.[135] In January of 1588, Felipe wrote, "It is necessary to increase the plantation of trees in Catalonia because the forests are going to disappear."[136]

As devastating as the enterprise of 1588 eventually was for Spain, it was not decisive in the long run.[137] Felipe and his government worked hard to rebuild Spanish naval forces, both to fight the then open warfare with England and to prevent continued attacks along coasts and trade routes. The king quickly received a lot of support from his officials and the nobility. In the months immediately after the disaster, Cristóbal de Barros was at work in Santander

133 Lynch, *Spain 1516–1598*, 246.

134 Cesáreo Fernández Duro, *La armada invencible* (Madrid: Sucs. De Rivadeneyra, 1884–5) v.i, 241–243; and Lynch, *Spain 1516–1598*, 440.

135 AGS, GA, leg. 195, fol. 136 (1586).

136 AGS, GA, leg. 220, fol. 211 (1588). "Es necesario incrementar la plantación de árboles en Cataluña porque se van acabando los bosques."

137 For the 1588 armada against England, see Felipe Fernández-Armesto, *The Spanish Armada: The Experience of War in 1588* (Oxford: Oxford University Press, 1988); and Colin Martin and Geoffrey Parker, *The Spanish Armada* (New York: Norton, 1988).

repairing the armada.[138] Felipe read reports on the wood taken from Ribadeo, on the Asturian-Galician border, for the repair of remaining ships from the Armada in Galicia.[139] In December, a nobleman near San Sebastián offered to cut down trees in his own forest to build replacements for ships lost in the "enterprise of England."[140] The king received a report describing types of trees growing near Barcelona that would be available for building galleys in 1589.[141] In February of 1589, the Duke of Medina Sidonia sent timber from southern Spain to El Ferrol, Santander, and Bilbao to build new galleons,[142] where at least twelve were under construction.[143]

This pressure on traditional sources of timber for shipbuilding sent Felipe in search of new sources. It is well known that in the period after the Armada, Spain was forced to import more materials and a greater number of ships. The fact that Felipe also searched for more sources of timber within Spain itself is often overlooked. While Spain's shipyards were busy rebuilding the fleet with new vessels using timber from the north coast, some royal officials were eager to offer their service to the king by sharing knowledge of potential timber outside the traditional supply zones, including reports on forests near Gibraltar and Málaga.[144] The *corregidor* of Antequera responded to the king's need for new vessels and made a case that his forests would be very useful to build ships "for the sea of England."[145] Since it was necessary to use the best timber possible at the lowest cost, the *corregidor* stated that there was very good timber near Antequera, which is only seven leagues by road from Málaga, where they could build the vessels.[146] In Aragón, deep into the Pyrenees near the French border, the crown built a road to the forests of the remote village of Broto to facilitate the extraction of timber for galleon construction, causing disquiet among the people of Broto.[147] Such roads opened municipal forests to naval access introducing greater demand upon the local population to cut, replant, and manage plantations.

---

138 AGS, GA, leg. 227, fol. 182, (Sept–nov., 1588).

139 AGS, GA, leg. 236, fol. 97 (1588).

140 AGS, GA, leg. 242, fol. 58 (1588).

141 AGS, GA, leg. 227, fol. 148 (Sept.–nov., 1588).

142 AGS, GA, leg. 281, fol. 27.

143 AGS, GA, leg. 285, fol. 48 (June 1589).

144 Regarding Gibraltar, AGS, GA, leg. 264, fol. 184 (1589).

145 Interest in a second armada against England was shared by many, including the king. Lynch, *Spain 1516–1598*, 462.

146 AGS, GA, leg. 347, fol. 88.

147 AGS, GA, leg. 284, fol. 51 (May 1589).

The utility of these forests for the crown was determined by a number of factors. Accessibility was defined first and foremost by physical geography. By 1589, Felipe II sought forests from all over Spain in order to recover from the failed Armada, so information about terrain, rainfall, and proximity to rivers proved crucial to the recovery. One example of the efforts to determine forest accessibility and utility comes from a reconnaissance mission carried out soon after the failure of the Great Armada. In December of 1588, the governor of Castelló de la Plana, Luis Castelvy, notified authorities in Barcelona of two forests potentially suitable for shipbuilding, including the forest of Valcanera in southern Catalonia, and the forest of Valivana just over the border in Valencia.[148] The Viceroy of Catalonia, Manrique de Lara, received the tip from Castelvy, and responded by appointing a naval captain with shipbuilding experience named Antonio de Alçatte to lead an expedition to inspect the forests. The rest of the Catalan group included a master shipbuilder from Barcelona and an engineer named Jorge Setara who would be responsible for making a map to accompany the report by Alçatte.[149] The Catalans paid a visit at once to the Viceroy of Valencia, the Marqués de Aytona, who added a royal bailiff and other Valencian officials to the entourage. By sending specialists who knew how to identify the type of timber needed for Spanish imperial defense needs, the crown could identify a new resource base and begin to enforce naval forest management regulations over the area.

The group aimed to do more than find trees. They were on the lookout for specific kinds of trees, mainly straight and tall pines for masts and well-aged oaks for hull construction. They were interested in the kinds of soils where these trees grew. Could they plant more trees there? Would the land support it? What kinds of trees grew best there? Property relations were also an important factor. To whom did the land belong: the crown, a municipality, the church, or a private individual? The owner would be the one responsible for planting new trees and obtaining licenses to fell old ones. How did people utilize the forest? Did goats, sheep, cattle or pigs forage there? These animals could cause severe damage to new growth if not closely monitored. Also, what kinds of transportation options existed? The forest had to be accessible from the coast or a navigable river so that the timber could be transported to a shipyard. Was there a major river or road near these forests? The forest expedition analyzed the regional geography and economy in order to produce a cost-benefit analysis for the viceroys of Catalonia and Valencia. With all of this information, the crown would know where the right kinds of timber sources were, who needed to manage them, and how accessible they were for the navy.

148 AGS, GA, leg. 246, fol. 283 (23 January 1589). Reproduced in Bauer, *Montes*, 561–562.

149 AGS, GA, fol. 281 (Alçatte's full report).

So when the forest inspection crew finally ventured into the Catalan forest of Valcanera, and found great quantities of pine trees, they had to qualify their findings. Alçatte reported that only the very best trees from these forests would be useful for naval shipbuilding, and they found many to be growing in rocky soil, creating timber that "it seems would be more appropriate for building houses and small boats than galleys." The forest technically belonged to the king, but use of it had been granted to the city of Tortosa and the towns of Asenia and Uldecona. Presumably, the crown could stake its claim once again as it had done elsewhere.[150]

For these forests, the key coastal access point was the town of Vinaroz, situated on the Mediterranean about halfway between Barcelona and Valencia. The forest of Valcanera was six Catalan leagues from Vinaroz.[151] Four leagues from Vinaroz on the same road was the Monastery of Benifasar.[152] Alçatte said that if the king was interested in using the pines of Valcanera, they would have to build a road extending two leagues from near the monastery, at a cost of 1,500 *ducados*. Vinaroz would also be the coastal port for the Valencian forest. There was no road to cover the six leagues from the forest of Valivana to Vinaroz, but Alçatte claimed that it would be easy to build one, since the land was entirely flat for four of those leagues.

From the Catalan forest of Valcanera the commissioners traveled five leagues to the Valivana forest in Valencia. In contrast to Valcanera, this forest had no pine trees, but it featured very abundant oak trees, mostly very old. The master shipbuilder from Barcelona believed the larger ones would be suitable for shipbuilding, but there were few that could be used for the joints and braces. The town of Morella owned the forest "where, in the summer, a great number of livestock enter the forest and forage there." As long as the animals could be prevented from damaging the saplings, the forest could be utilized by the navy in the winter when the crown preferred to fell its trees.

Similar forest reconnaissance missions would be repeated elsewhere in Spain in forests located outside the traditional supply zones of coastal shipyards. What made Alçatte's reconnaissance mission in Catalonia and Valencia extra instructive was the fact that his report was accompanied by a map of the region. Jorge Setara's map depicted the forests, mountains, rivers, roads,

150 See Goodman, 68–103, Chapter 3: "Counting the Trees: The Conservation of Spain's Forests."

151 A league was a unit of distance that a person or a horse could walk in about 1 hour, approximately 3.5 miles or 5.5 kilometers.

152 Alçatte also mentioned that the friars of Benifasar were selling a forest that contained trees well suited for smaller vessels.

bridges, towns, and the political border between Valencia and Catalonia, from the Ebro River and Tortosa in the north to Morella and Peñíscola in the south (Figure 3).[153] The map takes the viewers' eyes from the dark forests in the west, at the top of the map, down the rivers and roads to the Mediterranean Sea in the east. He included the port city of Vinaroz and the road that would pass beyond the Monastery of Benifasar. The map also indicates the location and accessibility of the crucial timber resources. With the map and the written report together, state officials in Valencia, Barcelona, and Madrid would know not only what kinds of trees these forests contained, but also exactly where they were and the best routes to travel there.

Setara's map is special because it does not fall easily into any category of maps from this period treated by historians of cartography. It is not strictly political, topographical, or topological, but it shows where the important towns and natural features were located in the borderlands of Valencia and Catalonia. It is not a map of a military campaign such as a battle, but it was produced by an expedition with a clear military purpose. It is not a map of cultural patrimony and historical wealth in the chorographic tradition. Rather, the map demonstrates the complementary roles of visual representation and descriptive accounts for enabling state officials to comprehend resource accessibility. It had a very practical use for the crown and played an important role in the process of territorialization of forests in Spain for state interests. Similar maps were produced elsewhere in Europe around this time, such as in Venice, as state governments sought new and useful information on the resources in their territories.[154]

Overall, the forests contained timber that could benefit the king, but the matter of connecting the timber source to a shipbuilding center remained a problem. Pedro de Isunza, a *veedor* who had experience inspecting forests all along the Catalan coast in 1586, expressed some doubts about the usefulness of these forests. Even before Alçatte left on his expedition, Isunza had listed several reasons that made the extraction of timber in these forests impractical for naval shipbuilding. Isunza said the transportation of timber from these mountains by land to Vinaroz or Tortosa would be costly. Even to get to Barcelona by sea from Vinaroz, although not as costly, was dangerous "*por causa de los moros*," because of the Muslims who navigated the waters around the *Alfaques* or the Ebro River delta and were experts at naval attacks. To build

153 AGS, Mapas, Planos y Dibujos, 05/079.

154 Appuhn, *A Forest on the Sea*, 224–247. See also David Woodward, ed., *The History of Cartography*, vol. 3, *Cartography in the European Renaissance* (Chicago: University of Chicago Press, 2007).

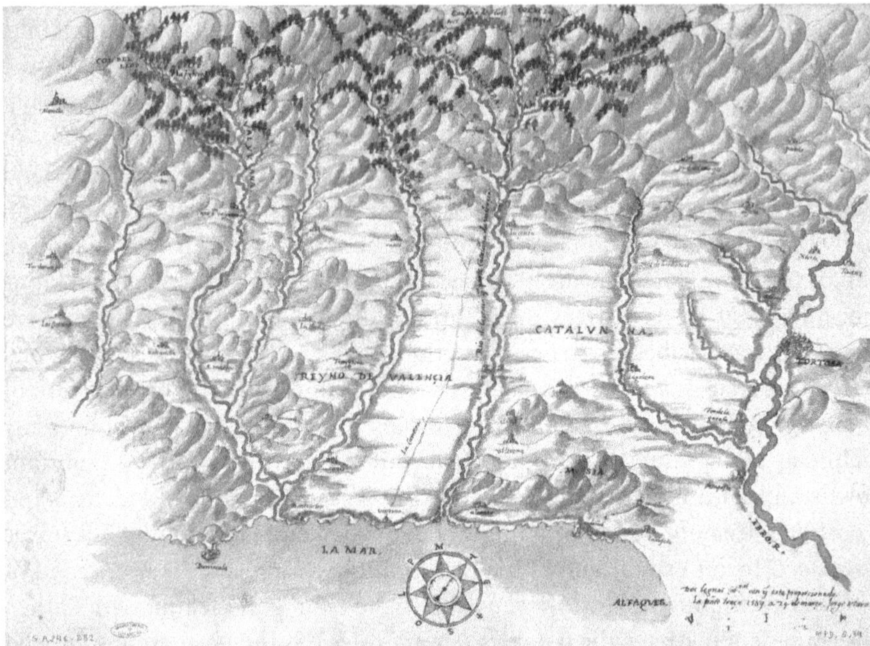

FIGURE 3 *Forests of Valivana and Valcanera – Jorge Setara produced this map of forests in the borderlands of Valencia and Catalonia in 1589, which were inspected by a group of naval shipbuilders and administrators. The map, coupled with a written report, informed state authorities of the region's terrain, transportation routes, distances from forests to the coast, types of trees in each forest, the quality of the timber, and forest ownership to serve as a cost-benefit analysis of its resources.*
SOURCE: ESPAÑA. MINISTERIO DE EDUCACIÓN, CULTURA Y DEPORTE. ARCHIVO GENERAL DE SIMANCAS. MPD,5,79.

vessels in the port of Vinaroz would make transportation less costly and less dangerous, but Isunza argued that it lacked the infrastructure and personnel of major shipbuilding sites such as Barcelona and Genoa. Roads would need to be built, supplies shipped in, and experts relocated for Vinaroz to be productive.[155] Alçatte, in a section tacked on at the end of his report, echoed Isunza's concerns. Nonetheless, extracting timber from Valencia and Catalonia still may have been preferable to relying on foreign suppliers. Despite the several challenges noted by Alçatte and Isunza, the forests of the Valencian-Catalan border region would prove to be a valuable source of naval timber for the crown. Particularly by the early eighteenth century, naval bureaucrats carried out more forest inspections, and timber from the region became an important source for a new arsenal at Cartagena.

155 AGS, GA, leg. 246, fol. 283 (January 23, 1589).

In a further indication of the importance of forests for the crown's naval goals, the superintendent of forests and plantations, Cristóbal de Barros, was named chief purveyor of shipping in Spain in 1591. In a *real cédula* written to all *contadores mayores*, Felipe II wrote that all assessments and audits of vessels in the service of the king had to go through Barros. Apparently, the differences in measurements and accounts in different ports had caused too much confusion. Felipe wrote that "to remedy this, all [accounts] pass through one hand, and that is to be of Cristóbal de Barros, my servant."[156] Barros would soon move to Seville, leaving behind his forest visitation duties. In 1594, however, he wrote a final report, which provided the groundwork for a smooth transition to a new superintendent, Hernando de la Riva Herrera, the author of the *cédula* from 1597 discussed at the beginning of this chapter.

This long history of forest regulation mostly in the interests of local communities met new challenges in the sixteenth century as Spain entered a period of global commitments, requiring the crown to reassess the value and functionality of its forests. Royal policies, then, played a significant role in shaping the landscape in the sixteenth century. Many of these policies emerged to address the crown's chronic shortage of money, including the granting of higher numbers of new town charters, the sale of common land for cultivation, and the general favoring of agricultural over pastoral interests. Also, the maintenance of a Castilian monopoly in supplying the New World helped to reshape the landscape, particularly of Andalusia, as greater demand for Spanish agricultural products created a new market-oriented agriculture based on olive groves and vineyards. Some policies were related to royal pursuits such as palace-building and hunting. Increasingly, foreign policy would start to guide the crown's management of forests as access to timber for shipbuilding took priority.

Jorge Setara's map, royal legislation for the forests of Guipúzcoa and Vizcaya, Galicia's new forest ordinances, road construction in Broto and near the Monastery of Benifasar, forest inspections in Catalonia, and other developments mentioned above, all contributed to the territorialization of naval forestry in Spain from the middle of the sixteenth century. Carlos I and Felipe II ruled Spain during a period characterized by religious warfare, dynastic conflict, and colonization on a scale not seen before. In order to gain access to resources for imperial defense, the crown began pursuing multiple approaches simultaneously to obtain 'actionable intelligence' of Spanish geography. Forest conservation efforts in sixteenth-century Spain served as valuable experiences

156 AGS, GA, leg. 326, fol. 183 (1591). "Para remedio de esto he mandado que todo ello pase por una mano y que esta sea por la de Cristóbal de Barros mi criado."

and historical precedents for colonial forest administration in the seventeenth century as well as for Spain's first national forestry code, issued by Fernando VI in 1748.

Fears of timber shortages, as articulated in numerous pieces of legislation issued by the crown, drove state officials to monitor forests more closely, but early modern shipbuilders avoided non-selective clear-cutting when procuring timber. The navy was not interested in using every single tree, and complete destruction of forests would have been calamitous for the local communities and the social order. The crown could not have abandoned its traditional concern for the general population's access to forest resources or dismissed the long established legal tradition that protected local access and usage rights. Historian David Goodman pointed out that Felipe II in fact praised his forest superintendents when they were able to take timber from dispersed sources so as not to place too large a burden on any one municipality.[157]

Rather, the crown regulated local behavior in certain territories without issuing a national forestry code. People in important shipbuilding regions risked being fined if they did not plant a quota of trees, and certain trees were off limits to needs other than major building projects, resulting in some examples of successful reforestation efforts by the early seventeenth century. The source of the laws was no longer the municipality, but the king and the Council of War. The crown's identification of valuable forest regions expanded during this period, beginning with Vizcaya and Guipúzcoa and eventually included the entire north coast and even parts of the southern and eastern coasts. These were not minor changes, but neither were they particularly severe or drastic. Due to a strong legal tradition that protected local rights, technological challenges to timely and reliable communication, lack of fully effective techniques to monitor and enforce royal policy, and deficiencies in geographic knowledge of its own territory, the early modern Spanish crown faced severe limitations to its ability to enforce royal regulations.

The unprecedented demands on Spanish government during the sixteenth century helped generate forest regulations for a new global age. Spain's population and economy grew for most of the sixteenth century, but by the end of the 1590s, a complex coincidence of economic crises, bad weather, overextension of resources, and a rise of powerful states in northern Europe posed a new set of challenges for Spain. The goals and form of state forestry would change little over the next one hundred years, but the real challenge would be the retention of state control over forests in a period of political and economic readjustment.

157 Goodman, 78.

CHAPTER 2

Forests of the Ultramar

Colonial Forest Territoriality in the Sixteenth Century

The Spanish monarchy's territories overseas, or the *ultramar*, opened up new and profitable lines of trade with Spain and the rest of Europe. Much of the wealth of the colonies went to pay the expenses of war and interest on loans from the crown's creditors. By the middle of the century, the merchant vessels and naval warships plying the Indies run provided the lifeblood of the empire. As the potential wealth of the New World became apparent to Spain's enemies drawing pirates and interlopers, there emerged a greater need to protect the treasure fleets and defend the trade ports of the Indies. The establishment of a Superintendent of Forests and Plantations in Spain in the sixteenth century marked an important step towards closer forest oversight in the Iberian Peninsula, where the need to regulate forests connected directly to the crown's interests in acquiring enough timber to supply Spain's shipyards. The Spanish exploited forest resources of the *ultramar* as they did in Spain, but the process of territorialization unfolded in different ways. Like in Spain, royal regulation of forests focused on acquiring timber for shipbuilding, but Spanish kings never permanently established anything like the Superintendent of Forests and Plantations in the colonies. Colonial forests never received the systematic annual inspections that Spanish forests received. Forest communities were not given plantation quotas, they were not required to keep plantation records, and they were not fined for failing to make new plantations. Colonial legislation appeared later than in Spain because in most cases restrictions on forest exploitation seemed unnecessary. The New World offered an abundance of natural resources in contrast to the shortages experienced in Spain, and it took some time before natural limits were apparent. Shortages near colonial shipbuilding centers eventually drew royal attention, and the restrictions on the use of trees suitable for naval interests increased. As in Spain, the crown supported a range of efforts to gather intelligence about strategic resources, and these efforts proved to be an essential component of establishing control over colonial forest access and use. Early settlers and explorers depended on indigenous support, but the use of forced labor to procure materials distinguished colonial forest exploitation from peninsular practices.

The delay in the emergence of forest regulations for naval timber conservation in the colonies could only partially be explained by the natural abundance

 | DOI 10.1163/9789004261372_004

found in the Indies. In addition, colonial communications with Madrid were slow. Communications within the peninsula were more rapid and reliable, thereby enabling closer crown involvement. Just as important, shipbuilding centers required supplies of associated materials, such as iron, sails, rope, and many other kinds of items.[1] The infrastructure of Spanish shipyards already had such related industries nearby or at least connected through commercial networks, greatly reducing the cost of the materials and their transportation. It took many years for such a complex of industries to develop in Spanish colonial settings, and in most cases colonial shipyards continued to rely on European imports, especially iron. Colonial shipbuilding became active in the sixteenth century where there were enough skilled laborers, materials, and demand for colonial-built vessels. Wherever such resources were exploited, royal officials asserted authority over forests and the behavior of able-bodied laborers. Colonial forest territorialization intensified from the later sixteenth century, characterized by searches for suitable woods in unfamiliar environments, greater efforts to enforce control over access, and closer regulation through royal legislation in regions where important naval shipyards existed. The rising costs of labor and materials beginning in the later sixteenth century in Spain also provided a strong impetus to protect access to colonial forests. By 1650, only one-third of Spanish vessels were built in Spain while one-third were built in America and one-third were purchased from abroad.[2]

This chapter analyzes the crown's overseas forest regulations during the sixteenth century identifying the parallels and differences with the forest territorialization process in Spain. It opens by presenting the scope and aims of geographic knowledge acquisition in environments unfamiliar to Europeans. The crown relied on first-hand accounts from explorers, merchants, priests, bureaucrats, and indigenous peoples to gain an understanding of the resources of the monarchy's *ultramar*. Because royal forest regulation increasingly aimed to protect the shipbuilding industry, the chapter then explains the shipbuilding process including some of the labor and material requirements of most shipyards, which remained dependent upon the methods and

1 José Alcalá-Zamora y Queipo de Llano, *Historia de una empresa siderúrgica española: Los altos hornos de Liérganes y La Cavada, 1622–1834* (Santander: Centro de Estudios Montañeses, 1974); José Alcalá-Zamora y Queipo de Llano, *Altos hornos y poder naval en la España de la edad moderna* (Madrid: Real Academia de la Historia, 1999); Carlo M. Cipolla, *Guns Sails, and Empires: Technological Innovation and the Early Phases of European Expansion 1400–1700* (London: Minerva Press, 1965).

2 J.H. Parry, *The Spanish Seaborne Empire* (Berkeley: University of California Press, 1990), 249 [First edition, New York: Knopf, 1966].

technologies of the Old World. Important colonial shipyards emerged in key strategic regions that owed much of their value to their proximity and control over suitable forests, so the chapter then explores the history of forest regulation near Realejo and Guayaquil on the Pacific coast, Havana in the Caribbean, and Manila in the Philippines. The chapter ends with a look at the ramifications of Iberian colonial success for Spain's imperial rivals, who devised their own policies toward colonial forests as they challenged Spanish supremacy.

Geographic Knowledge Channels across the Atlantic

The crown had been eager for information about the Indies beginning with Columbus's first voyage. The Spanish grew more familiar with New World peoples, climate, and topography by following sources of labor and mineral wealth. The crown's depth of knowledge about its colonies improved through additional institutionalized ways of collecting this information. Some important recent research has demonstrated how empirical methods of observing nature and geography in the New World carried out by merchants, explorers, and royal bureaucrats resulted in the collection of information for the crown about its realms.[3] Entrepreneurs requested royal grants to seek profitable commodities or natural curiosities and would in turn inform the *Casa de la Contratación* or the Council of the Indies about what they found. Simultaneously, as it would do in Spain, the crown commissioned questionnaires, mapping projects, and book publications, all of which were meant to provide information to aid in the good government of the empire and exploitation of its natural resources.

The first Europeans who wrote about the landscapes of the New World typically described the variety and abundance of trees and forests with great enthusiasm. Columbus described Cuba as being so "full of tall, cool trees that it was a glory to see," with "everything full of palm trees and groves."[4] La Española's mountains were "all accessible and full of trees of a thousand different kinds, so high that they seem to reach the sky."[5] In describing Cuba, Bartolomé de las Casas "affirmed that all was 'flat land full of woods and glades'

3 See for example, Antonio Barrera, *Experiencing Nature: The Spanish American Empire and the Early Scientific Revolution* (Austin, TX: University of Texas Press, 2006).

4 Reinaldo Funes Monzote, *From Rainforest to Canefield in Cuba: An Environmental History Since 1492* (Chapel Hill: University of North Carolina Press, 2008), 7.

5 Quoted from Jon Cowans, ed. *Early Modern Spain: A Documentary History* (Philadelphia: University of Pennsylvania Press, 2003), 29.

and one could walk for 300 leagues and remain under the trees."[6] The early explorers and conquistadors enticed their audiences with their descriptions, but in many ways were limited by their inabilities to identify accurately what they saw. The variety of trees seemed to confound Columbus, who wrote,

> I saw many trees very dissimilar to those of our country, and many of them had branches of different sorts upon the same trunk; and such diversity was among them that it was the greatest wonder in the world to behold. Thus, for instance, one branch of a tree bore leaves like those of a cane, another branch of the same tree, leaves similar to those of the lentisk [flat and leathery]. In this manner a single tree bears five or six different kinds. Nor is this done by grafting, for that is a work of art, whereas these trees grow wild, and the natives take no care about them.[7]

Of course, as Shawn Miller has written, the New World "is seen more accurately as just another old world."[8] The Spanish encountered people almost everywhere they went, and the millions of inhabitants who had long existed before the Spanish arrived had refashioned American nature through hunting, agriculture, agroforestry, irrigation, terracing, urbanization, road-building, warfare, and many other ways. Even in the supposed wilderness of Amazonia in the mid-sixteenth century, for example, the numbers of fruit-bearing trees could not have been as high as they were without human intervention.[9] Europeans depended on indigenous peoples' knowledge to familiarize them with nature in the Americas, because even the most learned among observational scientists of the Renaissance were "at best familiar with a very limited selection of Classical works, that were frequently cited only for effect, sometimes in the final stages of revision."[10] Nothing in Strabo, Pliny, or Ptolemy prepared European travelers to the New World for what they encountered on their journeys. José de Acosta wrote, "One thing is certain: those who have not seen and tasted these fruits, they will gain little by reading this, and the ear will grow

---

6 Funes Monzote, 7.

7 Cristóbal Colón, *Cartas y documentos completos* (Madrid: Alianza Universidad, 1984): 36–37; Miguel Jordan Reyes, "La deforestación de la isla de Cuba durante la dominación española, 1492–1898" (PhD diss., Universidad Politécnica de Madrid, 2006), 29.

8 Shawn W. Miller, *An Environmental History of Latin America* (New York: Cambridge University Press, 2007), 9.

9 Miller, *Latin America*, 18.

10 Karl W. Butzer, "From Columbus to Acosta: Science, Geography and the New World," *Annals of the Association of American Geographers* 82, no. 3 (Sep. 1992), 544.

tired, and I will get tired as well."[11] Diego Álvarez Chanca, the physician on Columbus's second voyage, wrote in a letter to the Cabildo of Seville in 1493, "On this island [Marigalante] were such dense groves, it was wonderful, and such unusual trees not known to anyone."[12]

To help convey what they saw, early explorers made comparisons and analogies to what was familiar to them in Spain and discussed the economic potential of natural resources. For many, it was an earthly paradise not far from El Dorado. At the same time, it was "a secular wonderland of untapped wealth."[13] They communicated back to Europe that despite all that appeared different, forests functioned pretty much as they did in Spain. Fray Bernardino de Sahagún explained how the trees grew roots, trunks, branches and leaves, just as they did in Spain.[14] Columbus described trees so green and beautiful, "like the trees in May in Spain," except these trees kept their foliage year-round.[15] He pointed out the "'great and marvelous pinewoods' from which one could build as many ships as one desired." Explorers of the Caribbean, along the Pacific coast, and even in the Philippines continued to anticipate the utility of what they encountered. Miguel López de Legazpi, for example, wrote to King Felipe II in 1568 praising the "abundance of timber" in the Philippines.[16] Crucially, therefore, the economic potential of the forests appeared limitless.

The necessity of maintaining the favor of the crown gave Columbus good reason to be enthusiastic and optimistic about the economic potential of what (and whom) he encountered. Of course, Columbus and other explorers wrote about much more than forests. They described the people they encountered, their material culture, and other aspects of the natural world, including climate, rivers, harbors, and precious metals. Further voyages led to new settlements and more extensive explorations. Ventures throughout mainland America resulted in many new reports in the 1520s and 1530s. In 1527, the crown took a more active role in the process by ordering pilots and ship masters to send reports of their voyages indicating their routes taken, islands and lands they found, and details of coastline geography to go towards producing or improving the master portolan chart, the *Padrón General*, held at the *Casa*

11 Jordan Reyes, 39.

12 Jordan Reyes, 31.

13 William D. Phillips, Jr. and Carla Rahn Phillips, *The Worlds of Christopher Columbus* (Cambridge: Cambridge University Press, 1992), 183.

14 Jordan Reyes, 41.

15 Cowans, 29.

16 Greg Bankoff, "'Tree as the Enemy of Man': Changing Attitudes of the Forests of the Philippines, 1565–1898," *Philippine Studies* 52, no. 3 (Sep. 2004), 321.

de la Contratación. Gonzalo Fernández de Oviedo wrote his *Sumario de la historia natural de las Indias* (published 1526) based on his own observations and experiences rather than on classical authorities. To help Oviedo publish a more complete natural history of the Indies and expand on his summary account, King Carlos I decreed in 1532 to all officials of the Indies to send back information about the land, islands, their people, plants, and animals. All contributors had to sign their names. In 1534, a *real cédula* told Fray Tomás de Berlanga, Bishop of Tierra Firme, to go personally to Peru for the purpose of informing the king of what he was to find, including ports, towns, rivers, *montes*, *dehesas*, animals, and "the qualities and strangeness of the province."[17]

Before he sent out questionnaires in Spain that would result in the *Relaciones topográficas de España*, Juan de Ovando carried out a *visita* of the Council of the Indies, an investigation of the council's operations, which led to important reforms. His work for the *visita* between 1567 and 1571 led him to seek all kinds of information from the New World that would enlighten the council on the region they administered but typically knew little about from first hand experience. Between 1569 and 1575, Ovando sent out questionnaires of varying lengths to jurisdictions in the New World. In June of 1573, Felipe II issued a *real cédula* ordering a compilation of this information into several books.[18] Ovando created a special post to help with the project, the *cosmógrafo-cronista*, to which he appointed his secretary and assistant in the *visita*, Juan López de Velasco. The work of López would result in an extensive collection of information on the geography and people of the New World. Responses to the questionnaires returned to Spain between 1578 and 1586, and were gathered together to form the *Relaciones geográficas de Indias*. Ovando also commissioned the scientific voyage of Francisco Hernández (1571–77). In 1570, Hernández, who was the court physician for Felipe II, was sent to "gather information generally about herbs, trees, and medicinal plants." His voyage resulted in the identification of over 3,000 new plants for Europeans, the names of which often retained the indigenous terms. Hernández produced a 16-book natural history and 2,071 pages of paintings of plants.[19]

Spanish exploration of the New World generated a unique system of knowledge production about unfamiliar geography, of which forests were only one component. The crown learned about its distant realms through a very

17 Archivo General de Indias (Hereafter, AGI), Lima, 565, legajo (Hereafter, leg.) 2, Folio 1 verso, Real Cédula, 19-07-1534.

18 Stafford Poole, *Juan de Ovando: Governing the Spanish Empire in the Reign of Felipe II* (Norman, OK: University of Oklahoma Press, 2004), 141.

19 Antonio Barrera, "Empire and Knowledge: Reporting from the New World," *Colonial Latin American Review* 15, no. 1 (June 2006), 49.

diverse group of individuals sharing lived first-hand accounts. The accepted wisdom of the natural world had surpassed its limits, so the practice of empirical observation was the only source of knowledge for Europeans about the New World.[20] Upon entering channels of highly regularized lines of communication geographic and natural historical reports made their way to the two most important administrative bodies dealing with the American kingdoms, the *Casa de la Contratación* and the Council of the Indies, shaping royal policy in the colonies including forest policy. In the colonies, exploration, commerce, and defense were closely interconnected, and each relied on the vessels built in peninsular and colonial shipyards.

Spanish Shipbuilding in the Old and New Worlds

For the most part, shipbuilding practiced by Europeans continued to rely on long-established methods. The rate and timing of construction for the Spanish fleets, whether carried out under the crown directly or through private contract, adhered to many of the same limitations as elsewhere in Europe. Adjustments in shipbuilding methods would have to be made in colonial settings where weather patterns, plant life, and sources of labor differed from Northern Spain, but not to the technology and methods of construction. Phases of the moon guided the timetables of construction. In Europe, master carpenters preferred to make their cuts in the forest sometime between November and February during periods of a waning moon in its last quarter. This practice was based on ancient astrological writings, particularly Pliny the Elder's *Natural History*, which advised not to cut during a waxing moon because the wood deteriorated quickly and decayed. Shipbuilders in all Europe's maritime powers shared this belief, and it was based on close observation of the swelling of plants during the waxing phases of the moon. It was important for master carpenters to fell at a time when the swelling of moisture or sap in the tree was at a minimum, because of the belief that this would produce more durable wood. The winter months were preferred because plants and trees were long dormant. Shipbuilders also believed the heat of the summer months would cause the sap to ferment and the wood to rot. The felling of spars was an exception; carpenters cut them in May seeking greater elasticity. Other benefits to logging in the winter included the absence of leaves, which allowed for unobstructed assessment of the deciduous trees' conditions and the greater availability of labor during the slow months of the agricultural calendar.

20 Barrera, "Empire and Knowledge," 46.

The major drawback in Spain, but not in more tropical colonial environments, was the winter weather, which could make roads impassable and forests inaccessible.[21]

Before the cutting and hauling of the timber down the mountain, the master carpenter inspected the forests for suitable trees. The type of terrain and climate influenced the characteristics of the wood, and carpenters as well as superintendents would know for example that wood is more porous and grows more rapidly in more humid areas, and in areas of harsh winters of ice, snow, and wind they would be more likely to see cracks and irregularities in the distribution of the fibers. They inspected the bark texture to look for signs of disease, cracks, or impaired growth. They took note of the distribution of branches, which determined the location of knots in the wood. Shipwrights were interested in the healthy parts of the trunk's heartwood for larger pieces of the ship's framework. Typically, the branches were cut off and left for local domestic use. Shipwrights were also interested in finding compass timber that grew naturally into the desired curved or bent pieces. To exploit the strength of the fibers of the wood to the fullest, the carpenter would cut along the grain, ideally in the shape of the piece that would be carved (Figure 4). The lumberjacks' work of stripping the bark, carving, and sawing was performed in the forest. Newly cut timber, called 'green' timber needed time to 'season' without rotting from the effects of aerial fungi. Storage aimed to prevent fungal attacks which occurred mostly on moist timber and in poorly ventilated spaces. Some pieces were stacked and rotated to allow air to pass between them. Some pieces, like the keels, were submerged in brackish water, which facilitated the exit of sap while protecting it against the fungi, and then later dried in the open air. The time for proper seasoning varied, but larger pieces of oak required about two years. Given the high demands for ships in this era, the employment of green timber in construction was more common than the crown would have wished.[22]

21 Isabel Vicente Maroto, "The Art of Shipbuilding in Spain's Golden Century," *History of Technology* 30 (2010), 79–94; José Luis Casado Soto, *Barcos y Astilleros: La Construcción Naval en Cantabria* (Santander: Puerto de Santander, 1993); Maria Lourdes Odriozola Oyarbide, *Construcción Naval en el País Vasco, siglos XVI–XIX: Evolución y Análysis Comparativo* (San Sebastián: Diputación Foral de Gipuzkoa, Departamento de Economía y Turismo, 2002); Carla Rahn Phillips, *Six Galleons for the King of Spain: Imperial Defense in the Early Seventeenth Century* (Baltimore: Johns Hopkins University, 1986), 19–77; David Goodman, *Spanish Naval Power, 1589–1665: Reconstruction and Defeat* (Cambridge: Cambridge University Press, 1997), 109–137.

22 Shipbuilders and the king alike also disliked prolonged delays in construction, especially after the cuts were made. The wood could rot and putrify in the winter rains.

The wood then had to be transported from the forest to the coast. After winter, rising water levels in rivers allowed an easier and more cost-effective transport option than over land, but this still required skill and strength while posing significant risks.

The practice of shipbuilding was an art that was passed on through experience. Attempts at improving measurement standards were made, but remained difficult and a subject of much dispute in the seventeenth century. The process of construction began on the beach or a riverbed. The proportions of the vessel were determined by the keel, which was laid down first. The traditional formula said "each unit of measurement of the beam corresponded with two units of keel and three units of length," known in Spain as the rule of "as, dos, tres."[23] This relationship evolved as the search for ideal proportions

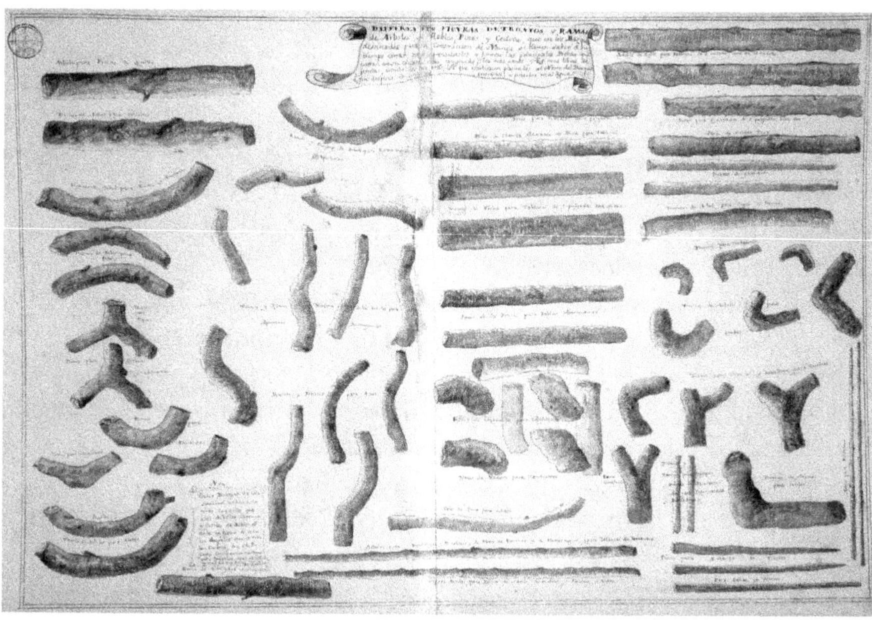

FIGURE 4 *Tree trunks and branches for shipbuilding – This page from the Marqués de la Victoria's* Dictionary of Naval Construction *shows drawings of trunks and branches to give an idea of what parts of trees were appropriated for the purposes of shipbuilding.*
SOURCE: ESPAÑA. MINISTERIO DE DEFENSA. ARCHIVO DEL MUSEO NAVAL. DICCIONARIO DE CONSTRUCCIÓN NAVAL DEL MARQUÉS DE LA VICTORIA. AMN. PL 235-1, HOJA 21. DIFERENTES FIGURAS DE TRONCOS Y RAMAS DE ÁRBOLES PARA DAR UNA IDEA DE SABERLOS APROPIAR A TODAS LAS PIEZAS QUE ENTRAN EN LA CONSTRUCCIÓN DE UN NAVÍO.

23 Vicente Maroto, "Art of Shipbuilding," 85.

continued and improvements to cargo capacity were made.[24] After the keel was laid, the stempost and sternpost were attached at either end and reinforced by sleepers, and the floor was reinforced by other timbers. Joined on top of the sleepers were the large curved pieces known as ribs, ideally made from holm oak. Horizontal planks were affixed as tightly together as possible. Caulkers would then fill the gaps with oakum (*estopa*) and tar (*betún*). The hull was then dragged to the water by hundreds of teams of oxen. Launching the hull into the water had to occur under ideal weather conditions. In northern Spain, this was limited to the summer months. The masts, spars, sails, and rigging were all added to the hull on the water. Sails were made of cotton canvas, and the rigging was most often made with hemp, but sometimes with esparto grass. After the rigging was completed, carpenters built the cabins, decks, and superstructures. The final step was the loading and securing of the artillery.

The costs and material demands of imperial defense were deemed necessary as long as New World treasure could continue to be carried safely. In the sixteenth century, the Spanish maritime presence was the largest the world had ever seen. Shipbuilding expert Tomé Cano wrote in 1608 that Spain had more than one thousand large ocean going vessels in about 1583.[25] At the time the Spanish king assumed the Portuguese throne in 1580, the Iberian powers together had about 300,000 tons of shipping, or nearly half of Europe's total. Only the Dutch came close with 232,000 tons.[26]

The two most prominent Spanish vessels of the sixteenth and seventeenth centuries were the *nao* and the galleon, and they continued to be built for the most part in peninsular shipyards. In the 1500s, 90% of the Spanish fleet was constructed in northern Spain, and the timber requirements were among the highest in the early modern world.[27] In this time, 190 *naos* and 181 galleons were built in the Basque Country, amounting to 68.1% of the total number of vessels built there. In the seventeenth century, 57 *naos* and 370 galleons were

24 Francisco Fernández-González, "The Spanish Regulations for Shipbuilding (Ordenanzas) of the Seventeenth Century," *International Journal of Naval History*, 8, no. 3 (December 2010), 1–30; Iván Valdez-Bubnov, "War, Trade, and Technology: The Politics of Spanish Shipbuilding Legislation, 1607–1728," *International Journal of Maritime History*, 21, no. 2 (December 2009), 75–102.

25 Tomé Cano, *Arte para fabricar y aparejar naos (1611)*, Enrique Marco Dorta, ed. (La Laguna, Canary Islands: Instituto de Estudios Canarios, 1964), 96.

26 Phillips, *Six Galleons*, 8, 22–23; Fernand Braudel, *The Structures of Everyday Life*, vol. 1 of *Civilization and Capitalism 15th–18th Century*, tr. Siân Reynolds (New York: Harper and Row, 1981), 363.

27 José Luis Casado Soto, *Los barcos españolas del siglo XVI y la Gran Armada de 1588* (Madrid: Editorial San Martín, 1988), 7–8.

made, making up 55.7% of the total.[28] The *nao* was principally used as a cargo carrier, but could also be used in warfare. It was one of the most-used ships of the sixteenth century, and was built mainly on the Cantabrian coast. Though quite variable in its typology, in general it was large with a carrying capacity between 100 and 600 *toneles*. The *nao* had three masts with rectangular sails on the mainmast and foremast, and a lateen sail on the mizzenmast. The galleon was of Iberian origin and was designed more explicitly for war, particularly to serve as defenders for the *Carrera de Indias*. They were typically heavily armed and were 300 *toneles* or more. By the 1700s, the age of the galleon had passed, and *navíos* and *fragatas* made up 25.2% of vessels built in the Basque country. However, production in this region had declined as arsenals elsewhere in Spain and the Americas became preferred, so only 82 of the 227 *navíos* built in the eighteenth century came from the Basque region.[29] *Navíos* were sometimes referred to ships-of-the-line by the eighteenth century, but the term came into use in the seventeenth century as a generic denomination for *naos* or similar vessels of small or middling tonnage.[30]

Spain's sixteenth-century naval fleet at the time of the Great Armada required an estimated 57,868 *toneladas* of wood.[31] For each *tonelada*, ten cubic meters of cut timber was required and about twenty cubic meters uncut, adding up to more than one million cubic meters of wood. Spanish shipyards consumed about six million trees, or 120,000 hectares of good forests if one hectare could hold about fifty good trees.[32] Two or three hectares of hardwoods and two hectares of pine were required to build a single large galley.[33] Galleons ranging from about 330 to 540 *toneladas* in the early seventeenth century required several thousand trees of moderate size each.[34] About forty pine trees

28 Odriozola Oyarbide, 201.

29 John D. Harbron, *Trafalgar and the Spanish Navy* (Annapolis: Naval Institute Press, 1988), 164–173.

30 Vicente Maroto, "Art of Shipbuilding," 80–85.

31 Erich Bauer Manderscheid, *Los Montes de España en la Historia* (Madrid: Fundación Conde de Valle de Salazar, 2003; First ed., 1980), 173.

32 Bauer, *Montes*, 170. The Spanish navy consumed 3 million trees in the era of the 1748 Forest Ordinances alone, María Valbuena-Carabaña, Unai López de Heredia, Pablo Fuentes-Utrilla, Inés González-Doncel, and Luis Gil, "Historical and recent changes in the Spanish forests: A socio-economic process," *Review of Paleobotany and Palynology* 162 (2010), 498, citing José Merino Navarro, *La Armada Española en el siglo XVIII* (Madrid: Fundación Universitaria Española, 1981).

33 Sam White, *The Climate of Rebellion in the Early Modern Ottoman Empire* (New York: Cambridge University Press, 2011), 16.

34 Phillips, *Six Galleons*, 60, 80, and 262 (n. 16).

were needed for the spars alone. Later in the century, a seventy-gun ship required 78,000 cubic feet of oak, or 2,800 mature trees. By the middle of the eighteenth century, a first-rate vessel built in Havana required more than twice that amount.[35] A 110-gun ship used about 4,000 oak trees, or the equivalent of 30–40 hectares of woodland.[36] Of the wood used in ship construction, 30–50% came from coniferous forests, while 50–70% came from deciduous forests. In general, the cost of wood was the greatest expense among shipbuilding materials, totaling about one-third of the final cost.

Despite changes in vessel types and typical sizes, shipbuilding practices remained relatively unchanged during the early modern period with respect to the general requirements for wood. As a result, superintendents and other representatives of the crown simply had to look for ways to increase the resource base through new plantings and new access to unexploited forests for the navy when there were greater demands for output. Nothing could be done about the restrictions on the timing of shipbuilding, which were guided by the weather, the phases of the moon, and customary or ancient wisdom.

Regulating Colonial Forest Use in the Sixteenth Century

Spanish colonists exploited forests of the *ultramar* from the very beginning. For the first half of the sixteenth century, other than designating municipal forests as commons, there was little effort from the crown to regulate forests restrictively, as it had already started to do in Spain. This held true for forest exploitation in general, not just naval shipbuilding. A recent study of the impacts of silver mining on forests in New Spain raised the point that the crown did not implement conservation regulations on wood as an important fuel source near silver mines because the crown did not wish to do anything that would curb the production of precious metals.[37] Castilians sought to

35 John Robert McNeill, *Atlantic Empires of France and Spain: Louisbourg and Havana, 1700–1763* (Chapel Hill: The University of North Carolina Press, 1985), 229 n. 66.

36 Michael Williams, *Deforesting the Earth: From Prehistory to Global Crisis, An Abridgment* (Chicago: University of Chicago Press, 2006), 193; cited in Bankoff, "Tree as the Enemy of Man," 324.

37 Daviken Studniki-Gizbert and David Schecter, "The Environmental Dynamics of a Colonial Fuel-rush: Silver Mining and Deforestation in New Spain, 1522–1810," *Environmental History*, 15, no. 1 (2010), 94–119. Historian Lawrence Clayton added "When extracting timber, one needed simply to observe some regulations directed at preventing the abuse of Indian labor. The forests themselves, as indeed most American resources in this period, were virtually free to be exploited." Lawrence A. Clayton, *Caulkers*

replicate their Old World landscapes in the new by bringing with them familiar crops like wheat, grapevines, and olive trees, and familiar animals like horses, cattle, pigs, and sheep.[38] They brought weeds, vermin, and diseases, too. Old World pathogens caused more indigenous deaths than European brutality did, facilitating European conquest. As Alfred Crosby has pointed out, any explanation of the demographic advance of Europeans in the New World has to include the demoralization and annihilation of the indigenous population as well as the successful establishment of European agricultural systems.[39] Settlements began on the islands of the Caribbean, but soon the Spanish established contact with more peoples, leading to the eventual conquest of the two largest and most powerful empires in the Americas, the Mexica (Aztecs) in the central Valley of Mexico and the Incas in the highlands of Peru. As during the Reconquest in Spain, all newly conquered land was declared *tierras de realengo*, posessed by the crown, and then the king granted land to settlers. Also as in Spain, forests were designated communal property. Some of the earliest legislation issued in the New World reserved the *montes* as commons. In 1510, Doña Juana sent a royal order from Monzón to the governor of La Española stipulating that fruit-bearing trees in the recently discovered forests are to be in common use by all residents.[40] Similar orders were made for forests elsewhere in the empire in 1532, 1541, 1543, and 1556.[41] In 1539, Carlos I decreed from Valladolid that *encomenderos* had to plant trees for firewood, including willows.[42] In 1559, Felipe II issued a law stipulating that Indians could cut wood from the *montes* for their own use.[43]

Shipbuilding in the Caribbean became important early on to carry out explorations of coastlines, and was aided by royal legislation. Between 1516 and 1518, the king issued three royal decrees for residents to build ships for

---

and Carpenters in a New World: The Shipyards of Colonial Guayaquil (Athens, Ohio: Ohio University Center for International Studies, 1980), 105.

38 Karl W. Butzer, "Ecology in the Long View: Settlement Histories, Agrosystemic Strategies and Ecological Performance," *Journal of Field Ecology* 23 (Summer 1996), 144–148.

39 Alfred Crosby, *Ecological Imperialism: The Biological Expansion of Europe, 900–1900* (New York: Cambridge University Press, 1986), 146–147.

40 AGI, Indiferente, 418, L.3, F.28, 15-06-1510.

41 AGI, Mexico, 1088, L.2, F.32r–46v (03-20-1532); AGI, Santo Domingo,868, L.2, F.80v (15-04-1541); AGI, Santo Domingo, 868, L.2, F.159v (01-05-1543); AGI, Santo Domingo, 899, L.1, F.32 (18-10-1556); *Leyes de Indias*, Libro IV, Titulo 17, Ley 5, 7, 8, 16.

42 *Leyes de Indias*, Libro IV, Titulo 17, Ley 16.

43 Gaspar de Aranda y Antón, *La Administración Forestal y los Montes de Ultramar durante el Siglo XIX* (Madrid: ICONA, 1995), 54.

regional trade and exploration.[44] In 1518, a royal decree authorized inhabitants to equip vessels with guns to discover and conquer new islands and territories.[45] In 1527, a *real cédula* sent to the governor and officials of Tierra Firme confirmed that colonists there could build *navíos* for explorations and rescue missions.[46] In the ensuing decades, shipbuilding expanded to produce merchant vessels for regional trade and warships for the convoy system linking the mines of Mexico and Peru with Spain. In the unfamiliar environments of the New World and the Philippines, shipbuilders adapted their work to the available materials when they could. To produce the oakum used for caulking, for example, the Spanish in Ecuador stripped the bark off coconuts. In the Philippines, Abacá, a species of banana, was used for the rigging. The Spanish would find species of pine trees, oak trees, as well as completely unfamiliar trees that possessed the qualities they were looking for, either for a sturdy hull, flexible masts, or resistance to dry rot and shipworm. As in Spain, most construction of larger vessels was carried out in one of two ways: under direct royal administration or through *asiento*, a contract between the crown and the shipbuilder.[47] At first, in either case, Spaniards performed the skilled labor of shipbuilding, including caulking, carpentry, rope-making, and metal-working, but over time this was done by *criollos*, or American-born Europeans. Indians and Africans provided skilled labor as well, but were mostly employed for manual labor at the shipyards or in the procurement of raw materials.[48] Shipyards emerged in several locations, anywhere there was the right combination of forest resources, suitable harbors, available labor, and economic activity. Their characteristics varied widely. They could be 'royal' shipyards, usually only limited to the time in which vessels were being built for the crown, or they could be private. They could be permanent installations, but most were temporary, small and situated on a riverbed.[49] For most of the century, shipbuilding occurred in many locations that were beyond any viceregal

44 Alejandro de la Fuente, *Havana and the Atlantic in the Sixteenth Century* (Chapel Hill: University of North Carolina Press, 2008), 128.

45 Funes Monzote, 20.

46 AGI, Panama, 233, L.2, F.242r–245v; Respuesta al gobernador y oficiales de Tierra Firme; 16-03-1527.

47 See I.A.A. Thompson, *War and Government in Habsburg Spain, 1560–1620* (London: Athlone Press, 1976). The debate carried on to the Philippines: AGI, Patronato, 262, N.1, R.3, 1587.

48 Marina Alfonso Mola and Carlos Martínez Shaw, "Los Astilleros de la América Colonial," in *Historia General de la América Latina*, v. 3, n. 1, Alfredo Castillero Calvo and Allan J. Kuethe, eds. (Madrid: Trotta, 2000), 280 and 282.

49 Alfonso Mola and Martínez Shaw, 281.

authority, and distance continued to be a formidable obstacle to territorial oversight.

However, by the end of the sixteenth century production had concentrated in a small number of larger and permanent sites, including Havana on the Atlantic side and Guayaquil on the Pacific, or *Mar del Sur* coast, both places maintaining the capacity to construct first-rate galleons. A similar process occurred in the Philippines in the later sixteenth century with the shipyard of Cavite near Manila building some of the largest galleons in the empire. Shipbuilding, then, developed in three major colonial zones, and a fourth zone emerged slowly in the Río de la Plata region centered on the city of Asunción del Paraguay, all situated near hardwood and pine forests capable of supplying most materials, with iron being the one major exception. Due to numerous first-hand accounts obtained by royal officials, it was clear enough in these early decades that where there were problems supporting a shipbuilding industry, the shortage of skilled and unskilled labor was more of an impediment to success than a shortage of timber.

Realejo, Guayaquil, and the Pacific Coastal Region

The Straits of Magellan posed too many dangers to be considered an operative link between the Atlantic and Pacific, so shipbuilding on the Pacific emerged out of necessity to explore the western coastline (Map 3). In 1518–19, Vasco Nuñez de Balboa was the first European to build vessels on the Mar del Sur, at the Gulf of San Miguel (Panama), but the resources he used were not local.[50] Rather, indigenous laborers carried materials overland across the isthmus, and around 2,000 died from the experience.[51] Several shipbuilding sites on the Pacific emerged between central Mexico and the Gulf of San Miguel. Maritime transport helped develop a Pacific coast trade system in the 1530s and 1540s in which goods entered Panama from Mexico and silver from Peru for the fairs at Nombre de Dios or Portobelo. Peruvian silver, the most valuable component of the Pacific coast trade, accounted for at least forty percent of the value of treasure that arrived in Spain from all American possessions. Upon the regular arrivals of the Manila galleons, Asian goods were reexported to Peru via Mexico for Peruvian silver, Chilean copper, and Ecuadorian cacao. Francis Drake's entry into the Pacific in 1578 and his raids along the coast through 1579 spurred the Spanish to improve security for the Peruvian silver on its voyage from Arica, the port of Potosí, to Callao, and on to Panama. To patrol the coast from

50 AGI, Patronato, 260, N.2, R.38.

51 William J. McCarthy, "The Yards at Cavite: Shipbuilding in the Early Colonial Philippines," *International Journal of Maritime History* 7, no. 2 (December 1995), 157.

Tierra del Fuego to the isthmus, the crown established the *Armada del Mar del Sur*. Typically, it was composed of two large galleons and four or five small roundships. It remained in operation into the eighteenth century.[52]

A mountainous spine runs down the western edge of Middle and South America, with smaller ranges and plains to the east. In the northern limits of the Spanish empire, Mexico, like Spain, exists on a high plateau, called the altiplano, and sits between the Sierra Madre Occidental and Sierra Madre Oriental mountain ranges, narrowing towards the southeast. The mountains descend to coastal plains, and the plateau ends at a formidable chain of volcanoes southeast of Mexico City. Mexico's western Sierra Madre receives plenty of rain, but blocks the weather from the altiplano. The vegetation of Mexico increases southward, going from sandy desert and arid scrubland in the widest parts of the altiplano, to grasslands and stunted trees, to denser woodland south of the Tropic of Cancer. The central highland of Mexico experiences reduced temperatures at tropical latitudes due to elevations of around 7,000 feet. The coasts, in contrast, become fully tropical south of the Tropic of Cancer at about 23 degrees latitude. Heat, high humidity, and low altitudes create the conditions for extensive tropical forests, with the wettest conditions occurring along the lower Gulf coast in the regions of Veracruz and Coatzacoalcos. The Pacific coast of Central America has a lighter covering of tropical forests, and in the higher altitudes of Guatemala, Honduras, and Nicaragua there are some mixed forests of conifers and broadleafed trees. These forests would serve the shipbuilding industry of Central America's Pacific coast. The forests on the Isthmus of Tehuantepec were used by Cortés and his men to build ships near the city of Tehuantepec on the Pacific coast. Soon after, in the 1530s, Pedro de Alvarado built three galleons in Iztapa, in Guatemala. Around the same time, shipyards in Acajutla, El Salvador, and Huatulco, in Mexico, became active. From the 1540s, the port of Huatulco participated in trade with Peru and built small vessels. Other early establishments were located at Alanje in Panama, the Gulf of Nicoya in Costa Rica, and Realejo in Nicaragua. In the second half of the century, the Mexican cities of La Navidad and Acapulco built galleons, some for the Manila trade. But it was Realejo from 1533 that was the first and most significant major shipbuilding site on the Central American coast. Demand for ships to transport Indian slaves from Nicaragua to Panama provided the impetus for shipbuilding there.[53] Shipbuilders produced vessels for coastwise trade to Panama as well as larger vessels for the Mexico-Peru

52 Goodman, 4.

53 David R. Radell and James J. Parsons, "Realejo: A Forgotten Colonial Port and Shipbuilding Center in Nicaragua," *Hispanic American Historical Review* 51, no. 2 (May 1971), 300.

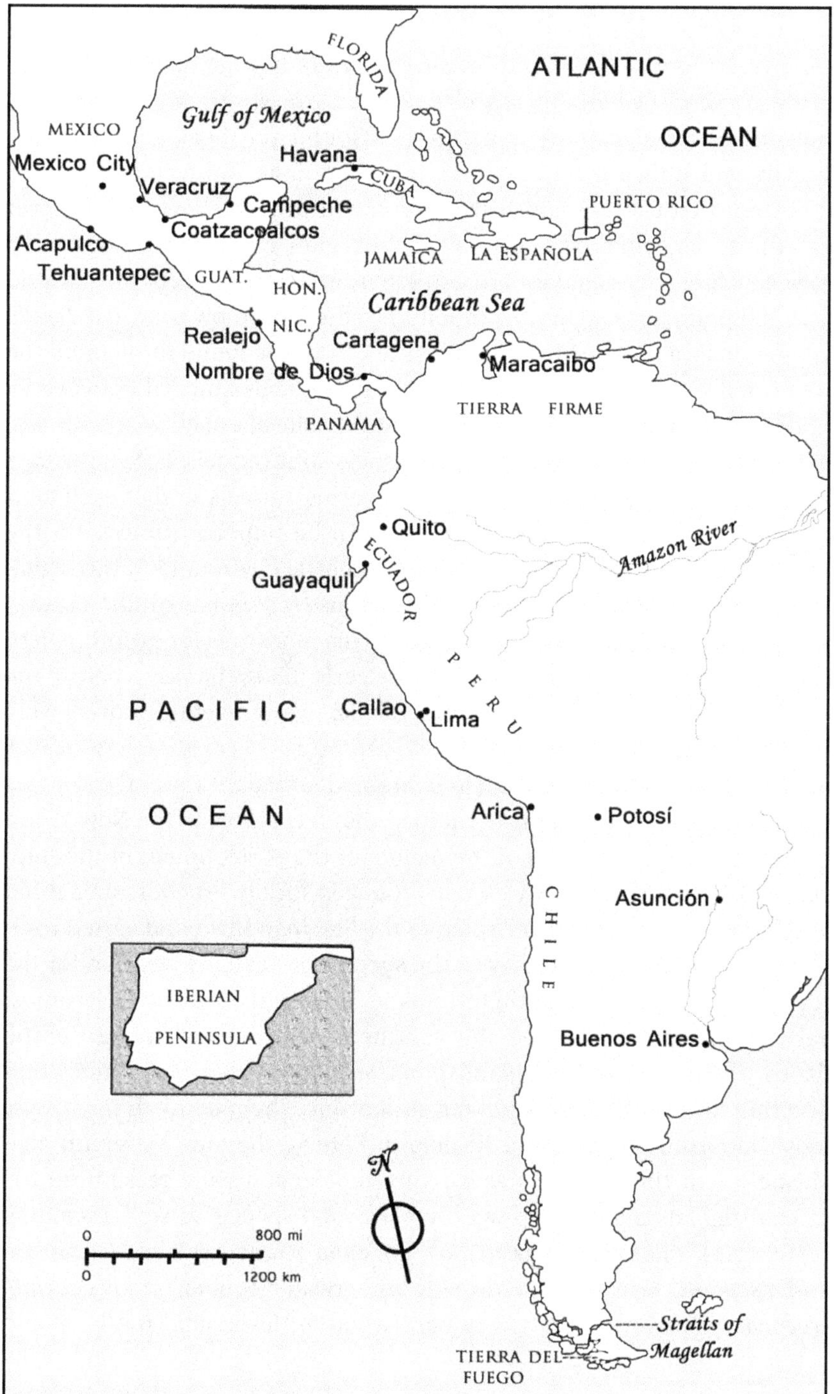

MAP 3 *The Americas*

trade, and even several galleons for the Manila trade. The port was situated on a mangrove estuary at the mouth of several rivers flowing across the León-Chinandega plain. The harbor was well protected from strong winds and swells by three islands, but settlers and travelers found the heat and humidity unpleasant and unhealthy. Coastal sailing conditions were unfamiliar and could be treacherous, so Spanish navigators had to learn how to manuever through the strong winds and currents of the region, especially in the Gulf of Papagayos, before they could establish trade from Realejo. Except in very cold waters, all shipbuilders had to worry about the effects of shipworms. The *Teredo navalis* is actually a marine bivalve mollusk and could be found throughout the world's littoral zones and mostly where water temperates ranged between 52 and 77 degrees Fahrenheit. It uses its hard ridged plate of a shell to burrow into the hull of a ship and then along the grain to make long tunnels. The shipworms do not break the surface or enter other shipworms' tunnels so that even after months the extensive tunnel-work would still be imperceptible from the outside.[54] Despite the challenges of Realejo's coastal environment, the region became an important port for colonial Nicaragua by supplying Indian slaves, timber, dyewood, and naval stores, and producing foodstuffs for export to Peru from the fertile plains around the city. By the early 1530s, thanks in part to the heavy-handed governorship of Pedrarias Dávila, nearly twenty caravels were transporting slaves from Nicaragua.

The forests and fields near Realejo contained abundant natural resources suitable for shipbuilding. Colonists claimed that few places in the New World could compare to the resources of the region for the development of the shipbuilding industry.[55] Only iron was not produced locally. Anchors and other iron products were shipped from Spain to the San Juan River and carted overland from Granada to Realejo. Key to the success of Realejo's position for the southward trade to Panama and Peru was its proximity to the southernmost natural stand of pine forest on the continent, which grew in the Central Highlands of Nicaragua. The upland pine forests provided masts and other construction timber, pitch for caulking, and resins. The pine pitch (*brea*) was the most important export from Realejo to Peru in the later sixteenth century. Realejo sent timber to Callao for ship repairs and wine casks from the 1540s. The Highlands grew ship-worthy hardwood species as well, including cedar (*Cedrela*), mahogany (*Swietenia*), *guácimo* (*Guazumo*), *madero negro* (*Gliricidia sepium*), which was a "durable 'underwater' timber," *palo cuadrado* (*Macrocnemum glabresceus*), "a very hard wood for levers and rudders," and

54 Miller, *Fruitless Trees*, 23.

55 Radell and Parsons, 301.

sapodilla (*Achras sapota*), "famed for its resistance to *teredo*."[56] The strong reputation of timber and rigging used in construction at Realejo was known all along the Pacific. The Marqués de Varinar wrote that

> this port of Realejo is best fitted for the purpose in all of Peru for the construction of ships for Your Majesty as well as for private purposes, on account of the excellence of the woods, as well as the abundance in which they occur, the availability of the Indians, many of whom are carpenters and sawyers: pitch and tar and much cordage, all from the land, are also found there.[57]

Indigenous laborers acquired and transported these forest products under harsh conditions. According to Bartolomé de Las Casas, the strenuous task of hauling timbers overland to the coast was one of the main causes of death and misery for the Indians of Nicaragua. Indians grew and wove cotton for textiles, and the Spanish redirected their cotton production to make sailcloth as well. Other fibrous plants cultivated locally were used for cordage, including maguey (*Agave*) and sisal hemp, or *cabuya* (*Furcraea cabuya*). The fibers of these plants were soaked in water then sun-dried before being twisted into lines and covered in a mixture of pitch and tar.[58] Parts of Nicaragua became renowned for plants that could make stronger cordage that those in Spain, along with superior cotton sailcloth, pitch, and planks.[59] The toll the Spanish presence took on the indigenous population was so severe that the Realejo shipyard recommended importing African slaves as early as 1578.

The shipyard at Realejo remained productive throughout the colonial period, but suffered from increased pirate attacks from the late sixteenth century. In 1579, Drake bypassed the port, but the threat of his return depressed commerce in the area. The harbor was not well defended and the city was easily accessible to raiding parties. Production also suffered from the increased construction of vessels in the Philippines for the Manila-Acapulco trade. The city of Realejo itself lost some of its relevance due to the relocation of León to within three leagues from the port. As the region's commercial and political center, León attracted Realejo's merchants and influential citizens away from the coast to the more hospitable inland location. There is evidence of gradual depletion of timber, but general mismanagement and shifts in the economy contributed more directly to the decline of the shipyard. By the seventeenth

56 Radell and Parsons, 302.
57 Clayton, 80.
58 Clayton, 85.
59 Clayton, 80.

century, the most important colonial shipyard on the Pacific was located at Guayaquil.

The reputation of Realejo would remain strong, mostly due to its forest resources, even by comparison to its Ecuadorian counterpart.

> In the vicinity of this port [Realejo], so says a report, there is an abundance of cedars, mulberrys, balsas, balsam of tolus, guachapelis and many others, being the best that has been discovered in the Indies to date. And, even given the case that the woods of Guayaquil lasted as long as those of Realejo, the latter are of a better type: Guayaquil woods split upon receiving the nail, and are more brittle: Realejo woods suck up the nail without splitting, and long experience with ships that have been broken up at each port proves this: Guayaquil woods are damaged where touched by the nail, because they are split by them: Realejo woods are undamaged.[60]

The woods of Realejo were understood to be "lighter, the type which made swifter ships," and shipbuilding was made easier for the fact that the wood did "not throw off splinters," and the timbers were larger than those of Guayaquil, allowing for Realejo's vessels to be stronger and built with fewer parts.

Guayaquil had its defenders, too, and benefitted from a more central role in the Pacific coast economy. Like Realejo, the success of Guayaquil can be attributed to a combination of superior natural resources, an indigenous labor supply, and an active port with extensive commercial ties to the region. In addition to shipbuilding and timber exporting, Guayaquil was the port of the Ecuadorian highlands making it the center of the export trade of cacao, as well as the exporter of gold and silver from Quito. Vessels built in Guayaquil formed part of the *Armada del Mar del Sur* and served the routes of Pacific intra-coastal trade. The exceptional characteristic feature of its location was that the port was situated in the nearest region north of Lima with the forest resources to support a local shipbuilding industry. The rains brought by the Equatorial Counter Current stop just to the south of Guayaquil, making the coast down to Chile, the Peruvian coastal desert, unsuitable for wooden ship construction. The resources of central Chile were comparable, but were often considered inferior to Guayaquil. The forests of Chile were south of Lima, away from the trunk routes of trade with Panama and New Spain. As it was, a roundtrip from Guayaquil to Callao took three to six months. The routes were defined by the Humboldt Current, which sweeps north up the coast of Peru, and southwesterly trade winds. The current turns west at 2 or 3 degrees south latitude, about

60 Clayton, 79; Radell and Parsons, 310; Ricardo Cappa, *Estudios críticos acerca de la dominación española en América* (Madrid, 1894), X, 100–101.

the same latitude as Guayaquil making travel south more difficult. Earliest evidence of ship construction appears in 1557 on the orders of the Viceroy of Peru, Andrés Hurtado de Mendoza, Marqués de Cañete on the island of Puná at the mouth of the Guayas River. The industry became established over the next decade and was able to supply warships to defend the silver ships against pirates in the late 1570s and 1580s, at a time when Guayaquil had about one hundred Spaniards, not including wives and children. The viceroys typically made the orders for ship construction and delegated authority to superintendents working for the crown. The decrease in viceregal spending slowed construction in the 1590s, but Guayaquil would thrive again in the seventeenth century as a production site of royal as well as private vessels.

Similar to the importance of Realejo's position near highland pine forests, Guayaquil could benefit from *palos maría*, strong and straight trees that could be used for masts and spars. Perhaps this was the primary reason for the establishment of ship construction in Guayaquil.[61] They were purported to be stronger but less flexible than European pines. For many other parts of the ship, the *guachapelí* tree was preferred. In the eighteenth century, Dionisio de Alsedo y Herrera wrote of *guachapelí* wood,

> Aside from its firm resistance to water rot and shipworm, it possesses the special quality of being easy to work with the axe and the adze, and the chisel, and the auger; from its trunk is made the keel, from its roots the crotches, and from its branches the stem, stern-post, knees, top timbers, futtocks, blocks and pulleys, and other smaller pieces; the brownish colored tree, called *guachapelí mulato* by the carpenters of the country, is the best.

Red and yellow oak were used for planking, but these oaks were not considered as strong as European oaks. Cinnamon laurel and mangrove were hard and heavy woods useful as keels and lower beams under the water line. Laurel was used for oars and sometimes masts when the more flexible *palos maría* was not available. *Cañafistola* was susceptible to wood rot and the *teredo navalis*, but was used above the water line. Most of these woods were found in the forest of Bulubulu, near the source of the river by that name, a tributary of the Guayas.[62]

In the sixteenth century, local Indian populations were a source of tribute or forced labor. As in Realejo, the Spanish reoriented some of the local handicrafts towards the shipbuilding industry. Cotton from the inland provinces of

61 Clayton, 85.

62 Clayton, 83–85.

Chachapoyas and Cajamarca was woven into sailcloth. Sisal hemp (*cabuya*) was worked into cordage. Cordage was produced on the island of Puná by Indians, and was even exported to Panama and Lima. Guayaquil had its own pitch lake nearby. One report noted, "In the town of Colonquillo there are pits from which a liquid or tar is extracted called copey, which is pitch." Still, pitch from New Spain was imported to dilute the tar. To produce the oakum used for caulking, the Spanish stripped the bark off coconuts, mashed it into a pulp, then separated the fibrous material and dried it out. It was then inserted into the seams of the ship. When it became wet, it expanded and then hardened producing a watertight seal. Hemp (*cáñamo*) was preferred as a source of oakum, and could be imported from Chile.

Superintendents were appointed by the viceroy and placed in charge of organizing the labor and acquiring the materials for ship construction. Under the authority of the viceroy, they could employ anyone they felt they needed and fine those who refused to comply. There was no Superintendent of Forests and Plantations as in Spain, but there was a chief lieutenant who operated as the liaison between the superintendent and the local craftsmen. As a trusted individual with an intimate knowledge of the forest and of the local shipbuilding industry, the chief lieutenant was usually a native or long-time resident. Other royally-appointed officials aiding the superintendent might include the *corregidor*, an accountant, treasurer, and a supply officer. Underneath this administrative level were enslaved and free laborers and craftsmen. The labor supply at Guayaquil was almost always sufficient due to the activity of private shipbuilding. By the early seventeenth century, private vessels made up most of the thirty to seventy vessels sailing the Pacific coast. Royal vessels were typically built in a flurry in response to threats of pirate attacks.

Havana and the Caribbean Region

While the colonies on the Pacific coast were markedly isolated from maritime trade on the Atlantic side, the resources of the forests along with the capacity to exploit Indian labor and skill enabled the Spaniards to build vessels on a par with those built in Spain. Colonists showed an ability to adapt to using local products as substitutes for European ones, which were sometimes deemed superior to the resources of Vizcaya or Cantabria. The emergence of a comparable shipbuilding industry on the Caribbean side came later and never became as self-sufficient; shipyards on the Atlantic side never needed to be, because of more reliable and affordable connections to Europe. Havana's success as the principal shipbuilding site in the Caribbean by the early seventeenth century was owed in part to its extensive forests of cedar, mahogany, oak, ebony, and pine, in part to its deep and easily-protected harbor, but mostly because of its role as the rendezvous point for the *Flota* and the *Galeones*.

Modern research seems to have confirmed the early European descriptions of Cuba as being almost entirely wooded. In the sixteenth century, forests probably covered nearly ninety percent of the island, of which 75–80% was tropical forest and the rest was composed of pinewoods or short forest. Five percent of the island was savanna and five percent was barren, coastal, or cleared. The numbers of early European settlers on the island rose after colonization began in 1510, but were reduced in the next decade as they moved on to explore the mainland. The Spanish population declined by as much as eighty percent between 1520 and 1540. The indigenous population also declined in the first decades of Spanish settlement, and no free Indians remained as early as 1517. The collapse of gold production in the 1530s played a factor in the island's stagnation during the early sixteenth century as well. By this time, though, merchants and navigators began to recognize Cuba's importance in Spain's Atlantic system. Havana emerged as the most important point of departure on the routes of maritime trade between America and Spain as they were being defined between the 1510s and the 1530s. After 1535, the Spanish had to deal with increasing corsair raids, first by the French and then by the English later in the century.[63] In 1538, the king ordered a fortress built at Havana. In 1553, the king ordered the governor of Cuba to move his permanent residence from Santiago to Havana, confirming the city's status. However, in 1555, the city was burned in an attack by the French corsair Jacques de Sores. A brief respite occurred when peace was declared with France in 1559, and the king halted the construction of two *navíos* on the island of La Española.[64] In the reconstruction of Havana, three forts were built and manned by a permanent garrison of 450. In 1561, Havana officially became the port of call for the Indies fleets, stimulating local commerce, agriculture, and livestock production.

Municipal councils in Cuba granted large estates with usufruct rights, but forests, streams, and pastures remained communal property. Pigs were raised in forests and cattle roamed in mixed forest and savanna. Settlers cleared some forests to expand pasture land. The sparse population during the early phase of colonization meant that disputes were comparatively rare. Increased settlement in the 1550s led to increased land grants and some early forest legislation, particularly around Havana. In 1550, black slaves were prohibited from cutting mahogany and cedar, which they used for troughs and boats but

63 John Lynch, *Spain 1516–1598: From Nation State to World Empire* (Oxford: Blackwell, 1991), 235.

64 AGI, Santo Domingo, 899, L.1, F.142 (09-09-1559).

were increasingly used for church and house construction, in a radius of 2 leagues, or about 8.5 km, of the city. If they were caught doing so, they would serve ten days in the stocks and receive three hundred lashes. If the slave acted on the orders of his master, the master would pay an additional twenty gold pesos. In 1557, Governor Francisco Carreño complained to the city council that people had to travel two or three leagues to find wood and suggested the creation of a municipal woodland. The council responded favorably to the governor, stipulating that no one was allowed to fell trees for any reason within a radius of one league from Havana so that residents would be able to access firewood and timber. The Spanish and other Europeans soon desired the precious and exotic woods from Cuba's forests for building projects across the Atlantic, but the Havana city council sought to prevent the cutting of Cuban trees for shipment to Castile in 1552. The crown would still have its way despite the wishes of the city council. A *consulta* from the Council of the Indies in 1576 discussed sending pine trees suitable for masts to San Juan de Ullúa and to Spain.[65] Cuban ebony, mahogany, ácana, and quiebrahacha were used for the construction of Felipe II's palace-monastery of San Lorenzo de El Escorial.[66]

By the time it designated Havana as the port of call for the Indies fleets, the crown had realized the potential for Havana's capacity to build vessels locally to defend and protect the area. In 1551, the first Cuban galleon was listed in the registry of the *Casa de la Contratación*. Governor Pedro Menéndez de Avilés boasted in a letter to the Council of the Indies in 1568 that he built six vessels for the purpose of conquering Florida that were as excellent as the vessels built in Vizcaya.[67] From mid-century, Cuba no longer had difficulty attracting the settlement of skilled craftsmen, who worked in several shipyards by 1600. In addition to the main shipyard, there were two other yards in the Bay of Havana, in the neighborhood of Campeche (1588) and across the bay on the 'pier Maria Melena' (1598). Another was east of the city at Puerto de Baracoa (1589), near stands of hardwoods, and another west of Havana on the Bay of Mariel. While some larger vessels were built in the sixteenth century, the average tonnage of vessels built in Cuba between 1590 and 1600 was 185 *toneladas*. Before 1590, most were fewer than 100 *toneladas*, and the largest rarely exceeded 300 *toneladas*. Cuban hardwoods suitable for shipbuilding included mahogany (*Swietenia mahagoni*), cedar (*Cedrela odorata*), and oak (*Tabebuia calcicola*). Cuban cedar was favored by shipbuilders because it was more resistant to dry rot than European oak and did not splinter as much

65 AGI, Indiferente, 738, N.232 (24-09-1576).

66 Funes Monzote, 19.

67 Funes Monzote, 21.

when hit by cannonballs.[68] Pine trees (*Pinus caribaea*) could be used for planks, pitch, and masts, but mainmasts and spars continued to be imported from northern Europe, principally Norway. Cordage was produced locally from the fibers of the majagua tree (*Hibiscus elatus*), but imports of rigging and cables came from Riga, England, and Flanders. Iron was also imported.

From 1576, the crown began regulating the use of forests around Havana to preserve the primary interests of the navy, the earliest such legislation in the Indies. That year a royal order to the governor of Cuba allowed for the cutting of pines within thirty leagues of Havana for the purpose of constructing *naos*.[69] Royal commissions for new vessels became more common in the later sixteenth century. In 1589, the king ordered up to eighteen *fragatas* built in the main Havana shipyard to help defend the Caribbean and sent 800,000 *reales* between 1589 and 1591 to build the initial six. In 1608, the king commissioned five galleons to become part of a permanent fleet to defend the Caribbean, the *Armada de Barlovento*. The mission of the fleet was in part to prevent Protestants from "planting their seed" in the Americas, but this fleet's operation was intermittent. In 1617, a new era began for Havana after a royal decree mandated the construction of an arsenal, confirming the city's privileged status as primary shipyard of the Indies.

Cavite and the Philippines

The Spanish never stopped their search for a western route to Asia in the sixteenth century. The settlement reached in the 1494 Treaty of Tordesillas demarcated a longitudinal line 370 leagues west of the Cape Verde Islands (46 degrees 30' W) allowing claims for lands discovered to the west for Spain and lands east for Portugal, but the treaty did not mark a similar line on the opposite side of the globe, and claims to the islands of southeast Asia remained open for dispute for years between the Iberian powers. At the time of Miguel López de Legazpi's voyage of 1564–5, the Philippines were not a unified polity or nation, but an archipelago of dozens of large islands and thousands of small ones. It is estimated that there are about 7,100 islands in the present-day nation. The island chain is part of an old longitudinal line of volcanic mountain range folds, and continues to experience volcanic and seismic activity. The islands are generally divided into three regions, Luzon and surrounding islands in the north, home to Manila and most of Spanish settlement activities, the Visayas in the middle, home to Cebu, where Magellan spent his final days, and Mindanao and surrounding islands in the south, largely controlled by Islamic

68 McNeill, *Atlantic Empires*, 133.

69 AGI, Patronato, 259, R.62 (1576).

powers with connections elsewhere in southeast Asia. The two largest islands, Luzon and Mindanao make up two-thirds of the total land area of the archipelago. The islands are mostly mountainous with narrow coastal plains and small but swift rivers. They are tropical and receive a lot of rain, which leaches much of the soil. Soil is renewed, though, by mountain streams and volcanic ash (Map 4).

Islands were inhabited by numerous groups of Malays who spoke eighty-eight different languages. Early settlers came chiefly from the south, but some people arrived from the Chinese mainland. Society was organized around small familial village groups called *barangays*, and led by chiefs, or *datus*. The arrival of the Portuguese in Southeast Asia dispersed many merchants away from Malacca and spurred Islamic political leaders to establish influence for themselves in other parts of the archipelago. One of these Muslims established a small sultanate on Mindanao. Among the sultanates in the region, the most powerful one was the Sultan of Brunei, on the island of Borneo, who had strong commercial and political ties to the Philippines. These island sultanates were conducive to the expansion of Islam and Islamic trade networks in the region, but Islam did not take root beyond the ruling elites in the Visayas and Luzon.

By 1569, Legazpi had to retreat north to Panay away from the Portuguese. In 1570, he sent a force of ninety Spaniards and three hundred Visayans to seize the *barangay* of Manila and defeated the Islamic ruler Raja Suleiman. Negotiations with local *datus* resulted in the establishment of a permanent Spanish settlement at Manila in 1571. Helping to ensure its success, Andrés de Urdaneta had established a viable return route to New Spain in 1565. Legazpi's settlement in Manila, then, put the Spanish out of the range of Portuguese and 'moorish' influence, but still within range of the Spice Islands.[70] Manila continued to be an important commercial center, with merchants from Japan, Malaysia, India, Sri Lanka, and Indochina regularly stopping by to trade. The Spanish also opened up and increased trade with the Chinese, who had already established relations with local merchants in Manila, a connection that would prove to be more meaningful for the Spanish in the long term than trade with the Moluccas. The Philippines remained an entrepôt for regional trade goods, so agriculture was largely neglected and the Spanish did not develop the

70 This was debated by Legazpi, who said in 1569 "The Philippines ought to be considered of little importance because at present the only article of profit that we can get from them is cinnamon." Quoted in Henry Kamen, *Empire: How Spain Became a World Power, 1492–1763* (New York: Harper Collins, 2003), 205.

MAP 4 *The Philippines*

islands along the lines of Latin American haciendas and plantations.[71] Spanish power on the islands was limited; their numbers were small, the indigenous population's regular contact with Old World diseases meant no demographic collapse ocurred as a result of the Spanish arrival, and the economy was not reshaped by cash crops or mining. Shipbuilding remained the most important colonial industry until sugar and tobacco production in the late eighteenth century.[72]

With such a small settler population, the Spanish employed shipwrights and other skilled personnel from the region, especially China. By 1586, the Spanish and mestizo population numbered about eight hundred, and were greatly outnumbered by the ten thousand Chinese merchants, artisans, and farmers in Manila at that time.[73] The Spanish, though, needed to rely on indigenous labor to support their nonagrarian settlements. In 1586, Santiago de Vera reported that the people of the Philippines proved to be "very skillful in making ships."[74] Cavite was the first important shipyard, and was operative by 1582.[75] When a fleet was being prepared in Cavite, about 1,400 carpenters were employed. By the early seventeenth century several other locations were active elsewhere on Luzon, including Pangasinán province north of Manila and the Camarines coast and Albay province in the Bicol Region of southeast Luzon. Shipbuilding also occurred on the islands of Mindoro and Marinduque, just to the south of Luzon, Masbate on the Visayan Sea, and at Otón and Arévalo, near Iloilo on Panay Island in the Visayas. In the early decades of the colony, vessels used by the Spanish in the Philippines were sometimes built elsewhere in the region, including Japan or Siam. In 1578, the captain general of the Philippines, Francisco de Sande, ordered the construction of galleys in Borneo.[76]

Early colonists, including Legazpi, wrote about the abundance and variety of the forests of the Philippines, and were quick to point out the potential of

71 John A. Larkin, "Philippine History Reconsidered: A Socioeconomic Perspective," *American Historical Review* 87, no. 3 (Jun. 1982), 598–601; McCarthy, 151.

72 McCarthy, 155.

73 Kamen, 207. "As late as 1637 Manila counted only a hundred and fifty Spanish households within its walls, an incredibly small number after some eighty years of colonization," 209.

74 Greg Bankoff, "'Tree as the Enemy of Man': Changing Attitudes of the Forests of the Philippines, 1565–1898," *Philippine Studies* 52, no. 3 (September 2004), 323; Greg Bankoff, "One Island too Many: Reappraising the Extent of Deforestation in the Philippines Prior to 1946," *Journal of Historical Geography* 33, no. 2 (April 2007), 318.

75 McCarthy, 154.

76 AGI, Patronato, 24, R.41.

the forests for naval purposes.[77] Juan Pacheco Maldonado estimated in a letter to Felipe II in 1575 that the Spanish could construct three to four galleons each year on the islands.[78] Some wrote about the "incorruptible woods" found throughout the region and an "abundance of wood for all kinds of vessels."[79] Antonio de Morga wrote in his *Sucesos de las Islas Filipinas* (1609),

> The land is everywhere well shaded by trees of different kinds, and there are fruit trees which beautify it all the year round, both along the shore and inland, on the plains and in the mountains. (It is well supplied with good fresh water rivers, both large and small, which flow into the sea. All of them are navigable and abound with every kind of fish, which are very tasty.) For this reason there is plenty of wood which is cut, sawn up, and then dragged to the rivers along which it is brought down. This wood is suitable for houses and buildings as well as for constructing large or small boats. There are in addition many stout, straight trees which are also light and pliant and can be used for making masts for ships or galleons. Thus any sort of vessel may be fitted with a mast made from a single trunk from one of these trees, without there being any need for splicing or fishing; or to make them up from different pieces. For the hulls of the ships, for keels, futtock- and top-timbers, and any other kinds of futtocks, breast-hooks, *puercas*, transoms, *llaves* and rudders, all sorts of good timber can be found easily. There is also good planking of quite suitable timber for the sides, decks, and upper-works.[80]

The importance of ships and shipbuilding in the colony was clear. Santiago de Vera wrote to King Felipe II in 1589, "It is most important that there be for this navigation plenty of ships, both for the emergencies of war which may arise, and for the preservation of these islands, which are supported by trade."[81] It did

---

77 Bankoff, "'Tree as the Enemy of Man'," 323.

78 Bankoff, "'Tree as the Enemy of Man'," 323.

79 Bankoff, "One Island Too Many," 318.

80 Antonio de Morga, *Sucesos de las Islas Filipinas*, tr. and ed. J.S. Cummins (London: Hakluyt Society and Cambridge University Press, 1971), 253–254. Indigenous laborers, not Spaniards, transported wood down the rivers.

81 Emma Helen Blair and James Alexander Robertson, eds. *The Philippine Islands, 1493–1898: explorations by early navigators, descriptions of the islands and their peoples, their history and records of the Catholic missions, as related in contemporaneous books and manuscripts, showing the political, economic, commercial, and religious conditions of those islands from their earliest relations with European nations to the beginning of the nineteenth century*. 55 vols. (Cleveland, OH: A.H. Clark Co., 1903–1909), v.VII, 78.

not take long for settlers to make the argument for building vessels for the Acapulco trade in the Philippines rather than in America. In 1582, Juan Bautista Román, *factor* and *veedor* of the Royal Hacienda of the Philippines, wrote "In the Philippines, *navíos* can be made very cheap to transport spices (cinnamon), and the proof is that they have already built two *naos* of 500 *toneladas*."[82] These substantial vessels were the *naos La Trinidad* and *Santa María de Jesus*, constructed in Manila in 1579 and Otón (Iloilo) in 1580, respectively.[83] Two more vessels over 500 *toneladas* were built within the decade.[84] Plenty of raw materials were nearby, except for iron, which was a problem shared by most Spanish colonies. Iron and lead were regularly imported from China, Japan, and India.[85] Members of the Real Hacienda of the Philippines argued that vessels were built cheaply there "because of the abundance of wood and because the *navíos* from China carry iron at a good price."[86] The Viceroy of New Spain, the Conde de Coruña, also claimed that it was more expensive to build vessels for the Manila trade in Mexico.[87]

The Spanish introduced the *repartimiento* to extract forced labor from the indigenous population. Filipino "Indians" were typically carpenters and laborers. They were paid a meager amount and many were required to travel long distances for the *corte de madera*, a labor draft that required municipalities to provide people to work in the yards or procure raw materials. The Tagalogs and Pampangos near Cavite were affected most by the *corte de madera*, but drafts also occurred in Leyte, Samar, and Bicol. The labor was organized seasonally and into mobile camps. Workmen were removed from their family and land for extended periods of time. The work and conditions were terrible and some colonists spoke out against the timber labor draft, including Hernando de los Ríos Coronel, who wrote in 1621 that 6,000 Filipinos worked for 3 months to transport masts of one galleon.[88] The staunchest opponents among the Spanish were most often members of religious orders. Fray Gaspar de San Agustin wrote in a letter from about 1690,

---

82 AGI, Filipinas, 29, N.38 "Relación de Juan Bautista Román, factor y veedor de la Real Hacienda de Filipinas, sobre importancia del Maluco," 12-06-1582.

83 AGI, Filipinas, 29, N.32, 10-06-1579; AGI Filipinas, 29, N.33: "Despacho a Nueva España del navío Espíritu Santo en 1578 y de la nao La Trinidad en 1579, y de la nao Santa María de Jesus en éste de 1580, las dos últimas fabricadas en esas islas, resultando muy baratas por la abundancia de maderas y porque los navíos de China traen hierro a buen precio."

84 AGI, Filipinas, 339, L.1, F.354v–355v, 03-07-1588.

85 Alfonso Mola and Martínez Shaw, 287.

86 AGI, Filipinas, 29, N.33, 1580.

87 AGI, Filipinas, 339, L.1, F.305v–306r, 09-06-1584.

88 Bankoff, "'Tree as the Enemy of Man'," 324.

> For it is a fact that if one consider the life and lot of most of them [the indigenous Filipino population], they resemble that merchant in the gospel of Matthew (chapter 13), who gave all that he had for the precious pearl; for it costs them more than is apparent to become Christians, with so much cutting of timber, and many personal services; and thus God gives them the true rest of death, as to poor and needy ones.[89]

Payments, when made, were inadequate. Forty *reales* per month was considered a livable wage, but unskilled labor received four *reales* before wages were raised to eight in 1619. Masters, those who fashioned masts and rigging, received between 24 and 32 *reales*. Filipino smiths earned 12 *reales* per month, but Chinese smiths earned 28 *reales*.[90]

At the time of Legazpi's arrival, the Philippines contained six different types of forests covering about 92% of the islands.[91] Large evergreens comprised mostly of the dipterocarps family were found throughout the islands up to altitudes of 800 meters. Valuable hardwood species included amugis (*Koordersiodendron pinnaturo*), guijo (*Shorea guiso*), kalumpit (*Terminalia micocarpa*), and yakal (*Hopea malibato*). Lesser quality hardwoods included apitong (*Dipterocarpus grandiflora*), lauan (*Shorea contorta*), and tanguile (*Shorea polysperma*). Non-dipterocarp hardwood forests, known as molave forests, were also found in numerous locations and contained durable species such as balete (*Ficus balete*), ipil (*Leucana leocephala*), molave (*Vitex parviflora*), and narra (*Pterocarpus indicus*). Pine forests grew in northern and central Luzon as well as on the island Mindoro, just south of Luzon. The wet high-mountain regions of about 1200 meters altitude supported mossy-forests. Mangrove and beach (not beech) forests grew in coastal areas. Quality hardwoods that were comparatively impervious to the teredo ship worms allowed for the construction of sturdy galleons.[92] Captain Sebastián de Pineda related information to the crown in 1619 on specific species utilized in shipbuilding, along with active shipbuilding regions and the personnel involved with the industry. Maria (palo maría or *Calophyllum inophyllum*) were plentiful in the Visayas' coastal and beach forests and were useful for compass timber like knees and futtocks since they are sturdy but grow short and not straight. They were also resistant to rot. Today, they are decorative and ornamental for their flowers and fragrances. Guijo, common in Mindoro, grew tall and straight,

89 Blair and Robertson, v.XL, 261–262.

90 McCarthy, 159.

91 Bankoff, "One Island Too Many," 324.

92 McCarthy, 155.

was also very hard and durable, so it was used for keels, beams, mastheads, and oars. Lauan was used for planking, masts, topmasts, and yards. The banabá colorado (*Lagerstrœmia speciosa*) was used for sheathing and planking due to its resistance to shipworms. Dongon is strong, but is not resistant to the shipworm, so was most often used for roof-timbers. The strong yakal and ipil were used for keels and sternposts. Molave was used for the keel, rudder, and frames. Non-timber materials were found mostly nearby. Abacá, a species of banana, was used for the rigging. Its fibers contributed to another common name, Manila hemp, which is best know today for being pulped to make paper products such as manila envelopes. Canvas for the sails was made of cotton from the province of Ilocos.[93]

Naval Forestry and Imperial Rivalries

The success of the Iberian empires in the sixteenth century attracted rivals to the East and West Indies. Spain's annual convoys of treasure ships were increasingly harassed by pirates who challenged Castile's declared monopoly over the Indies trade. With escalating maritime presences around the globe, Spain's European imperial rivals quickly viewed the forests of the Americas and Asia as potential naval supply zones. The English, Dutch, and French each founded companies to help capitalize long-distance trading ventures.[94] The English in North America encouraged shipbuilding, and by 1686 about half the trade between New and Old England was in American vessels with Boston and Philadelphia emerging as leading shipbuilding centers by the early eighteenth century. The English crown set up regulations in the woodlands, such as the Broad Arrow policy, to prioritize the needs of the royal navy.[95] The French

93 Blair and Robertson, v.XVIII, 160–180; Alfonso Mola and Martínez Shaw, 287.

94 James D. Tracy, ed. *The Rise of Merchant Empires: Long-distance Trade in the Early Modern World, 1350–1750* (Cambridge: Cambridge University Press, 1990) and James D. Tracy, ed. *The Political Economy of Merchant Empires: State Power and World Trade, 1350–1750* (Cambridge: Cambridge University Press, 1991).

95 Robert Albion, *Forests and Sea Power: The Timber Problem of the Royal Navy, 1652–1862* (Cambridge, MA: Harvard University Press, 1926), 231–280; William Cronon, *Changes in the Land: Indians, Colonists, and the Ecology of New England* (New York: Hill and Wang, 1983), 110–111; Jennifer L. Anderson, "Nature's Currency:The Atlantic Mahogany Trade and the Commodification of Nature in the Eighteenth Century," *Early American Studies* 2, no. 1 (Spring 2004): 49; Williams, *Deforesting the Earth*, 213–216; John Perlin, *A Forest Journey: The Story of Wood and Civilization* (Woodstock, VT: The Countryman Press, 1989 and 2005), 263–286; Y. Eyüp Özveren, "Shipbuilding, 1590–1790," *Review (Fernand*

likewise tried to extend royal control over the forests of New France. Towards the end of the seventeenth century, they established sawmills along the St. Lawrence and explored the region to search for good shipbuilding sites. The French also established sawmills in Louisiana in the early eighteenth century. While the masts cut in New France may have been much less expensive than those in the Pyrenees, France's attempts in Canada were suspended after 1700 due to transport costs and lack of trans-Atlantic vessels. After 1731, only the French merchant marine was using Canadian masts.[96] Portugal experimented with building in its colonies in India and Brazil due to the depletion of supplies and increasing costs in Lisbon and Porto. A royal order from 1585 encouraged building in Goa, where strong Indian teak could be found and labor costs were one-third the Portuguese price. By 1620, Lisbon still employed 1,500 men in the royal dockyard, but the advantages of building stronger, more affordable, and longer-lasting vessels in India were clear. In Brazil, labor was unpaid, but transport costs and the lack of skilled labor delayed significant production in Bahia and Rio until after 1650.[97] Portuguese colonial policies were perhaps the most restrictive. Lisbon claimed exclusive control over any and all of Brazil's trees with timber that was potentially of use for shipbuilding. Shawn Miller has argued that the exclusive right policy was ultimately wasteful and a disincentive to other kinds of investment in forest resources for the colony.[98] Ultimately, Portuguese colonial policy promoted greater destruction of Brazil's forests. The Dutch took measures to control access to teak in Java from 1651, and by the early eighteenth century had water and wind-powered sawmills operating on the north coast of the island. The Dutch East India Company had to negotiate with

---

Braudel Center) v. 23, n. 1, Commodity Chains in the World-Economy, 1590–1790 (2000): 38, 42; Geoffrey Scammell, "British Merchant Shipbuilding, c. 1500–1750," *International Journal of Maritime History* v. XI, n. 1 (June 1999), 41–42.

96 Paul W. Bamford, *Forests and French Sea Power, 1660–1789* (Toronto: University of Toronto Press, 1956), 21; Özveren, "Shipbuilding," 56, 60–61; McNeill, *Atlantic Empires*, 66–67.

97 Özveren, "Shipbuilding," 24; Xavier M. Martins, "Some Aspects of Portuguese Shipbuilding in India During 16th–17th Centuries," *Indica* v. 41, n. 2 (Sept. 2004), 142; F.W.O Morton, "The Royal Timber in Late Colonial Bahia," *Hispanic American Historical Review* 58, no. 1 (1978), 41–44; Lucy Maffei Hutter, "A madeira do Brasil na construção e reparos de embarcações," *Revista de Instituto de Estudios Brasileiros* 26 (1986), 47–64; Miller, *Fruitless Trees*, 50–51 and 79; A.J.R. Russell-Wood, *The Portuguese Empire, 1415–1808: A World on the Move* (Baltimore: The Johns Hopkins University Press, 1998), 28, 86; A.R. Disney, *A History of Portugal and the Portuguese Empire: From Beginnings to 1807, Volume 2: The Portuguese Empire* (Cambridge: Cambridge University Press, 2009), 232, 234.

98 Miller, *Fruitless Trees*, 46–47, 61–62.

and provide lavish gifts to Javanese sovereigns to hire labor and establish annual teak quotas in treaties, which were drawn up in 1677, 1705, and 1733.[99]

Nearly all maritime empires of the time sought out new sources of timber beyond traditional shipbuilding centers, including empires on the Mediterranean Sea. For example, the Venetian Republic had put tremendous timber demands on its small territory to support its urban population and its Mediterranean fleet. Concerns that Venice could effectively be isolated from mainland resources rose as a result of the War of Chiogga against Genoa in 1378, leading the Venetians to expand into the Veneto in the early fifteenth century.[100] The needs for shipbuilding timber also sent them across the Adriatic, and even in raiding parties into the Ottoman-controlled Balkans.[101] The Ottoman Empire emerged as a naval power in the sixteenth century, and the need for new ships led the government to explore the empire for suitable timber. The search for large trees radiated out from the imperial arsenal in Istanbul into the Balkans, western Anatolia, along the Mediterranean, and along the Black Sea coast.[102]

By the end of the sixteenth century, Spain had developed more systematic ways to exploit forests of the *ultramar* for naval purposes. Forests continued to be exploited for other industrial and local uses, of course, but growing reports of shortages around the major shipyards of Havana, Guayaquil, and Manila spurred more restrictive measures of forest use and renewed efforts to locate new stands of usable naval timber. Colonial forestry relied on forced labor, operated sporadically as shipbuilding needs fluctuated, and did not operate

99 Nancy Lee Peluso, "The History of State Forest Management in Colonial Java," *Forest and Conservation History* v. 35, n. 2 (Apr. 1991): 66–67. For more on Dutch colonial forestry, see Peter Boomgaard, "The VOC trade in forest products in the seventeenth century," in Richard Grove, Vinita Damodaran, and Satpal Sangwan (eds.), *Nature and the Orient: The Environmental History of South and Southeast Asia* (Delhi: Oxford University Press, 1998) 375–395; Peter Boomgaard, "Exploitation and Management of the Surinam Forests, 1600–1975," in *Changing Tropical Forests: Historical Perspectives on Today's Challenges in Central and South America*, eds. Harold K. Steen and Richard P. Tucker (Durham, NC: Duke University Press, 1992), 252–264. For Dutch forest history, see Jaap Buis, *Historia Forestis: Nederlandse bosgeschiedenis*, 2 vols. (Utrecht: Landbouwhogeschool-Wageningen, 1985).

100 Karl Appuhn, *A Forest on the Sea: Environmental Expertise in Renaissance Venice* (Baltimore: The Johns Hopkins University Press, 2009), 16 and 32.

101 J.R. McNeill, "Woods and Warfare in World History," *Environmental History* 9 (July 2004), 397.

102 Selçuk Dursun, "Forest and the State: History of forestry and forest administration in the Ottoman Empire" (PhD diss., Sabanci University, 2007), 50; Colin Imber, "The Navy of Süleyman the Magnificent," *Archivum Ottomanicum* v. VI (1980), 228–231.

with any superintendents of forests and plantations as in Spain. Increasingly, Spain's imperial rivals expanded their own control over colonial forests as the struggle to control the world's sea lanes intensified. Soon, the pressures of competing maritime empires would encourage the Spanish government to orient strategy more definitively away from the Mediterranean to the Atlantic, resulting in more attention paid to its colonial forests. The wars of the early seventeenth century would push the government to the limits of its resources, and cause the crown to promote further measures of observing, conserving, and exploiting forests in the monarchy.

CHAPTER 3

The Struggle to Stay Afloat in the Seventeenth Century

Transition and Persistence in Spanish Forests under the Later Habsburgs

Fed up with years of non-compliance, the Superintendent of Forests and Plantations of Guipúzcoa, Domingo de Idiáquez, a former soldier of the Army of Flanders, ordered the seizure and imprisonment of the mayor of Asteasu, Domingo de Lizárraga in 1609. Idiáquez incarcerated Lizárraga for refusing to submit testimonies of tree plantings made by his townspeople. Such village testimonies from towns along Spain's north coast contributed to the crown's knowledge of local forest conditions and shaped its evolving policy of naval forest management. Instead of submitting the testimonies to Idiáquez, however, Lizárraga submitted them to the local *junta* of administrators. By sending the testimonies to the *junta*, the mayor resisted more royal oversight and asserted local authority in matters of forest conservation. Unimpressed, the royal superintendent did not release the mayor until the town relinquished the testimonies fourteen days later.[1]

This example shows one way in which royal officials met local resistance to state forestry policies. Guipúzcoans, in this case, simply ignored the orders from the forest superintendent for several years. By the early seventeenth century, the crown enforced its policies of forest conservation through a network of forest superintendents rather than relying solely on new legislation as had been done multiple times since 1547. The superintendents inspected the towns' written testimonies to know if the towns had reached their planting quotas or had fallen short, thus incurring a fine. The seizure of the mayor of Asteasu was a turning point in the superintendent's authority in Guipúzcoa. Such blatant resistance from Guipúzcoan towns would not be as widespread after 1609.

In the seventeenth century, the king placed superintendents in multiple regions, including Galicia, Asturias-Cuatro Villas, Vizcaya, Guipúzcoa, and Catalonia. Each region responded to the encroachment of royal power into the forests in different ways. The division and multiplication of the office of the

1 Ricardo Gómez Rivero, "La superintendencia de construcción naval y fomento forestal en Guipúzcoa," *Anuario de historia del derecho español* 56 (1986), 617.

 | DOI 10.1163/9789004261372_005

forest superintendent extended state power to increasingly local levels. More bureaucrats meant more inspections and more information for the crown. The proliferation of forest superintendents and the establishment of family lines in each region to inherit the office became a key component of Habsburg state forestry in the century. The effectiveness of these bureaucrats depended greatly on the personal character of the superintendent, the traditional experiences and expectations of local forest communities, and the evolving broader historical context of the economy: international politics and naval power. Similar factors shaped colonial forestry, but the colonies lacked a specialized forestry bureaucracy. Instead, the crown relied on legislation, naval officials, local indigenous or Creole experts, and forced labor practices to extract and conserve timber for the navy. The Spanish became more concerned about their defenses in the Caribbean in the early seventeenth century as their enemies from England, France and the Netherlands became more entrenched in the region. Forest protection remained anathema on Barbados, St. Kitts, Martinique, and other islands that were converted to sugar production, so Spanish protectionist efforts for naval interests were unusual among the colonial powers of the Atlantic at the time. As in Spain, however, naval forest regulations led to periodical conflicts with colonial land holders.

This chapter analyzes territorialization of forests by the Spanish Habsburg state in the forms of bureaucratic administration and forest legislation from 1598 to 1700. For the first half of the seventeenth century, Spain struggled to maintain its primacy in European affairs by defending its interests and possessions in Europe and around the globe.[2] The demands of these wars continued to place great strains on Spanish forests, which were managed closer than ever by maritime experts with experience in matters such as ship construction and naval service. The transition of power from Cristóbal de Barros, Felipe II's representative in charge of overseeing forest conditions, to his successors in the late sixteenth and early seventeenth centuries was not always smooth, but the crown eventually established a greater and more regular presence on the north coast. The lives and careers of several bureaucrats from the early seventeenth century are examined below to understand better their policies, the operation of forest conservation in this period, and the establishment of veritable dynasties of superintendents.

2 Relevant secondary literature on this period is extensive and I have drawn heavily from R.A. Stradling, *Europe and the Decline of Spain* (London: George Allen & Unwin, 1981), J.H. Elliott, *Spain and its World, 1500–1700* (New Haven: Yale University Press, 1989), David Goodman, *Spanish Naval Power, 1589–1665: Reconstruction and Defeat* (Cambridge: Cambridge University Press, 1997), and Carla Rahn Phillips, *Six Galleons for the King of Spain: Imperial Defense in the Early Seventeenth Century* (Baltimore: Johns Hopkins University Press, 1986).

As Spain's imperial rivals asserted stronger presences in Asia and the Americas forest conservation became a pressing concern of the government, as it had been in Spain for the past century, but each state approached the challenges of resource management in different ways. Studying resource control strategies is particularly well suited for Spain in the later seventeenth century. Some authors have remarked on how population decline and a depressed shipbuilding sector meant less demand for timber. The naval decline of the period has even caused some scholars to view state forestry under the later Habsburgs as virtually non-existent.[3] Logically, if the crown did not use timber to build many ships and the population of its monarchy did not need to extend arable land into upland forests due to demographic stagnation, the state would not require a forest administration system. However, the final section of the chapter examines the persistence of Habsburg state forestry in an era not noted for Spanish naval activity. By studying Spanish attempts to affect, influence, and control forest communities' relationships to forest resources, state forestry policies become something that did not necessarily involve resource extraction and deforestation. In other words, this chapter is not concerned with the changing forest cover in Spain, but rather with forest management and the politics of control over forest access. Just because the crown built fewer ships did not mean that it reduced concern for its forest resources. In fact, it is argued that the Habsburg crown took some very important steps in the later seventeenth century that in many ways helped enable a revival of shipbuilding under the Bourbon dynasty in the eighteenth century. After 1630, in an era frequently described as one of naval decline, the Habsburgs carried out resource control strategies aimed to maintain and improve access to forests for crown interests. They maintained the bureaucratic network of forest superintendents, issued new forest regulations based on local-level experience, decreed more stringent forest legislation that enhanced royal control of behavior in forests, and commissioned more forest reconnaissance missions to locate new domestic and accessible timber areas in which the crown could exercise a greater influence.

State Forestry in Times of Peace and Global Conflict in the Early 1600s

Toward the end of Felipe II's reign, Spain entered a prolonged period of demographic and economic crisis, made worse by the increasing expenses

3 Erich Bauer Mandersheid, *Los montes de España en la historia* (Madrid: Fundación Conde de Valle de Salazar, 2003), 174.

of maintaining its possessions in Europe and defending its empire against its rivals.[4] During the 1590s, Spain was facing war on all fronts in Felipe II's attempt to prevent English encroachment in the Americas, Dutch independence, and a Calvinist-ruled France. Military victory remained elusive, and the cost of maintaining the Army of Flanders, along with maritime defenses in the North Sea and the Atlantic, increased Spain's debt to eighty-five million *ducados*, nearly ten times its annual revenue.[5] The crisis led the king to scale back and seek peace with France. In 1598, Felipe II signed the Treaty of Vervins with France and accepted Henri of Bourbon's claims to the French throne, after Henri returned to the Catholic fold.[6]

Felipe III (r.1598–1621) continued with his father's overall strategy of defending the empire against rivals and heretics. Initially, he increased Spanish military involvement in Europe by landing a force in Ireland and increasing the tempo of war in Flanders. He also renewed Spanish commitment to fighting Islamic forces based in North Africa and sent a force against Algiers in 1601. However, victories were few and Felipe III soon began to seek peace as well. The death of Queen Elizabeth I in 1603 opened the way to peace with England under her successor, James I, in 1604. In the Netherlands, the Spanish siege of Ostend in 1604 came at a great cost, leading to devastating mutinies from unpaid soldiers. The Dutch and the Spanish signed an armistice in 1607 and a twelve-year truce in 1609. The search for peace in the early seventeenth century resulted from Spain's financial, economic, and psychological exhaustion.[7]

The era of peace did not result in a complete suspension of hostilities outside of Europe, however, nor in a cessation of naval construction.[8] Spanish forces fought the Dutch in the Caribbean and in East Asia during the twelve-year truce. The crown also turned its attention to the Mediterranean and its problems with piracy and the presence of Islam in the Iberian Peninsula. The expulsion of the Moriscos from Spain in the years from 1609 to 1614 required heavy logistical support from the army and navy.[9] While the crown succeeded in its goals of removing the Moriscos, some of the exiled population exacted revenge against Spain by joining the corsairing forces of Algiers and Salé. A special naval force, the *Armada de la Guardia del Estrecho*, served to protect

4 Phillips, *Six Galleons*, 8.

5 Paul Allen, *Philip III and the Pax Hispanica, 1598–1621: The Failure of Grand Strategy* (New Haven: Yale University Press, 2000), 2.

6 Stradling, 31.

7 Elliott, *Spain and its World*, 116.

8 Goodman, 16–17.

9 *Moriscos* were Muslims who had been baptized and outwardly converted to Christianity.

Spanish waters and the Strait of Gibraltar in this period. Beginning around 1617, in the midst of the twelve-year truce with the Netherlands, Spain began an active shipbuilding program. Felipe III did not neglect naval forces during this era of European peace, and remained committed to overseeing his forest resources to ensure access to shipbuilding materials.

Many authors have argued that the architects of peace wanted to use the time to place Spain in a better position vis-à-vis the Dutch once hostilities in Europe restarted, which seemed inevitable.[10] Spanish ambassadors in central Europe had reactivated close relations with the Austrian branch of the Habsburg family in the early seventeenth century.[11] After much debate, Spain promised to help Germany fight its rebels, hoping to receive the favor in turn once the truce with the Dutch ran out. War broke out in 1618 in central Europe and in the Netherlands in 1621. For much of the next decade, Spain experienced a string of military and naval successes.[12] Spanish armies helped defeat a Protestant army at White Mountain in 1620, secured key Alpine passes, and occupied strategic territory in Alsace and the Lower Palatinate. Ambrosio Spínola captured the important town of Breda in the Netherlands in 1625, immortalized in the famous painting by Diego de Velázquez. Naval victories that same year included the recovery of Portuguese Bahía in Brazil from the Dutch and the defense of Cádiz, both led by Fadrique de Toledo, and memorialized in paintings by Juan Bautista Maíno and Francisco de Zurbarán, which were hung in the Prado Palace of Felipe IV (r.1621–1665). Spain also sent aid to the Portuguese in the Persian Gulf and the Indian Ocean, and sent multiple fleets to protect the Philippines.[13] The global successes in this decade would stand as the high point of Spanish power in the century.

One of the boldest elements of Spanish strategy in the 1620's, championed by the Count-Duke Olivares, but developed before his time, was an economic blockade of the Dutch through the control of the North Sea and the Baltic.[14] Control of this area would also guarantee access to Baltic pine and spruce trees. Western European powers relied on pines from the Baltic region for masts, typically called Flanders pines by the Spanish, because they arrived via Dutch middlemen. The Baltic trade operated on a much larger scale than the

---

10 For example, Allen, 238.

11 J.H. Elliott, *The Count-Duke Olivares: The Statesman in an Age of Decline* (New Haven: Yale University Press, 1986), 56.

12 Stradling treats these successes together in his chapter, "From little wars to total war, 1610–1628," 50–84.

13 Goodman, 15.

14 Stradling, 62–64.

extraction of timber in Asia and the Americas by European empires. The Baltic region's role in the sixteenth and seventeenth centuries as a supply zone of bulk goods such as grain and timber for Western merchants meant it operated as a "collective home market," or a semi-colonial zone within Europe.[15] Towards the end of the eighteenth century, the average import of all colonial timber to Great Britain amounted to 1% of the volume of Baltic timber imports.[16] Before the American Revolution, it was still only about 20%. The benefits of the Baltic for shipbuilding materials included numerous navigable rivers, dense forests, and closer proximity to European shipping centers than the colonies. The Dutch controlled the carrying trade of Baltic goods, including Norwegian timber, from the late sixteenth century to the early eighteenth with their specialized *fluyts*. These vessels were medium-sized, lightweight, and required comparatively little wood to build. They were not suitable for warfare, as Spanish vessels were required to be, but they were cheap to build, inexpensive to maintain, and were useful for bringing bulky goods to the shipbuilding centers of Holland, particularly Zaandam.[17] By 1700, most of the carrying trade of northwest Europe was in the hands of the Dutch, who had 900,000 tons of total shipping on the water. Soon, however, Dutch shipbuilding would decline and the British would overtake the Baltic trade. The British were not alone, however, because powerful Baltic states emerged in the later seventeenth century, including Sweden, Denmark, and Russia, who took measures to preserve the best timber for themselves.[18] During times of war, Spain's geographic position made access to Baltic resources very difficult. Spain was at the greatest distance from these forests among the Atlantic empires and either had to make costly purchases through Dutch merchants or risk sailing past England, France,

---

15 Jason W. Moore, "Amsterdam is Standing on Norway Part II: The Global North Atlantic in the Ecological Revolution of the Long Seventeenth Century," *Journal of Agrarian Change* 10, no. 2 (April 2010), 193 and 196.

16 Y. Eyüp Özveren, "Shipbuilding, 1590–1790," *Review* (*Fernand Braudel Center*) v. 23, n. 1, Commodity Chains in the World-Economy, 1590–1790 (2000): 61.

17 Phillips, *Six Galleons*, 23; Özveren, "Shipbuilding," 17; E.W. Petrejus, "The Dutch Flute," in Joseph Jobé, ed. *The Great Age of Sail*, tr. Michael Kelly (Lausanne: Edita, 1967), 81.

18 Per Eliasson and Sven G. Nilsson, "'You Should Hate Young Oaks and Young Noblemen': The Environmental History of Oaks in Eighteenth-and Nineteenth-Century Sweden," *Environmental History* 7, no. 4 (Oct. 2002), 659–677; Bo Fritzbøger, "Environmental Agency: Forest Histories from Denmark c.1500–2000," in *Transcending Boundaries: Environmental histories from the Øresund region*, vol. 9 of *Skrifter med historiska perspektiv*, eds. Fredrik Björk, Per Eliasson, and Bo Poulsen (Malmö, Sweden: Malmö University, 2009): 74–87; Aleksey V. Postnikov, "The Russian Navy as Chartmaker in the Eighteenth Century," *Imago Mundi* v. 52 (2000): 79–95.

and the Dutch to get to the source. Spain was at war with at least one of these states for most of the early modern period, but still had to resort to purchasing timber from Poland and Riga or import foreign-built vessels. In short, Spain's geopolitical situation made the use of Baltic pine more costly, dangerous, and unreliable than it was for any of its western European rivals. Access to Baltic timber was so important, however, that Felipe IV relaxed prohibitions against Dutch shipping in Spanish ports solely for importing masts.[19] Such a grand strategy to control the North Sea indicates the great lengths to which the crown was willing to go to ensure access to forests. Changes in fortune for the Spanish navy and financial troubles in the late 1620's and 1630's meant the plan never succeeded, which reinforced the crown's desire to open access to greater stands of domestic timber, and added greater motivation to developing mechanisms of procurement in colonial forests.

The kings of Spain continued to rely on the bureaucracy to govern and administer the empire. Forest superintendents acted as purveyors of shipbuilding as well, joining both sets of responsibilities. Ultimately, this insured that the crown's forest and naval policies remained closely aligned. During the 1620s and 1630s, the crown relied on the zeal and dedication of bureaucrats, nobles, and military officers when finances and supplies fell short.[20] This strain could only be withstood in the short term, but despite the defeats suffered after about 1630, Spanish forces were able to retain most of the empire. Throughout, the bureaucracy, including the forest superintendents, continued to operate and supply the crown with valuable information and resources.

The Spanish monarchy remained a collection of political units that maintained separate governing structures, currencies, and law courts. At times, the bureaucracy faced severe resistance to crown policies in territories with strong *fueros* that protected local rights. For a century, Castile had been bearing a disproportionate share of the burdens and costs of empire compared to the Crown of Aragón, and Olivares sought reforms to the tax structure of the monarchy to receive a more even contribution from the regions. Olivares's Union of Arms of 1626 was an ambitious attempt at a new territoriality of Castilian power. The logic behind the attempt was that by unifying the laws of the empire, removing internal customs barriers, and receiving equal tax and military contributions from all regions, the king could deploy his bureaucrats anywhere and monarchical power and resources would greatly increase. Spain could then afford to face all of its imperial commitments. Such notions were too radical for the constitutional diversity of the monarchy, however, and the

19 Goodman, 139.

20 Phillips, *Six Galleons*, 220.

reforms met with considerable resistance.[21] One example of local resistance to policies formulated in Madrid arose in response to attempts to institute an obligatory naval registration program in the Basque region in the early seventeenth century. Traditionally, naval enlistment was voluntary, with the exception of rare levies in wartime, but in the early seventeenth century a number of decrees made naval registration mandatory so that sailors could be recruited more efficiently in times of conflict. A decree for Guipúzcoa in 1605 stated that only registered sailors could continue to fish and make a living from the sea. Local authorities in Guipúzcoa responded that such a decree violated traditional rights and liberties, which had been protected by royal decrees in the past. After years of correspondence, the crown realized the region would continue to resist with "respectful non-compliance."[22] Outbreak of war in 1618 and 1621 renewed the crown's efforts to register sailors, particularly in the Basque provinces of Guipúzcoa and Vizcaya, and Felipe IV issued a new decree in 1625. Again, Guipúzcoans resisted the orders. The Superintendent of the General Registry on the north coast, Martín de Aróstegui, himself a Basque, recommended that local men assist in the recruitment efforts because of the delicate relations of the region with the crown. The government's need for Basque mariners meant the king had to continue with traditional methods and with the help of local officials and naval experts.[23] Forest superintendents would face similar resistance when they implemented royal policy, but the crown in the early seventeenth century expressed a vigorous commitment to greater control over forest resources to help in its wars and overall strategy in the Atlantic, Mediterranean, and North Sea. After 1598, Felipe III and Felipe IV installed a greater number of forest superintendents on the north coast and divided jurisdictions to enable better coverage and increase the number and rate of forest visitations.

To a New Generation of Forest Inspectors: Barros's Letter to Riva Herrera

Spanish commitments around the globe ensured that the need for naval resources only escalated after the failed armada against England in 1588, and a period of rapid recovery and frantic shipbuilding in the 1590s unfolded in

21 Elliott, *Spain and its World*, 179.

22 Carla Rahn Phillips, "Naval Recruitment and Basque Resistance in Early Modern Times," *itsas Memoria: Revista de estudios marítimos del País Vasco* no. 5 (2006) 175.

23 Phillips, "Naval Recruitment," 173–176.

many regions in Spain. In this context, a new generation of forest superintendents gained their experience in naval or military matters, and their descendents would oversee Spanish state forestry for most of the next century.

By 1592, Cristóbal de Barros had left his position as Superintendent of Forests and Plantations for a post as *proveedor* (purveyor) in Seville. He was succeeded as superintendent by Hernando de la Riva Herrera, who had begun inspecting forests in Asturias and around Santander at this time.[24] Riva Herrera observed during his forest inspections in the summer of 1594 that towns failed to achieve their planting quotas.[25] Probably confused about how he was to be effective at overseeing forest conservation in the thousands of towns in his jurisdiction, two leagues from the sea along the coast from Portugal to France, he contacted the crown and his predecessor Barros for some direction. There is no known copy of an appointment letter for Riva Herrera, which would have outlined his duties. Most likely, they were not very different from Barros's, but the uncertainty of the specifics of the job left some unanswered questions and required clarification. To help with the transition, Barros wrote a letter in October of 1594 to report on the duties of the forest superintendent and told Riva Herrera what he should expect while on the job.[26] The letter articulated lessons learned from years of experience by the first forest superintendent.

In his letter, Barros began by stating generally how important forests were for shipbuilding, house construction, foraging animals, and hunting, demonstrating a clear awareness of forests as sites of multiple uses. In the regions of Guipúzcoa and Vizcaya, shipbuilding and iron making had put a great strain on forests, so the crown felt a special urgency to conserve forests in these regions, although Riva Herrera's total jurisdiction was far larger. Barros told his successor that he had only tried to persuade each resident to plant two oak trees every year and had refrained from forceful methods. He said that he was not obligated to visit all of the plantations within this jurisdiction, nor did he need to make specific visits at specific times, "as Riva Herrera supposed." Barros worked without taking money from anyone and estimated that his work had cost him 3,000 *ducados*. Above all, his job was to persuade people to plant trees and make the orders of the king clear, while local justices had the responsibility for collecting fines, although Barros admitted that "the desire of most was not to plant to manage and clear the forests." Barros told Riva Herrera to ensure that other personnel involved with forest conservation performed their

24 Archivo General de Simancas (Hereafter, AGS), Guerra Antigua (Hereafter, GA), Legajo (Hereafter, leg.) 403, Folio (Hereafter, fol.) 99.

25 Goodman, 75.

26 AGS, GA, leg. 403, fol. 102.

duties responsibly, including the local justices. Riva Herrera or other officials needed to continue visiting forests "from France to Portugal," and not just around Guarnizo (Santander), one of the most active shipbuilding centers on the north coast. Justices who did not conform to the king's demands were to be punished, typically with a fine.

Barros also instructed Riva Herrera on how to make proper plantations. For new plantings, a hole two feet in circumference was required so that grass and other weeds would not prevent the sapling from taking root. The hole also collected water, allowing for healthy growth. For these young trees, Barros liked to plant thornbushes around them to protect them from livestock. The young trees should be planted in February or March in the low, humid lands so that winter rains did not rot the roots. Then they would grow in the moderate moisture of the summer. The trees planted in the wetter lowlands grew thicker than in the highlands. As a result, Barros wrote, trees in the highlands needed to be planted farther away from one another, and he warned that they grew more slowly. Vizcaya was the area best suited for shipbuilding, but some of its best trees were being used for other things, such as making charcoal. Barros noted some of the biggest trees came from this region, including some that grew four *brazas* wide,[27] which could produce one hundred cartloads of firewood. Barros implied that such trees should be reserved for naval use.

Barros conceded that "it is almost impossible [for the superintendent] to visit every plantation."[28] Knowing that Riva Herrera would have to prioritize his visitations, Barros made clear that Vizcaya and Cuatro Villas were the most important areas, requiring the most attention. Barros noted that when he did carry out visitations, his precise planting directions worked. The territory he could effectively control, however, was clearly far smaller than the territory of his full jurisdiction, and he was unable to carry out frequent visits to observe the growth of new plantations.

Since Barros only attempted to persuade local populations to plant trees and left the enforcement up to the local justices, most towns had simply ignored his instructions. Still, Barros's approach may have been part of a conscious strategy, since he presumably knew that harsher tactics would have led to stronger resistance and social unrest. Given that background, it is not surprising that in the short-term Riva Herrera witnessed non-compliance in his inspections of 1594. Barros's approach did not result in the kind of resistance discussed below in later years under superintendents who used

---

27 A *braza* was about 1.67 meters.

28 AGS, GA, leg. 403, fol. 102. "Esto bien que pensar que el superintendente lo a de visitar todo, es casi imposible."

stronger methods. In fact, town councils in Cuatro Villas in 1594 "said they were won over by the leniency of the superintendent who, instead of imposing heavy fines, asked them to make good the planting quotas they had failed to fulfill."[29]

The crown's response to Riva Herrera's concerns over jurisdiction in the 1590's eventually led to a new era of state forestry territorialization. Barros was as able a bureaucrat as any serving Spain at this time, but he had been unable to visit everywhere in his jurisdiction to carry out his duties. Beginning in 1598, the job he once held was divided into smaller territorial units in order for more efficient forest conservation. In the seventeenth century, there were separate forest superintendents for Galicia, Asturias and Cuatro Villas, Vizcaya, Guipúzcoa, and Catalonia.[30] Riva Herrera became the superintendent of Asturias and Cuatro Villas, a region in which he had shipbuilding experience.[31] The letter Barros wrote in 1594 remained influential for the methods of forest inspection and new plantations, but later superintendents would employ their own methods of enforcement.

The Formation of Forest Superintendent Dynasties

Cristóbal de Barros had been named the first forest superintendent by extension of his expertise in ship construction and other naval matters. His promotion to *proveedor* for the entire Armada is one indication that the position of forest superintendent was a highly regarded and important office. It should not be a surprise, therefore, that his successors were for the most part accomplished servants of the state with similar naval expertise. One that stands out above the rest, however, was the superintendent of forests in the shipbuilding region of Vizcaya, Agustín de Ojeda.

29 AGS, GA, leg. 403, fol. 98, Gonzalo del Rio, testimony of inspection of forests, 16 July 1594, Santander; translation to English from Goodman, 104.

30 Whereas the superintendents of the coast from Portugal to France combined the duties of ship construction and forest conservation, Catalonia had its own forest *conservador*, who worked closely with the superintendent of the *atarazanas*, the shipyards of Barcelona. Goodman, 74.

31 Goodman says Hernando and Fernando were two different people, father and son, but I have not found any evidence that they were two people. In fact, in a list of superintendents from Cuatro Villas in the Museo Naval, Fernando is listed right after Barros, starting in 1593. There is no Hernando in the list. (Museo Naval, Ms.1294) The names Hernando and Fernando were often interchanged; for example, Christopher Columbus's son's name has been written as both.

Ojeda was one of the most important and most prolific shipbuilders of the early modern era.[32] Between 1589 and 1598, he built thirty galleons and two *galizabras* for the crown. His appointment in 1598 as "*Superintendente de fábricas y arqueamiento de navíos, y conservación de montes y plantíos en el Señorío de Vizcaya*" was the culmination of a long and distinguished career that totaled fifty-six years of royal service. Kings and councils sought his expertise until his death in 1619. In January of the following year, his widow wrote to Felipe III to ask for some financial assistance because their family had fallen on hard times. To make her case, she reported on the distinguished years of service performed by her late husband, providing us a good sense of the career of Agustín de Ojeda, which set him on the path to being forest superintendent of Vizcaya.[33]

Before he built galleons, Ojeda worked on the Indies run and helped prepare for the Armada against England. In 1578, he commanded a *flota* carrying silver, cochineal, and over 20,000 *ducados* worth of cargo from Santo Domingo back to Spain. On his way home, however, he was captured in the Azores, which had entered into rebellion against Felipe II's claim to the Portuguese throne.[34] Ojeda managed to escape to Lisbon, and he informed the king about what was occurring in the islands. In 1589, Ojeda's shipbuilding career began when Cristóbal de Barros sent orders from the king to build twelve galleons to help with the recovery after the loss of the Armada the year before. Six were to be built by Hernando de la Riva Herrera, the future Superintendent of Forests in Cuatro Villas, and the other six by Ojeda. In 1594, in another example of how some forest superintendents in this period knew each other well, Ojeda worked alongside Antonio de Urquiola, a future superintendent of Guipúzcoa, in preparing galleons in Pasajes, near San Sebastián. Over the next four years, the crown commissioned Ojeda to build twenty-four more galleons, an exceptionally high number for such a brief period. During Felipe II's urgent program to rebuild the navy after

---

32 María Isabel Vicente Maroto, "Agustín de Ojeda y la construcción de navíos a finales del siglo XVI," in María Isabel Vicente Maroto y Mariano Esteban Piñeiro (coords.) *La ciencia y el mar* (Valladolid: Sever-Cuesta, 2006), 311–344.

33 AGS, GA, leg. 1003, "El 18 de agosto de 1629, el Consejo de Guerra hace recuerdo a V.Mgd. volviendo a sus reales manos la inclusa Consulta sobre la pretensión de Doña Antonia de Velasco, viuda del capitán Agustín de Ojeda, Superintendente de fábricas y plantíos en Vizcaya." Discussed in Vicente, "Agustín de Ojeda," 313–314.

34 The Portuguese succession crisis began after the death of King Sebastian I at the Battle of al Qasr al Kabir in Morocco in 1578. Felipe II of Spain claimed the throne, but "Most of the Azores archipelago refused to recognize the Spanish succession, instead acknowledging the claim of Dom Antonio, prior of Crato, Sebastian's illegitimate cousin." In 1582 and 1583, two Spanish fleets were sent to end the resistance. From Geoffrey Parker, *The Grand Strategy of Felipe II* (New Haven: Yale University Press, 1998), 167.

the Armada of 1588, three future forest superintendents received formative experience in royal naval service.

There is evidence that Ojeda assumed Barros's job in Vizcaya by 1596, but he was not appointed superintendent of the *señorío* officially until 1598.[35] His early inspections revealed that forests were being consumed rapidly and there had been no planting for many years. Over the course of Ojeda's career, we can see that he effectively increased the timber stocks in Vizcaya. Between 1602 and 1615, the people of Vizcaya planted 467,036 trees in the region.[36] In 1603, the king commissioned Ojeda to serve as superintendent in Galicia as well, which had not had a forest superintendent up to this point, to oversee a contract with Isidro Sánchez de la Mota, who supplied timber for shipbuilding.[37] In 1612, Ojeda returned to Galicia to visit the forests there and established conservation practices similar to those outlined in Barros's letter to Riva Herrera. An inspection by Ojeda's successor in Galicia, Bernabé Márquez, in 1624 indicated that timber stocks had increased since his 1612 visit.[38]

Ojeda succeeded at his forest superintendent duties without having forestry experience, but he was an able administrator and had a great deal of knowledge about the kind of timber that naval shipbuilding required. His real expertise was in ship construction, and the crown sought his advice in 1607 and 1610 when the top mariners in Spain came together to discuss ordinances for regulating ship design. Scholars have shown that Felipe III's reign was a turning point in naval policy.[39] During his reign, policy-makers sought ways of perfecting ship design, and in 1605 the king called an assembly of Spain's maritime experts to establish optimum design specifications. In 1607, eleven experts, including Ojeda, met in Madrid to debate recommendations for the best length to beam ratios and establish a formula for calculating a ship's capacity from its dimensions.[40] This was not the last of such meetings during the reign; other groups of experts continued to debate the matter before the crown issued ordinances in 1613 and 1618.

35 Vicente, "Agustín de Ojeda," 316.

36 Goodman, 105.

37 Vicente, "Agustín de Ojeda," 318.

38 Goodman, 104. The comparatively late arrival of a full-time forest superintendent in Galicia meant that episodes of resistance did not occur until the 1620's and 1630's. Ofelia Rey Castelao wrote that the first formal protest against royal forest inspectors in Galicia occurred in 1629, in Ofelia Rey Castelao, *Montes y política forestal en la Galicia del Antiguo Régimen* (Santiago de Compostela: Universidad de Santiago de Compostela: Servicio de Publicacións e Intercambio Científico, 1995), 171.

39 Phillips, *Six Galleons*, 28.

40 Vicente, "Agustín de Ojeda," 319; Goodman, 116.

Ojeda by this time was one of the most accomplished shipbuilders and naval administrators in the monarchy. The king wanted to make sure the job of forest superintendent of Vizcaya stayed in Ojeda's family.[41] However, his two sons died before he did, serving in the navy. As a result, the king said that whoever married his eldest daughter would succeed Ojeda, if approved by the *Junta del Señorío de Vizcaya*. It turned out that both his daughters married highly accomplished mariners. One daughter married Martín de Aróstegui, who was the *veedor general* of the entire Atlantic fleet and mentioned above as part of the crown's naval recruitment efforts. Ojeda's other daughter, María, married her father's successor, Martín de Vallecilla y Ochoa del Casal. Vallecilla descended from a well-established family from Portugalete and was admiral of the Indies escort squadron. The daughter of María and Martín would marry Francisco Díaz y Pimienta, who would become captain general of the Indies escort squadron and then of the Atlantic fleet.[42] The responsibility for conserving Vizcaya's forests remained in Ojeda's family for at least sixty-two years, not just because of strategic arrangements made within the family, but because of direct royal involvement.

During his career, Ojeda faced severe money shortages, skyrocketing timber prices, and personnel shortages. However, he asserted his authority as superintendent in both Vizcaya and Galicia and was able to improve the forests and fulfill his duties in ship construction, helping to set the standards in ship design of the day.[43] Vallecilla succeeded Ojeda and, according to a detailed study of his tenure,[44] was very diligent in recording the successful plantings in Vizcaya. Although Vallecilla often had to face the competing interests of the region's flourishing iron industry, the establishment of the forest superintendent's territorial control in Vizcaya occurred rather smoothly compared to other regions.

The effective establishment of the superintendent's office in Guipúzcoa took longer. In 1598, the king appointed Antonio de Urquiola, a Basque who directed the arsenal of Pasajes, built ships for the king alongside Ojeda, and served as captain general of the squadron of Cantabria. However, as superintendent he spent little time in Guipúzcoa, and when he did, the towns rarely complied with the planting quotas. The king and Urquiola both took a moderate approach to the situation, aware of the region's sensitivity to royal intervention, and they said they

---

41 AGS, GA, leg. 1003.

42 Vicente, "Agustín de Ojeda," 315.

43 For details on the evolution of Spanish ship design, see José Luis Rubio Serrano, *Arquitectura de las naos y galeones de las flotas de Indias, vol. I (1492–1590), vol. II (1590–1690)* (Málaga: Ediciones Seyer, 1991).

44 Goodman, 105–108.

wanted to persuade people to make the plantations, but did not want to violate the region's *fueros*. Such an approach produced very little new forest growth, but neither did it promote serious conflicts between crown and province.

Urquiola died in Madrid in 1600, and the province thereupon declared a return to their observance of the forest conservation ordinance of 1548, no longer recognizing the office of the superintendent.[45] The king looked for a replacement for Urquiola anyway, and appointed Domingo de Idiáquez Goicoechea in a letter dated July 21, 1601.[46] Since we do not have a copy of Urquiola's appointment, Idiáquez's appointment letter is the earliest we have for Guipúzcoa; most likely, the letters were very similar. Contrary to what Cristóbal de Barros had written to Riva Herrera, Idiáquez's appointment letter said he needed to make visits at least twice a year to all towns and villages in his jurisdiction two leagues from the sea or a navigable river, in the company of a local justice. Idiáquez's appointment letter enhanced the authority of, and added new responsibilities to, the office of forest superintendent.

Idiáquez was ordered to signify which trees were best suited for shipbuilding and to allocate the quantity of new trees to plant in each region, bearing in mind the suitability of the terrain. During his visitation, he was required to take testimony from each location specifying the quantity of trees planted, so that on his return trip he could take an account from the justices and *regidores* to see if they had complied with the stated quota. For each tree that they were short, the justice, mayor, or other responsible parties would be fined one *real*. The total fines collected by the superintendent would be divided three ways, with one-third going to the person who called attention to the infraction, one-third to pay for the superintendent's visits, and the remaining third to the crown. This three-part division of fines followed standard Habsburg practice. Each year, Idiáquez needed to make a full report of the quotas he set and of the testimonies from the towns. Finally, like Barros and Urquiola before him, Idiáquez had to oversee the disbursement of loans to shipbuilders working for the crown to build vessels of at least 300 *toneladas*.[47] His salary amounted to 800 *ducados* per year. This salary would remain unchanged for Idiáquez's successor, and the letter of appointment stipulated that anyone who performed the job of interim superintendent would be paid 400 *ducados* per year.

45 Ricardo Gómez Rivero, "La superintendencia de construcción naval y fomento forestal en Guipúzcoa, 1598–1611," *Anuario de historia del derecho español*, 56 (1986), 606.

46 Museo Naval, Colección Vargas Ponce, Tomo 31, Documento 31, Folios 170–174. A copy of the full text is in Gómez Rivero, 626–630.

47 For details on how the tonnage of a Spanish ship was calculated, see Phillips, *Six Galleons*, 60–61 and 259, footnote 50.

Idiáquez's appointment letter indicated that the superintendent had a tremendous amount of control over people's behavior in the forests, at least on paper. The superintendents monitored forests and forest communities more closely than ever before, and the fact that a crown official (essentially someone viewed as an outsider by the forest communities of Guipúzcoa) assigned tree plantings had unsettling effects in some areas. In an act of territoriality, a superintendent identified an area and signified to the community the number of trees they needed to plant in a certain amount of time. He brought with him the authority of the King of Castile and the power to fine those who did not comply. Such methods were an unwelcome change from Barros's and Urquiola's techniques. Much like royal decrees on naval recruitment, the king's assertion of rights to control local forests was challenged by Guipúzcoans' claims to autonomy, based on *fueros* and traditions. It is well to remember that all of the royal superintendents in Vizcaya and Guipúzcoa were Basque by birth, but this did not seem to enhance their authority in the eyes of local officials.

In September of his first year on the job, Idiáquez sent letters to town councils urging them to plant before his visitations in the spring.[48] He said that for each tree cut from common, as well as private land, two more needed to be planted. For new plantations, trees needed to be spaced eight *codos* from each other. Before the month was over, Idiáquez received word from the towns of Tolosa, Oyarzún, and San Sebastián that they did not want a superintendent and they planned to submit forest testimonies to the local *junta*, as they had been doing, not to Idiáquez.[49] In November, after complaining to the king of such resistance, Idiáquez received a *real cédula* confirming his role as superintendent, which he took with him on his forest inspections. Nonetheless, by the spring, Idiáquez had not received any testimonies, though the towns had sent them to the local *juntas*.

In July, Idiáquez, an upright former soldier not tolerant of such disobedience, decided to fine officials in Azpeitia one thousand *reales* for not planting.[50] Idiáquez may have intended to make an example of Azpeitia rather than fine them for not planting one thousand trees, but this failed to set other towns straight, and they continued to send their testimonies to local *juntas*. At the end of 1602, Idiáquez wrote to the king that he could not perform his duties. The king once again reconfirmed his authority as superintendent in 1603 and

---

48 Gómez Rivero, 607.

49 The town of San Sebastián added that they would rather plant trees 10 *codos* from each other, not 8 as ordered by Idiáquez. Gómez Rivero, 609. The *codo real* used in ship construction was 22 inches or 565 millimeters long. See Phillips, *Six Galleons*, 228.

50 Gómez Rivero, 612.

said simply that he must make the people obey his orders. Meanwhile, the province continued to ignore the superintendent. Over the next several years, Idiáquez continued to invoke the king's orders, tried to make inspections, and issued new quotas, but the towns of Guipúzcoa continued to flout his authority.

Idiáquez had run out of options by 1609. His seizure and imprisonment of Domingo de Lizárraga, the mayor of Asteasu, for not submitting testimonies of plantings made in his jurisdiction was a desperate effort to establish his authority in the region. Idiáquez had gone to great lengths to assert that authority, and it appears that his tactics finally had the desired effect. After the episode in Asteasu, most towns began cooperating with the superintendent. Even the most resistant town, San Sebastián, submitted testimonies withheld since 1605.[51] Not surprisingly, Idiáquez was not a popular figure in much of the region. The town council of Zumaya expressed its delight when Idiáquez left for his appointment as governor of Melilla in 1611 and asked the crown not to name a new superintendent.[52]

Idiáquez's tenure in Guipúzcoa demonstrated that the crown's forest conservation bureaucracy did not immediately establish a secure position of authority, though his high-handed tactics did much to overcome resistance from the town councils. As with Agustín de Ojeda in Vizcaya, Idiáquez's replacement would stay in the family, but his son Alonso was away serving in the navy in 1611. Domingo de Echeverri served as interim superintendent until 1618, receiving half salary. Echeverri managed to continue cooperating with Guipúzcoa and received testimonies directly from the mayors and *regidores*, although not all the time. Alonso de Idiáquez would eventually serve a long and distinguished career as forest superintendent from 1618 to 1645.[53]

In Catalonia, the job of building galleys and inspecting forests went to two different people, but they frequently worked together closely on forest matters.[54] Antonio de Alçatte, who carried out an inspection of forests in Catalonia and Valencia in 1589, served as the director of Barcelona's atarazanas (shipyards) into the early seventeenth century. The city's atarazanas were the most important for galley building in Mediterranean Spain. Alçatte's son, Carlos de Alçatte, replaced his father and served from 1607 until 1623.[55] The Alçatte family and multiple forest superintendents had trouble finding adequate supplies

51 Gómez Rivero, 618.
52 Gómez Rivero, 619.
53 Goodman, 262, Appendix A: "Superintendents of Forest Plantations, 1574–1660."
54 Goodman, 74.
55 Goodman, 125.

of timber near Barcelona; this had been the problem that led to Alçatte's forest inspection of 1589. The difficulties securing timber arose for other reasons besides deforestation. Timber sellers would refuse to sell to the crown, which offered less money and was less reliable in making payments than private buyers. As in Guipúzcoa, the crown urged superintendents in Catalonia not to harass local populations about tree plantings, but to try to persuade them to plant more trees. The royal forest of Tortosa became a valuable exception, since it both belonged to the king and was near Barcelona. The region continued to devise and administer local forest laws, including a landmark set of ordinances in the 1620s.

About the time when the crown's strategy to control the North Sea and access Baltic pine forests began to falter, the crown investigated ways to gain access to the great pine forests of Navarre via the Ebro River. If the crown could find a way to use the pines of Navarre, perhaps it could stop relying on Baltic timber for masts. The plan to build a road to one of the Ebro's tributaries in Navarre came from Pascual de Atocha, a shipbuilder from San Sebastián.[56] King Felipe IV approved a reconnaissance mission in 1637, adding, "This is of greater importance than anything in Spain."[57] Despite the great distance of 160 leagues, ninety days of travel, the employment of several skilled raftsmen, and great costs to build the roads and clear the river of obstacles, the possibility of domestic self-sufficiency led to a full inspection of the river network and supply system. Unfortunately, after the sample of pines arrived in Barcelona in 1639, authorities decided they were of inferior quality and too short to serve as masts. Nonetheless, the chance of finding better trees nearby encouraged the crown to attempt similar reconnaissance missions multiple times over the next century.

On the north coast, in the region known as Cuatro Villas and Asturias, Hernando de la Riva Herrera continued to serve until 1605, when his son Fernando possibly took over and served until 1632. As in other regions, the job remained in the family, and Fernando's nephew, Fernando de la Riva Herrera y González de Acevedo, served until 1647. It took a couple of generations, but after the death of Cristóbal de Barros, the Spanish forest inspection bureaucracy had grown to include powerful bureaucratic dynasties with broad powers to monitor and control forest management from Galicia to Catalonia. The reign of Felipe III and the early years of Felipe IV's reign essentially created a new Habsburg forest territoriality. Beginning in about 1598, the crown gradually established, formalized, and centralized control over extensive forest

56 AGS, GA, leg. 3168, August 12, 1637.

57 Quoted in Goodman, 139.

resources for the interests of the navy through its bureaucracy in the north of Spain.

The Waning of Spanish Imperial Might

Forest historian Erich Bauer Mandersheid wrote, "The one hundred years until approximately 1730 signifies a certain interval of tranquility for the forests of the country."[58] He says that during a period of naval shipbuilding decline, the crown placed a lot less pressure on Spain's forests, allowing them to recover from the decades of deforestation of the sixteenth century. One could add to this argument that population decline, or at least a plateau of growth, in this period also reduced the rate of deforestation, since the extension of agriculture had traditionally been a greater threat to forests than shipbuilding. Guillermo Muñoz Goyanes, who has also written on forests and forest legislation in Spain, has noted that measures for defending forest resources during a period of civil conflict after 1640 were notably absent.[59] With such contributions to our knowledge of forests in the seventeenth century, why bother studying state forestry in an era of naval decline, that is, from about 1630 to 1730?

The answer is clear. By exploring the idea of territoriality in the context of Spain in the later seventeenth century, we can study an aspect of forestry history that does not necessarily involve resource extraction and deforestation. Instead, the resource control strategies of the crown aimed to maintain and improve access to forests for crown interests. Such strategies included the continuation of the bureaucratic network of forest superintendents, the production of forestry advice based on local-level experience, more stringent forest legislation that enhanced royal control over behavior in forests, and more forest reconnaissance missions aimed to locate new domestic and accessible timber areas into which the crown could project greater authority. Such activity is reason enough to study state forestry in an era of naval decline and not to dismiss the period as one of tranquil forest recovery lacking royal involvement.

By the time war broke out in 1618 and 1621 after a decade of peace, the forest superintendents had established their authority in northern Spain. The high demands of war for the next forty years put great strains on financial, material and human resources, and the crown relied on the bureaucratic network of forest inspectors to help ensure access to shipbuilding timber. Olivares held naval

58 Bauer, *Montes*, 174.

59 Guillermo Muñuz Goyanes, *Crónica sobre bosques y montes de la Península Hispánica* (Madrid: Fundación Conde del Valle de Salazar, 1983), 112.

power in high esteem and worked to override regional liberties, the same ones that had been giving superintendents such as Antonio de Urquiola and Domingo de Idiáquez trouble, in order to help meet the crown's need for resources.

After a decade of military and naval success, Spain's fortunes began to wane after 1627. That year, Felipe IV declared bankruptcy and rescheduled his debts. The following year, at this crucial financial juncture, the Dutch captured an entire treasure fleet in Matanzas Bay, Cuba. In 1629, Olivares gambled on a quick intervention in the succession of Mantua to help prevent the title going to the French-born Duke of Nevers. At the time, France, under the leadership of Louis XIII and Cardinal Richelieu, was busy fighting the Huguenot-controlled city of La Rochelle. Olivares believed France with Richelieu's leadership posed Spain's greatest long-term threat and thus wanted to make a preemptive strike against French interests in Italy, while France was busy with internal affairs.[60] However, the Spanish siege of Casale in Mantua took longer than planned, and France sent forces to Italy to prolong the struggle for three years. The War of the Mantuan Succession was a costly distraction for Spain following the loss of the 1628 treasure fleet, and it unfolded at a time of poor harvests and rising food prices. John Elliott wrote "In retrospect, the years 1627–1629 seem to be the turning-point in the history not only of the Olivares administration but also of the attempts of Habsburg Spain to retain its European primacy."[61] However, at the time, this was not clear. Spain would continue to pursue its ambitious foreign policies and, at the very least, survive in its wars against the Dutch, the English, and the French. Nor did the crown abandon its commitment to conserve timber resources through the efforts of its forest superintendents and forest legislation.

While Spain suffered many losses after 1630, its military and naval forces could still mobilize effectively at crucial moments. Its ability to repeat extraordinary feats against all odds eventually waned, but the process unfolded slowly. In 1630, the Dutch returned to Brazil, and Spain sent another relief force to repel them, but the Dutch eventually established a long-term foothold in northeast Brazil. Dutch and English encroachment in the Caribbean also accelerated in this period, eroding the Spanish monopoly of American trade. The Dutch took Curaçao and the English established control in Barbados, Bermuda, St. Kitts' and Nevis. The 1630's also witnessed the beginnings of a serious naval threat from France. Richelieu had forty warships by the middle of the decade.[62] War broke out with France in 1635, which at the same time

60 Elliott, *Spain and its World*, 126–127.

61 Elliott, *Spain and its World*, 125.

62 Goodman, 24.

threatened Spain's most important forest regions in the borderlands of the Basque country and Catalonia.

In 1636, the Habsburgs of Germany and Spain launched a joint invasion of France, nearly reaching Paris. Over the next four years, however, Spain suffered some of its most demoralizing setbacks. The Habsburg invaders were soon expelled from France, and the Catalan army collapsed under desertions and conflicts within its ranks.[63] Richelieu opted to invade the Basque region rather than take advantage of the growing tensions in Catalonia. In 1638, the French blockaded the Guipúzcoan coast, attempted a siege of Fuenterrabía, destroyed ships under construction in Pasajes, and defeated a Spanish fleet off the coast near Guetaria.[64] Spain suffered even greater defeats in 1639 when a fleet sent to recover Brazil failed, and Antonio de Oquendo lost thirty-two of his forty-seven combat ships at the Battle of the Downs.[65]

In 1640, Catalonia and Portugal revolted against Castile. Castile could not face both fronts at once and gave Catalonia primary attention. Olivares could not overcome such disasters, and Felipe IV removed him from power in 1643. Spain sought peace with the Dutch to face the rebellions and the invading French forces in Catalonia. In 1648, Spanish representatives signed the Treaty of Münster, in which Spain finally recognized Dutch independence. Spain allowed the Dutch to provide shipping for Spanish commerce and the Dutch promised to end interference with Spanish trade in America. Fortunately for Spain, France descended into the civil conflict known as the Fronde later in 1648, allowing Castile a chance to recover Catalonia. In 1652, rebel authorities in Barcelona capitulated to Juan José, an illegitimate son of the king, ending the Catalan revolt. The recapture of Dunkirk that same year indicated that Spanish power had not completely disappeared. In fact, victories against France at Pavia in 1655 and in Flanders in 1656 brought France's Cardinal Mazarin to Madrid for peace talks.[66]

In the 1650s, however, the rising power of England during the Protectorate of Oliver Cromwell aimed to destroy Spanish sea power by gaining greater influence in America, weakening the Spanish position there. England combined forces with France, leading to another series of setbacks for Spain, including the definitive loss of Dunkirk (1658), and finally leading to the Peace of the Pyrenees in 1659. For Spain, peace provided an opportunity to deal with the Portuguese rebels, but support from England and France gave Portugal the

63 Stradling, 94.

64 Goodman, 24.

65 Phillips, *Six Galleons*, 218–219.

66 Stradling, 123.

means necessary to resist Spanish invasions in 1663 and 1665. Spain finally made peace with the rebels and recognized Portuguese independence in 1668. Spain made some commercial and colonial concessions to England as well, laying the groundwork for peace until the next century.[67]

Toribio Pérez de Bustamante's *Instrucción* of 1650

The Spanish population failed to grow in this period, and its economy continued to stagnate, restricting supplies of men and money for its military forces. Concerns for supplies, such as shipbuilding timber, also continued throughout this period. The crown made incredible demands on its bureaucracy, but it managed to find loyal subjects willing to serve the crown, often at great personal cost. Clearly, Spain remained quite active militarily after the "turning-point" of 1627–9, and it achieved many significant naval victories. What was also clear, however, was that by 1668, Spain had fallen to second-rate power status behind France and England. Nonetheless, throughout this period of economic depression and political realignment, the crown's concern for its forest resources remained at a high level. In 1656, for example, at a time when Spanish naval power was on the wane, Felipe IV confirmed the most important forestry document of the seventeenth century – the regulations written by Toribio Pérez de Bustamante.

Bureaucratic experience in dealing with the local forest communities had already resulted in various innovations and new tactics in forest management for state interests. The most influential contribution of forest administration expertise in the entire century, however, came in 1650 from the forest superintendent of Cuatro Villas, Pérez de Bustamante, as a result of two years of service in the region.[68] King Felipe IV confirmed Pérez de Bustamante's forestry instruction, originally meant for Cuatro Villas alone, for all of Spain and its empire in 1656.[69] While policies from the center often changed local level forest management, sometimes local conditions directly influenced centralized policies.[70]

---

67 Stradling, 144–147.

68 AGS, GA, leg. 3309, February 15, 1650, Santander. A copy exists in AGS, Secretaría del Despacho de la Marina (Hereafter, Marina), leg. 571.

69 Muñoz Goyanes, 113; Gaspar de Aranda y Antón, *La Administración Forestal y los Montes de Ultramar durante el Siglo XIX* (Madrid: ICONA, 1995), 54.

70 For another example, in Rentería in 1618, a parish priest presented a series of ordinances aimed to conserve the local forests. One section of the forest would be reserved for the

In 1648, Pérez de Bustamante, a purveyor of armadas, became Superintendent of Forests and Plantations in Asturias and Cuatro Villas. Two years later, he wrote an *Instrucción* on the conservation of forests located two leagues from the sea or a navigable river. In it, he wrote that copies of the instruction "should be made and distributed to the justice and *procurador* of each town council so that nobody could claim ignorance [of these topics]."[71] He ordered the towns to read the instruction publicly in churches when all the townspeople were together inside. Pérez de Bustamante's instruction would be the most important forestry document until the 1748 forest conservation ordinance, and would be referred to in court cases and forest policy discussions in government until the 1748 ordinance superseded it.[72] While the frequent appearance of a forestry ordinance might indicate continued noncompliance, the fact that other legislation did not replace the 1650 *Instrucción* for nearly a century warrants a closer examination of its contents.

At its heart, the instruction was a message to forest communities about the benefits that forest conservation gave to everyone, from His Majesty the king to the average peasant. If the crown could convince local populations that obeying the superintendent was good for them, the king would face less resistance and improve his timber stocks. This general approach was not new. However, Pérez de Bustamante took it to new levels of sophistication. As he noted, it was important to conserve the forests because, "if those who came before us did not conserve them, they could not be enjoyed by us today. And if we do not conserve them, they will not be available to future generations." He went on to say, "I do not want to order you to plant more than your customary ordinances demand, but if you do, you will see benefits for yourselves." He then provided a financial argument to address those concerned with the cost of planting. This was a new approach. He said,

> It costs one half a *real* to plant one tree, but after twenty years, without giving it much thought nor requiring any extra exploitation, other than

---

iron works. The rest would be divided into twelve equal parts, identified with numbers and a name of an apostle on stone posts. In any one year, the town could only take from one of the twelve sections, leaving the other sections alone. Philip III approved the ordinance and ordered them to be applied to all municipal forests. The king appreciated the effectiveness of this local plan and hoped it would be effective elsewhere as well. Goodman, 95.

71 AGS, GA, leg. 3309, Santander, February 15, 1650. "Se da copia aunténtica a la justicia y procurador de cada consejo para que no puedan alegar ignorancia."

72 For a court case example, in the Archivo Real Chacilleria, Valladolid, Pl Civiles. Fernando Alonso (F) Caja 2719.0003/2720.0001. For forest policy discussion, see AGS, Marina, leg. 571, 1743.

> leaving it up to God and the weather, it will be worth 15, 20, or 30 *reales*, plus you would be enjoying the acorns, leaves, and firewood the entire time.

Finally, he added,

> Although the trees are planted for His Majesty, the fruit of the trees are to be enjoyed by everyone. When the king needs wood for his vessels – they are his forests, after all – he pays a reasonable price to encourage vassals to plant, and looking at it this way, although the plantations are the king's in form, in substance they belong to the vassals.

As a result, he said, everyone should take great care in observing the laws and planting new trees.

Pérez de Bustamante urged all types of forest owners to conserve and enhance their forests. Those who cut down trees without permission would be "castigated with rigor." Such a strong-armed approach came as a response to people abusing permission to cut down trees for various construction projects and personal needs by cutting more than they were granted and leaving many trees cut to the trunks. With the authority given to him by the king in his jurisdiction, Pérez de Bustamante ordered people to obtain a license before making cuts, to gain consent of the town council, and to make cuts in the presence of officials or reliable townspeople to reduce damages to trees. When the council provided a license to cut trees, it was of "urgent necessity" to plant two, three, or more for every one tree removed.

He noted that the proper time to make cuts was from the middle of December to the middle of February. Trees cut in these months during a waning moon would be less likely to rot and putrefy than those cut in the hot summer months. Also, more laborers were available in the winter months than during the harvest season of summer.[73] Pérez de Bustamante wrote that it was worthwhile to locate terrain suitable for plantations, or people would waste time planting trees that would fail to grow. He also noted that in the same winter months, but during a waxing moon, people should plant thornbushes to protect saplings from livestock. To help protect the saplings against strong winds, he told people to tie them to stakes and plant them close together, not far apart, so they would help shelter one another. He also urged people to clear the underbrush annually during the two months between the middle of December and February to help prevent forest fires, but not to clear near new saplings, because

73 Vicente Maroto, "Agustín de Ojeda," 323.

the underbrush also helped protect the saplings against winds and livestock. Pérez de Bustamante noted that livestock posed a much greater threat than fires, since animals passed through forests nearly every day, whereas fires occurred more rarely. However, he added that justices and officials needed to remain vigilant against the outbreak of fire and they needed to reprimand severely those who caused the fires, or, he said, "notify me and I will do it."[74]

Pérez de Bustamante stated that each citizen needed to plant two sessile oaks every year. He specifically forbade planting chestnut trees, which were not as highly valued by the navy, next to or within oak plantations and gave three insightful reasons why. First, the chestnut trees would become the dominant species and prevent the oaks from growing. Interestingly, Jean-Baptiste Colbert's famous French forest legislation from 1669 did not include such a clause, and eventually French officials complained of faster growing trees preventing the growth of young oaks.[75] Second, Pérez de Bustamante argued that owners of the chestnut trees would take the fruit from their trees and then inevitably take material from the municipality's oak trees as well, resulting in fights and disputes. Finally, if people were allowed to plant chestnut trees and oak trees together, people with "bad souls" who sought better terrain for their trees would not hesitate to remove the oak trees to replace them with chestnut trees. In order to avoid such problems, Pérez de Bustamante notified owners of chestnut trees to uproot or cut down their trees within ten days of notification if their trees were next to the oak plantations, or else face a penalty of 10,000 *maravedís*.

Pérez de Bustamante also identified goats as serious threats to forest plantations, as they caused much damage, especially to small trees. He ordered the owners of goats to take them to graze in the high sierras where the animals would cause less harm. Goat owners whose herds continued to damage forests would be fined 2,000 *maravedís* the first time, 4,000 the second time, and 10,000 for the third time.

Finally, Pérez de Bustamante reminded people in his jurisdiction to keep a book of accounts and record the oaks that they planted and the ones that they cut down every year, including testimonies regarding the purposes for cutting down trees. The town councils were to place the books in safekeeping for the justices and *regidores* to inspect, because damages to the forests were ultimately their responsibility.

---

74 "Las justicias y oficiales han de estar vigilantes y castigar rigurosamente a los que los causen, o darme noticia para que yo lo haga."

75 Paul W. Bamford, *Forests and French Sea Power, 1660–1789* (Toronto: University of Toronto Press, 1956), 83–84.

The crown relied on the obedience of its subjects to comply with royal policies, and Pérez de Bustamante worked toward this end by making some strong arguments about the long-term benefits of forest conservation for everyone. For as long as the crown had written forest legislation, it had conveyed a message that forest conservation was important to the population as a whole, but some of the arguments Pérez de Bustamante stated clearly for the first time. He also made clear that those who did not comply would be fined severely. He anticipated future local level disputes, ordering people to keep oak tree plantations for the king separated from other private plantations. Although he referred to naval shipbuilding specifically only once, the importance of these new forests for the king's navy underlies each point of the Instruction.

Colonial Forest Territoriality in the Seventeenth Century

Havana and the Caribbean Region

The shipbuilding program initiated in the reign of Felipe III during the late 1610s extended across the Atlantic to colonial shipyards as well, especially Havana, thereby introducing a new concern for protecting colonial forests. The Dutch began pressuring Spanish trade in the Caribbean from the early seventeenth century, and with the outbreak of the Thirty Years' War in 1618 and the renewal of war with the Dutch in 1621 at the expiration of the Twelve Years' Truce, the crown expressed a greater concern for protecting treasure fleets in American waters against pirates and corsairs. The idea of an *Armada de Barlovento* stationed in the Caribbean to defend the fleet until it had safely sailed past the Bahamas into the open Atlantic had emerged in the late sixteenth century, but had floundered due to a lack of resources. The crown revived the *Armada de Barlovento* in 1605. However, Spain diverted its forces to Europe. It was revived again in 1627 and was paid for by the *Real Audiencia de Santo Domingo*, but operated intermittently. In the 1640s, the *Audiencia* asked for financial support from New Spain, but the fleet was again used to serve on the transatlantic convoy. Attempts to establish a permanent fleet in the Caribbean did not succeed until the late 1660s. With Spain's resources strained to the maximum during this era, its priorities remained in Europe, allowing for successful incursions by the Dutch, English, and French in the Caribbean. Planters from these nations on non-Spanish islands of the Caribbean were eager to cultivate sugar, and enthusiastically removed dense forests at a rapid pace.[76]

76 Shawn W. Miller, *An Environmental History of Latin America* (New York: Cambridge University Press, 2007), 84–85.

Protection of the Indies was also made more difficult by a reticence on the part of the government in Spain to commit support for an autonomous fleet in America. Olivares was concerned with the potential weakening of Castile's authority in the region if the crown granted an independent creole armada.[77] Throughout the century, even when American vessels started to make up a significant percentage of those involved in the Indies trade and defense, royal councils continued to express their preference for Cantabrian and Basque vessels for the *capitanas*, *almirantas*, and *galeones de la plata*.[78] In addition, total initial costs for vessels in America appeared higher than in Spain, 24 *ducados* per *tonelada* in Spain versus 36 *ducados* per *tonelada* in America.[79] Still, the crown could adapt to enemy encroachment to some degree as long as the Indies trade remained well-protected. To this end, the crown decreed on March 2, 1617, to establish an arsenal in Havana to serve as a permanent royal shipyard. This did not signal the beginning of major shipbuilding in Havana; the port had been building a varied typology of vessels for private and crown interest for long time. However, it did help establish a preeminence of naval interests in the area.[80]

As shipbuilding in the Caribbean became more concentrated in Havana, the crown asserted greater control over the forests of Cuba. This assertion of naval interests in Havana soon brought the manager of the shipyards, Captain Juan Pérez de Oporto, into conflict with the regional livestock owners, who had entered into litigation by 1620. Pérez de Oporto oversaw the felling of trees in estates near the arsenal site, and the landowners in the area protested the new demands of the navy. The crown came to the support of Pérez that year with a *real cédula* decreeing that shipbuilders had the right to take the trees they needed.[81] Disputes continued, to the detriment of Cuban forests, and in 1622 Felipe IV decreed from Madrid, "Considering the woods of mahogany, oak, and cedar are of the greatest importance for the *navíos* made on the island of Havana: we command governors and captain generals not to consent or permit

77 Goodman, 23.

78 José Luis Casado Soto, "Barcos Para la Guerra: soporte de la monarquía hispánica," *Cuadernos de Historia Moderna* v. 5 (2006), 43.

79 Marina Alfonso Mola and Carlos Martínez Shaw, "Los Astilleros de la América Colonial," in *Historia General de la América Latina*, v. 3, n. 1, Alfredo Castillero Calvo and Allan J. Kuethe, eds. (Madrid: Trotta, 2000), 288–289. The authors add that colonial construction was profitable in the long run due to American vessels' endurance and infrequent careening.

80 Alfonso Mola and Martínez Shaw, 289.

81 Miguel Jordan Reyes, "La deforestación de la isla de Cuba durante la dominación española, 1492–1898" (PhD diss., Universidad Politécnica de Madrid, 2006), 47–48.

the cutting of these save for the construction of vessels for our fleet."[82] This decree to conserve mahogany, oak, and cedar arrived in Havana on May 24, 1623. The next year, the king decreed that no cutting of any kind could be carried out for ten leagues along the coast in either direction from the city of Havana without a license from the governor of Cuba.[83] Restrictions on woodcutting were strongest at times of crisis. In 1629, one year after the Matanzas disaster, the crown advised Lorenzo de Cabrera y Corbera, governor of Havana and captain general of Cuba, that the Dutch were on their way back and to prevent any cutting of timber in the area of La Chorerra.[84] There were no superintendents of forests in the colonies as there were in Spain, but the shipbuilder, whether appointed by the governor or hired by private bid, oversaw the procurement of materials. In Cuba, wood was cut by slaves and transported by oxen to the river banks.

Havana was not the only Caribbean port building vessels for Spain in the seventeenth century, but it was the only site in the region that received special regulation of its forests for naval shipbuilding. Other productive sites included Campeche, Veracruz, Panama, Cartagena de Indias, and Maracaibo. Out of thirty-five large vessels built in the Caribbean between 1618 and 1648, though, twenty-one of them were built in Havana. The resources for vessels built in Cuba and other sites came from many places in the region as well as from Spain and the Baltic. The most important local source was, of course, timber. Most other items could be imported relatively easily. Rigging was sometimes made of henequen from Campeche. In 1624 and 1625, the king ordered rigging, nails, iron, tar, sails, and any other necessary items, even yards, to be sent from Cádiz to Havana for two galleons under contract with the Captain Francisco Díaz Pimienta.[85]

Increasingly, American-built vessels were allowed to become part of the treasure fleets. Ship builders in Havana had lobbied the crown in 1621 for their vessels to be admitted in the *flotas*. In 1629, by a *real orden*, this was accepted.[86] A *real cédula* from 1638 allowed for *navíos* built in Havana, Campeche, Santo

---

82 Jordan Reyes, 45.

83 Reinaldo Funes Monzote, *From Rainforest to Canefield in Cuba: An Environmental History Since 1492* (Chapel Hill: University of North Carolina Press, 2008), 22; author equates 10 leagues with 42.4 km.

84 AGI, Santo Domingo, 870, L.8, F.132v–133r, Aviso de llegada de enemigo a La Habana (1629-01-11)

85 AGI, Santo Domingo, 869, L.7, F.257r; AGI, Santo Domingo, 869, L.7, F.258v–260r.

86 Ovidio Ortega Pereyra, *El Real Arsenal de La Habana: la construcción naval en La Habana bajo la dominación colonial española* (La Habana, Cuba: Editorial Letras Cubanas, 1998), 34.

Domingo, Puerto Rico, and Jamaica to be in the *flotas*. A *real cédula* from 1648 added *navíos* built in Maracaibo as well as those built on the Pacific side at Realejo and Guayaquil.[87] By the end of the century, American vessels made up about half of the active Indies fleet.[88] There was an increasing prestige (and affordability) of creole vessels, as they became known for their longevity by contemporaries, who attributed the strength of the timbers as one of the main reasons.[89]

The crisis in the Indies trade, particularly the middle decades of the century, meant fewer vessels were sailing for Spain. Naval shipbuilding dwindled in the second half of the seventeenth century. In Havana, two *galeotas* were built between 1680 and 1685, and one galleon was ordered built for Governor Severino de Manzaneda. Two galleons were built in the later seventeenth century in Cartagena de Indias. The American contribution even remained small in comparison to the twenty *buques* built in the same period in Basque shipyards.[90] As historian Ovidio Ortega Pereyra has pointed out, shipbuilding in the seventeenth-century Caribbean was sporadic. In the 1650s and 1660s, Cuban naval authorities began facing off with the growing sugar industry, but a decline in both the sugar and shipbuilding industries in Cuba in the last decades of the century delayed conflict.[91] By the late seventeenth century, Spain's transatlantic convoys were less frequent. The numbers of vessels involved in the Indies trade dropped from 6,573 in 1600–50 to 1,835 in 1650–99.[92] The crisis had more to do with American economies developing and breaking away from dependence on the mother country rather than colonial economic collapse. As in Spain, though, the crown remained concerned with shipbuilding's access to timber resources. Timber shortages near shipyards spurred the crown to look for potential new sites that possessed the natural resources as well as the potential to support the required skilled and unskilled

87 Ortega Pereyra, *El Real Arsenal de La Habana*, 34.

88 John Robert McNeill, *Atlantic Empires of France and Spain: Louisbourg and Havana, 1700–1763* (Chapel Hill: The University of North Carolina Press, 1985), 132.

89 Alfonso Mola and Martínez Shaw, 289.

90 Alfonso Mola and Martínez Shaw, 288.

91 Funes Monzote, 31–32.

92 John Lynch, *Spain under the Habsburgs, Volume 2: Spain and America, 1598–1700* (New York: New York University Press, 1984): 208–209. This drop left "the last fifty years with only 22 percent of the century's traffic...In the period 1650–99 national vessels in the *carrera* numbered 289, or 31 percent of the whole; Spain's American shipyards supplied a further 211 vessels (22.6 percent); 275 (29.5 percent) were foreign-built, and 155 of unknown origin. The use of Spanish vessels rose especially in the 1670s and 1680s, coinciding with a period of improved export performance."

labor to produce affordable vessels. In 1665, for example, Captain Sebastián Fernández Gamboa, *guardia mayor* of the port of Cartagena, and *alguacil mayor de la Real Hacienda*, submitted a report on the installation of a shipyard at Jagua on the southern coast of Cuba because of its natural resources and its large harbor, as well as the insufficiency of those conditions in Cantabria.[93] Officials in America and Spain discussed the shipyard at Jagua well into the eighteenth century, but it was far from a populated site (Trinidad was 14 leagues away) and the lowlands surrounding the harbor were not very protective (Figure 5).[94]

Guayaquil and the Pacific Coastal Region

As seventeenth-century shipbuilding in the Caribbean concentrated in Havana, shipbuilding in the Pacific increasingly centered on Guayaquil. Like in the Caribbean, the main shipyard was supported by smaller ones in the region and was supplied with resources from a wide area. Guayaquil experienced its golden age of shipbuilding in the century, supplying large royal defense vessels for the *Armada del Mar del Sur* as well as numerous private merchant vessels for regional inter-colonial trade. However, it was also a century of increasing threats as Spain's enemies made new forays into the Pacific. In the early seventeenth century, that threat came in the form of large and well-organized Dutch fleets that aimed to destroy Spanish control of the region and even establish their own colonies nearby. They were often on their way across the Pacific to continue their efforts in Southeast Asia. The Dutch encountered the South Seas fleet off the coast of South America in 1615, 1624, and 1643. After this, the Dutch focused their attention on Brazil and the Caribbean. The threat to Spain's colonial interests in the Pacific changed character in the later seventeenth century, as pirates from the Caribbean made their way across the Isthmus of Panama or around the southern tip of South America to harass the merchant vessels, particularly the ones loaded with silver from the Potosí mines, for the last twenty years of the century. In 1687, the city of Guayaquil was destroyed and was forced to rebuild in a new location.

The crown urged the *Armada del Mar del Sur* to keep at least five seaworthy ships in operation from the early decades of the century. Construction in Guayaquil was even more sporadic than in Havana, and occurred mostly at a frenetic pace in response to apparent threats. The city became a designated royal shipyard only when actively constructing for the crown. This occurred in the

---

93 AGI, Santo Domingo, MP, 61.

94 Antonio Béthencourt Massieu, "El real astillero de Coatzacoalcos (1720–1735)," *Anuario de Estudios Americanos* 15 (1958), 374–377.

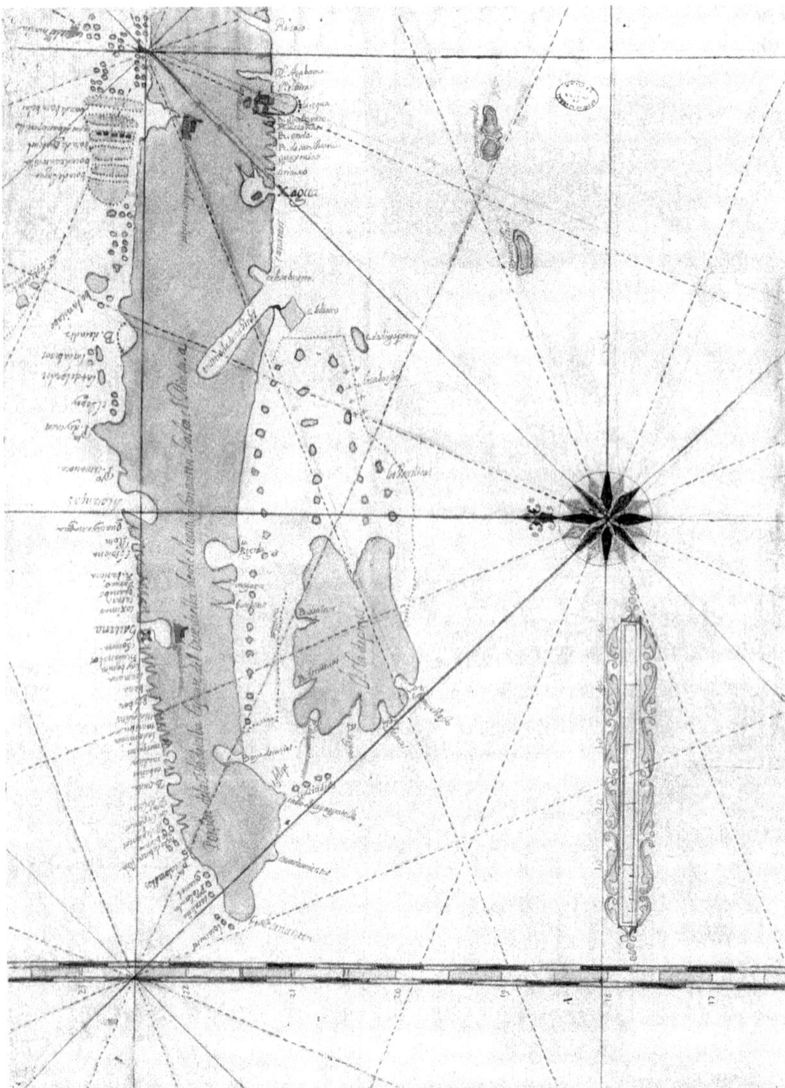

FIGURE 5 *Map of Western Cuba including Jagua – Concerns for timber shortages near Havana in the middle of the seventeenth century motivated searches for new shipbuilding sites in the Caribbean. The site of Jagua on the southern coast of Cuba (spelled "Xagua" on the above map from 1665) possessed timber resources and a large harbor, but was distant from labor sources and was not well-protected.*

SOURCE: ESPAÑA. MINISTERIO DE EDUCACIÓN, CULTURA Y DEPORTE. ARCHIVO GENERAL DE INDIAS. AGI,MP-SANTO DOMINGO,61 – PARTE DE LA ISLA DE CUBA.

seventeenth century at an average of once every ten years.[95] Guayaquil, then, never became a permanent royal shipyard like Havana would become by the early eighteenth century. The city was "a rough, ramshackle sort of frontier town during the sixteenth century, and remained temporary and sad in appearance throughout the seventeenth."[96] Economically, it still benefitted from timber exports, private shipbuilding, cacao exports, and remained the main port for the Ecuadorian highlands. Royal shipbuilding in the seventeenth century became increasingly under the direction of the viceroy in Lima. The viceroy initiated construction either through direct contract or by appointment of a lieutenant who contracted out the necessary stages of production. The superintendent held sweeping powers related to finance, supply, labor management, and judicial oversight, and superseded the authority of the *corregidor* in the seventeenth century. Orders from the viceroy outlined the details of the construction, including methods of procurement of resources. As in Spain, matters related to timber identification and extraction were left to men with extensive maritime experience, usually with the Spanish navy. However, as in other colonial contexts, there was no superintendent of forests and plantations designated for long-term forest conservation. As described above, one of the most important officials under the authority of the shipbuilding superintendent was the chief lieutenant. In the middle of the seventeenth century, this position was twice offered to Lorenzo de Bances León. In 1640, Martín de Valenzegui was appointed by the viceroy to build two galleons for the South Seas fleet, and completed his task in 1644. He chose Bances to serve him because Bances possessed "a good knowledge of the forests for he was a shipbuilder himself."[97] In 1656, Bances was reappointed by a new superintendent, Cristóbal de Mello, who was appointed to build another two galleons. Later that year, Bances was promoted to sergeant major and chief scribe of the treasury of Guayaquil. His commission as chief lieutenant stipulated his role in carrying out long visitations in the forests to oversee the cutting of timbers. His duties required knowledge of the appropriate woods for different parts of the vessels under contract as well as skill as a manager of labor. He had to negotiate contracts with the local population, known as the montañeros.[98]

95 Alfonso Mola and Martínez Shaw, 291.

96 Lawrence A. Clayton, *Caulkers and Carpenters in a New World: The Shipyards of Colonial Guayaquil* (Athens, Ohio: Ohio University Center for International Studies, 1980), 9.

97 Clayton, 29; from "Título de Teniente de la Superintendencia de la Real Fábrica de Galeones, a favor del Sargento Mayor don Lorenzo de Bances León," June 28, 1656, ACG, III, 327–330 (ACG, III, 327).

98 Clayton, 32. "Middle men such as Indian caciques (chiefs), local shipbuilders, and montañeros all delivered timber to the yards at one time or another under contracts specifying types, amounts, prices, and deadlines."

Other personnel involved with procuring timber were members of the *cabildo*, or city council, who received royal commissions to cut and deliver timber to the royal shipyard.

Spanish-creoles moved away from the manual labor of ship construction during the seventeenth century, but retained positions of authority in the administrative and financial sectors of the industry. Indian labor tribute continued to be directed to felling and transporting timber, as well as other unskilled tasks. Black slaves and mulattoes gradually replaced Spanish-creoles and Indian laborers at the shipyards. According to Lawrence Clayton, the early seventeenth century was a period of transition from a workforce of Spanish craftsmen, their black slave apprentices, and Indian laborers, to a "multi-hued" labor force composed mostly of black and mixed races in both the skilled and unskilled sectors by the late seventeenth century.[99]

Seventeenth-century construction was characterized by greater financial accountability, and greater concern for uniformity in design. Early in the century, galleons built in Guayaquil tended to be about 400 *toneladas*, but the galleons built by de Mello after 1656 were 825 *toneladas*, and were criticized by the crown for being too large. The king expressed a desire to keep the vessels at or under 500 *toneladas* to allow for greater maneuverability along the American coasts. Larger vessels may have offered the benefits of cheaper freight rates, but also brought with them greater risks. Overall, the vessels built at Guayaquil left a record of sturdy construction. Dionisio de Alsedo y Herrera wrote in the 1730s that the ships were "so durable that many have lasted sixty and seventy years notwithstanding the bad manner in which they are sailed."[100]

Recognition of depleting local resources for Guayaquil's shipyard came early in the seventeenth century, but legislation aiming to curb deforestation did not emerge until 1650. Antonio de Morga, the President of the *Audiencia* of Quito, remarked in 1615 that shipbuilding had greatly reduced nearby forests and had come at great cost to the indigenous population. People procuring timber "have to search very far and many Indians are lost in these endeavors."[101] The experience of Bances working as chief lieutenant from 1640 may have led to new regulations, because Bances reportedly encountered shortages wherever he went. Eight months into his commission he had been unable to acquire the necessary timber for one frigate.[102] In 1650, the *cabildo*

99 Clayton, 125–127.

100 Clayton, 72.

101 Clayton, 105.

102 Clayton, 106.

of Guayaquil and *corregidor* Francisco Vásquez de Silva passed legislation regulating the cutting and commerce of timber. Wood deemed essential for naval shipbuilding, including *guachapeli* and *palos amarillos*, were prohibited from being used in the construction of buildings, furniture, or any other non-marine items. Ship timbers, including "*guachapeli* masts, sternposts, stems, brace pendants, inner posts, knees, finished or unfinished futtocks," could not be exported. Violators of these regulations "faced the forfeiture of all illegal material found in their possession; half of the seized lot went to the royal treasury and half to the *cabildo*." In 1660, the *cabildo* passed additional regulations. The extraction of shipbuilding woods, which now also included *árboles de maría*, required a special license. Violators, in addition to giving up the material, had to pay fines of 50 *pesos* for first offenders, and 100 *pesos* for second offenders. Flagrant offenders faced exile. Enforcement of these regulations was extremely difficult because average citizens as well as royal officials were guilty of breaking these rules. Failure to enforce the regulations, as well as the increasing use of green timbers in the later seventeenth century, led to the reissue of the 1660 regulations in 1684 and again in 1689.

Piratical activity intensified in the waters along the western coasts of America in the 1680s. Even though shipbuilding costs were coming down in the later seventeenth century from a peak in the middle of the century, vice-regal treasuries could no longer support a fully-operational *Armada del Mar del Sur*. As a result, the *consulado* (merchants' guild) of Guayaquil funded the arming of merchant vessels in 1686 to defend the city. After 1689, the merchants' guild struck an agreement with the crown to pay off some of its debts by building royal warships. The final decade turned out to be more productive for the Spanish American Pacific region. In the years 1689–95, 38 vessels were built; 21 were built in Guayaquil, including 6 navíos, 6 fragatas, and 10 other ships. Central American yards built 15 (1 navío, 7 fragatas, and 7 other ships), and 2 were built in Chile (1 navío and 1 fragata). However, naval forestry remained ineffective in the forests of Guayaquil and the Pacific.

Cavite and the Philippines

Spain's conflict with the Dutch intensified in Southeast Asia during the early seventeenth century. The Dutch East India Company, chartered in 1602, aimed to keep Spain out of the Spice Islands to the south. The Dutch also hoped to disrupt the trade in and out of Manila, especially the Acapulco-bound galleons. No fewer than sixteen Dutch East India Company fleets sailed in the region before 1625. There were about forty or fifty armed Dutch vessels in the waters of the Phillipines in any one year, and they engaged in at least four major naval engagements.[103] The Spanish were also concerned about raids from the indigenous populations, especially the Islamic powers to the south.[104] Effective colonial control did not reach Mindanao or even the highlands of Luzon.

The great distance from the court in Madrid hindered aid and communications in the Philippines, and prevented the access to skilled personnel and essential goods from Spain that other colonial ports enjoyed. Authorities in the islands repeatedly complained of a shortage of Spanish shipwrights. The crown posted several notices in the major shipyards of Spain, including a *real cédula* for the *corregidores* of Guipúzcoa and Vizcaya in 1605 looking for three master shipwrights. Hernando Guerra was given the job, but by 1611 had never shown up. New orders were sent by the crown to the *Casa de la Contratación* in 1638, 1645, and 1647, and to the viceroy of New Spain in 1660. By 1670, the crown offered to pay for the full journey to the Philippines if anyone wanted to go. In November of 1671, a *real cédula* sent to the Marqués de Mancera, the viceroy of New Spain, stated that searches for master shipwrights to send to the Philippines failed in Vizcaya, Guipúzcoa, Cádiz, and Sanlúcar, but ordered further searches in the *Audiencia* of Santo Domingo, including the ports of Havana, Yucatán, and Cartagena. If they found anyone, they were to be sent to México and then directly to the Philippines. In May of 1672, the *Junta de Guerra* believed they had finally found their man, Bernardo de la Cruz of Triana, the mariners' enclave across the Guadalquivir River from Seville. However, by September, the king gave an order to renew the search for a master shipwright, because upon Bernardo de la Cruz's arrival, he was excluded for being too old and unqualified (*por exceso de edad y falta de inteligencia*).[105]

Despite the nearly continuous shortage of skilled personnel from Spain, colonists in the Philippines did manage to exploit the local forests for shipbuilding and defenses. In the first decades of the colony during the sixteenth century, most of the vessels operating in the Acapulco trade were built in Realejo, Guayaquil, or some other American shipyard. By 1619, though, Captain Sebastián de Pineda could state that six out of seven galleons were built in the Philippines. The shipyards benefitted from skilled shipbuilders from the region, especially China, and the employment of forced labor. When Cavite was active, it employed up to about 1,400 people.[106] Some of the largest vessels

103 Greg Bankoff, "'Deep Forestry': Shapers of the Philippine Forests," *Environmental History* 18, no. 3 (July 2013): 537–538.

104 AGI, Filipinas, 27, n. 52, July, 1605.

105 AGI, Filipinas, 331, L.7, F.24r–24v, September 13, 1672.

106 William J. McCarthy, "The Yards at Cavite: Shipbuilding in the Early Colonial Philippines," *International Journal of Maritime History* 7, no. 2 (December 1995), 155.

of the century were built in these islands. Timber came from nearby, while other essential materials were imported from elsewhere in the region. Local timber shortages were apparent by 1621 when *procurador general* Hernando de los Rios Coronel wrote of trouble finding suitable timber near shipyards, and only "by penetrating the thicker recesses of the woods" could adequate supplies be found.[107] Later on, other officials expressed concern about timber shortages, writing of mountains "almost everywhere destitute of forests," and "cleared and despoiled of trees."[108] In general, however, throughout the century, the Spanish were able to meet their timber demands. This required building in multiple sites, not only in Cavite. Vessels were constructed on the beaches of Camarines, the provinces of Albay and Cagayán, the coast of Pangasinán, the islands of Marinduque, Mindoro, and Masbate. At times, the Spanish obtained vessels from elsewhere in the region, including Siam and Japan, but by 1679 the crown forbade the use of foreign vessels.[109] In 1627, the *provisor general* of the Philippine province of Rosario, Dominican friar Melchor Manzaon, wrote a discourse to the king about the importance of the Spanish port in Taiwan, then called Isla Hermosa, for its "fertility and abundance of woods," at a time when the Spanish had just established a colonial presence in the northern tip of the island.[110]

The crown had also grown concerned about the treatment of the indigenous laborers, calling on the governor of the Philippines to pay them fairly for their work in 1679. The *corte de madera* remained the method of procuring labor throughout the century, prompting local officials to remark on its destructive qualities. In 1608, Rios Coronel suggested purchasing ships built in Bengal and Cochin to avoid the elevated costs of construction in the Philippines and the "excessive" work demanded from the indigenous populations. Sebastián de Pineda claimed in 1619 that "if shipbuilding continues, no natives will be left."[111] In 1621, Rios Coronel recorded that the masts of one galleon required six thousand Filipinos for three months to transport them.[112] Labor conditions at the shipyards were not as bad as those in the forests, but overexploitation is evident in Cavite by the first decade of the century. In 1608, Rios Coronel established a

107 Bankoff, "Deep Forestry," 538–539.

108 Greg Bankoff, "'Tree as the Enemy of Man': Changing Attitudes of the Forests of the Philippines, 1565–1898," *Philippine Studies* 52, no. 3 (September 2004), 321–322.

109 Alfonso Mola and Martínez Shaw, 287.

110 Martín Fernández de Navarrete, *Biblioteca Marítima Española* (Madrid, 1851): 483. The Spanish would lose their colony there to the Dutch in 1642.

111 McCarthy, 157.

112 Bankoff, "Trees Enemy of Man," 324.

sawmill, which he requested from the crown because of the inconveniences and delays associated with employing indigenous workers.

Revolts against Spanish colonial rule occurred numerous times during the century. Rebellions against the imposition of Spanish Catholicism occurred in Bohol in 1621 and Panay 1663. Rebellions against the imposition of labor demands, particularly the *corte de madera*, broke out in 1649 in Samar and lasted for over a year. Also in 1649, the extraction of laborers from Nueva Segovia (northwest Luzon) caused the resistance of religious members of the Dominican order in that region. In 1660–1, revolts occurred in Pangasinán, Ilocos, and Cagayán against labor and tax demands from the Spanish government, during which monks would again join the protest and purposefully set fires in the forest.[113]

The galleons built in the Philippines were reputedly very sturdy and durable because of the high quality timbers used in their construction. The woods were impervious to the teredo and they were expected to last thirty to fifty years, but the average life of a Manila galleon was about six years due to losses in storms and other wrecks as well as negligence of required maintenance. Diego de Villatoro, the *procurador general* of Manila, complained about the short lives of the vessels, blaming the quality of the wood and the high costs of maintaining them.[114] As was often the case, shipbuilding in this century used timber that did not have the time to season the desired amount. The use of green timber caused Pineda to remark "one must tear up the decks every two years and put down new ones."[115]

Despite the strategic importance of the islands, the Philippines were low on the king's list of priorities for most of the century. As in other colonies, the Philippines had no dedicated forest bureaucracy. The islands were unique, however, in that no local forest regulation directives were issued by the crown. The population was sparse, the forests were plentiful, and the islands offered enough coastlines and riverbanks to allow for periodic shipyards to appear in numerous places while Cavite remained the principal port for the galleons.

Habsburg State Forestry in an Era of Naval Decline

The reign of Spain's Carlos II (r.1665–1700) was characterized by the growing militarism of France under Louis XIV and struggles over influence at the court

113 John A. Larkin, "Philippine History Reconsidered: A Socioeconomic Perspective," *American Historical Review* 87, no. 3 (Jun. 1982), 604–606; McCarthy, 158.

114 AGI Filipinas, 28, n. 96.

115 Bankoff, "Tree as the Enemy of Man," 324.

of the feeble king in Madrid. Carlos was only four years old when his father died in 1665. His mother, Mariana of Austria served as Regent until he came of age in 1675. However, Carlos was the product of generations of inbreeding, and he had shown few signs of being capable of ruling alone by that time. The committee (*junta*) of officials set up by Felipe IV, along with Mariana and her favored ministers, continued to rule. Juan José, an illegitimate son of Felipe IV, had many years of dedicated service to Spain, including leading the recapture of Barcelona in 1652 and taking part in the setback at the Battle of the Dunes in 1658. It was no secret that he had pretensions to play a central role in government under the new king, his half-brother. His father had excluded him from the governing *junta*, and he believed Carlos had become too dependent on Mariana, leading him to attempt a coup in 1669, which led to the ouster of Mariana's favorite, Father Nithard and Juan José's retreat to Aragón. Mariana and her new favorite, Fernando de Valenzuela began to alienate the nobility and Juan José attempted another coup, this time with success, in 1676. Juan José had proven to be a skilled leader, and Spaniards placed high expectations on his arrival in Madrid. It was a rather difficult time to be entering royal government, however, since the demographic stagnation of the time made it difficult to supply men for the military, and economic stagnation coupled with famine intensified in the 1670's.[116]

After four years of peace, Spain entered war against France in 1672. Louis XIV had invaded the Spanish Netherlands in 1667, because he believed some territory there had "devolved" to France as a result of the non-payment of the dowry of Louis's Spanish wife.[117] Moreover, Louis had felt betrayed by a guarantee of protection to Spain from the United Provinces (Netherlands) and England against further French aggression in Flanders in 1668. Two years later, though, Charles II of England defected to support France, ending the Anglo-Dutch guarantee. In 1672, Louis attacked the Dutch, and the Count of Monterrey, the governor of the Spanish Netherlands, decided on his own to come to the aid of the United Provinces. French victories in parts of the Low Countries, Catalonia, and Sicily, as well as the capture of the Habsburg homeland, the Franche-Compté, led to the peace of Nijmegen in 1678. Juan José felt strongly about the need to make peace with Louis, believing it was in Spain's long-term interest, and he followed up the peace by arranging a marriage between Carlos II and a Bourbon princess, María Louisa of Orleans. Juan José did not live to see the union, however, and died in 1679.

116 See Henry Kamen, *Spain in the Later Seventeenth Century, 1665–1700* (New York: Longman, 1980).

117 Stradling, 147.

The 1680's was a peaceful decade for Spain, save for a short war against France in 1683–4, and witnessed effective financial reform and population growth that helped set the country on a road to economic recovery. The war had been a response to continued French bullying of frontier populations of Flanders. France occupied Strasbourg in Alsace and Casale in Mantua in 1681 and invaded Luxembourg in 1683. Spain responded by declaring war without the aid of its former allies. France invaded Catalonia and Flanders and punished Genoa for aiding Spain in the Mediterranean. The French had demonstrated that they had the capacity to wage war on several fronts at once, but when Luxembourg conceded defeat in 1684, France and Spain agreed to seek peace.[118]

The next time Louis XIV started a war in 1688, Spain tried to stay out of it. William III's succession to the English throne, and the Dutch and Austrian appeals for help, eventually led Madrid to enter the war in 1690. In the Nine Years' War (1688–97), France invaded Catalonia, Italy, and Flanders again. While France demonstrated a great deal of military power, it was unable to achieve any real victory. By combining naval forces and essentially surrounding France, the allies worked together to achieve peace at Rijswick in 1697. France maintained its prestige and some territorial gains but restored Barcelona and Luxembourg, hoping to preserve favorable relations with Spain for the impending succession crisis. Carlos's marriages with María Luisa of Orleans and then with Maria Ana of Neuburg had failed to provide an heir, and the king's declining health had captured the attention of the rest of Europe.

Even though the reign of Carlos II was a difficult time for Spanish forces, the regency council and the king himself maintained an active involvement in the conservation and management of Spanish forests. The bureaucracy continued to inspect trees and send reports to the crown, the crown issued new forest legislation, and it ordered new forest reconnaissance missions to find and gain access to new supplies of domestic timber, even in an era of naval decline.

The appointment letter issued to Domingo de Idiáquez in 1601 turned out to be the model for any seventeenth-century forest superintendent of Guipúzcoa. Unfortunately for them, so was the salary. The 800 *ducados* a year remained unchanged for the entire century. An appointment letter from January of 1690 given to Francisco José de Nicolalde for the job of forest superintendent in Guipúzcoa sounds strikingly familiar to Idiáquez's letter from eighty-nine years earlier.[119] Nicolalde needed to reside in the province, visit twice each year

118 Stradling, 176.

119 AGS, Libros de Registros, Libro 414 (1689–1692), fol. 63–67, January 28, 1690, "don Francisco Joseph de Nicolalde, Superintendente de Guipúzcoa."

every location two leagues from the sea and along navigable rivers, and inspect the forests with local justices. He was responsible for identifying the locations of new plantations and allocating to different areas the number of trees to plant. He also instructed the people in his jurisdiction to avoid damage from livestock by planting thornbushes around saplings, an idea explained by Barros in 1594. When people cut down timber to build houses, churches, or mills, they had to plant two or three new trees in the same location for each felled tree. Nicolalde, like his predecessors, had to gather and record testimonies of plantings made, so the next time he went out for a visit, he took account of whether the justices had complied with the quota. For each tree not planted, the fine remained one *real*. Finally, the superintendent continued to be responsible for issuing subsidies to encourage shipbuilders to produce vessels suitable for the crown to rent in wartime.[120]

Nicolalde replaced his father, who had served for fifty-two years.[121] The continuity and stability of this bureaucratic network was an essential part of Habsburg state forestry in the later seventeenth century. Several letters from bureaucrats in the 1720's, after the Bourbons had replaced the Habsburgs, complained that oversight of forests had become ineffective and indicated that the lull in shipbuilding likely contributed to greater absenteeism in the offices of forest superintendents. It is probable that some superintendents devoted their efforts to other jobs elsewhere beginning in the later seventeenth century, but it is possible that the War of Succession (1701–14), which established the Bourbons on the throne, was more damaging to the management of forests in Spain than the preceding era of naval decline. During the late seventeenth century, the king at least continued to receive reports on forest conditions from his forest superintendents.

In fact, in Galicia, royal oversight even intensified. Bureaucrats wrote in the 1720's that Galicia had the most effective forest administration of all, and this was arguably due to forest legislation issued in the later seventeenth century. Galicia had been at the margins of royal forest territoriality in earlier times, being far from the active shipbuilding centers in and around Santander, Bilbao, and San Sebastián. It was the last region in the north to get its very own forest superintendent.[122] A *real cédula* written in the name of the Queen-mother

120 See Kamen, *Spain in the Later Seventeenth Century, 1665–1700,* 115, for evidence of continued ship construction in the Basque regions in the later seventeenth century.

121 AGS, Libros de Registros, Libro 414, fol.63.

122 Ojeda served for both Vizcaya and Galicia in 1612, then Juan de Morales replaced Ojeda in Galicia. By then, all other provinces had established superintendents. See Vicente, "Agustín de Ojeda," 343, and Goodman, 262.

Mariana of Austria in 1675 to Juan de Paredes, the forest superintendent in Galicia, placed the forests of his jurisdiction under the direct authority of the Council of War and the *Junta de Armadas*.[123] This meant that officials in Madrid, rather than local justices, would now settle disputes over fines and plantations. This was the closest royal involvement in forest management anywhere in Spain up to this point. In order to improve forest conservation in the region, the queen-mother declared that the superintendent had "absolutely independent administrative and judicial power and was subordinate only and directly to the Council of War and *Junta de Armadas* in Madrid."[124] The rationale for this change in royal tactics was that heavy deforestation in Galicia had been reported, and forest conservation techniques were not succeeding. The 1675 *cédula* stated that,

> The forests are found greatly reduced today…Seeing that this resource is for the common good of the monarchy (for the wood enables the construction of the warships in royal service), this *cédula* gives the necessary measures to achieve the best conservation and expansion of royal forests.

If employed well, the queen believed, the actions outlined in the *cédula* would make local residents stop destroying trees suitable for shipbuilding.

Mariana specified methods of visitations and plantations and the types of cuttings that were acceptable, drawing heavily on precedents set a century before. The *jueces de montes* should visit forests and plantations every year to examine conditions and provide reasonable remedies for conservation problems. Mariana ordered the superintendent to identify the location and allocate the quantity of new trees to plant. The local population was required to have a license issued by the superintendent to undertake any kind of timber extraction. If they damaged trees without a license, they would receive a fine that corresponded to the crime. Mariana ordered the superintendent to carry out visitations in her name in the spring and summer each year. Galicia was larger than the other superintendents' jurisdictions, so the *cédula* ordered half the

123 I have only seen a 1743 copy from AGS, Marina, leg. 571, dated February 10, 1675, placed directly after a copy of Bustamante's 1650 Instrucción. It is also mentioned in Muñoz Goyanes, 114.

124 AGS, Marina, leg. 571, February 10, 1675, "Declaro que el officio de Juez de Montes y Plantíos de dicho Reino tiene jurisdicción privativa absolutamente independiente en lo formal y judicial y solo es subordinado y dependiente immediatamente del Consejo de Guerra y Junta de Armadas."

forests of Galicia to be inspected one year and half the next. This was a method unique among the provinces of the forest superintendents. "Otherwise," the queen said, "quality visitations would be impossible and towns would not have time to execute what you order." She wanted the superintendent to make his visits during times that were convenient for the farmers, so that he could get the full attention of the towns. The superintendent was to bring a bailiff and a scribe on each inspection to prevent disputes.

Perhaps the most important task ordered was the proper maintenance of the testimony books. In addition to containing information on any new plantations and tree cuttings, Mariana wanted the books to contain information on the forests and districts the superintendent visited, their distances to ports and the closest harbor, navigable rivers nearby, and the towns that were part of the forests. In other words, she wanted testimonies with richer geographic information than had been typical in other regions in the past. The books should also contain a record of the penalties made by the justices in the locations where they worked. The superintendent had the power to issue fines to those illegally cutting trees or setting fires. Finally, Mariana reiterated that any problems with the orders would be dealt with by the *Junta de Armadas* and its treasurer, Pedro de Guzmán.

This *cédula* offered some innovations to Spain's forest territoriality. Galicia for the first time became the most closely managed region. By the mid-eighteenth century, Galicia was the most important and productive shipbuilding area in the Iberian Peninsula. Perhaps the Habsburgs understood the potential role an area like El Ferrol could play as a strategic naval shipbuilding center. Perhaps the closer management of forests in Galicia in the later seventeenth century contributed to its selection as an important shipbuilding center under the Bourbons. In any case, we see a greater interest in the region from the crown at this time. The queen-mother asked for more information than her predecessors had, including detailed information regarding the surrounding geography of each forest. Knowledge of nearby ports and navigable rivers could help a superintendent decide if building a road from a certain forest was an economical option, thereby making the process of finding new timber stands more efficient. The general goal remained the same as it had been for the rest of the century: the crown wanted to gain access to the greatest number of forests as quickly as possible and at the least possible cost. Mariana also asked for a report from the head officials of shipyards on the types and sizes of timber stored in arsenals in Galicia. She encouraged the shipyards' officials to visit and examine the forests that supplied their timber, probably because their expertise would help the crown locate more usable timber in the vicinity.

Many of the rules and directives in the document originated earlier in the century, but its focus on Galicia can be seen as an extension of forest territoriality. As in other regions, in Galicia the crown expressed its dependence on local residents to help conserve forests and make plantations. The legislation also revealed firmer methods to control the population to improve compliance, including oversight by the combined authority of the superintendent and local officials, more complete written testimonies, and a broader knowledge of suitable forest locations. This occurred at a time when Spanish naval power was in decline, but the crown understood the importance of maintaining access to the supply of a crucial natural resource in bad times as well as good.

Additional legislation from Carlos II in 1694 and 1695 reinforced the jurisdiction of the *Junta de Armadas* over forests with trees destined for royal arsenals.[125] The king issued a royal decree in December of 1694 specifying that all forests with trees destined for shipbuilding in Galicia within the usual two leagues from the sea or a navigable river fell within the purview of the Council of War and the *Junta de Armadas* directly. About two months later, the king issued a royal order, addressed specifically to the justices of Galicia. It clarified that the districts in which the superintendent had authority corresponded legally to the jurisdiction of the Council of War and *Junta de Armadas*.

Finally, in addition to the continued interest in reports from the bureaucracy and in stronger forest legislation, the crown continued to search for more supplies of domestic timber, particularly pine from Navarre. In 1677, Carlos II approved a plan very similar to the one attempted in 1637 to gain access to the pine trees of Navarre via roads that would take the trees to a tributary of the Ebro River, which would carry them to the Mediterranean Sea and to the arsenal of Barcelona.[126] The earlier venture had failed, but the potential of finding better forests in Navarre or Aragón had led to another expedition. The urgency to remain self-sufficient in shipbuilding materials is apparent. Forest reconnaissance missions that aimed to extend the resources of the navy, and therefore to extend the territorial range of state forestry, had been carried out multiple times before and would be carried out multiple times in the eighteenth century as well. Notable past ventures included explorations of forests in northern Valencia in 1589 during a frantic period of rebuilding after the Armada against England; notable future ventures included similar

125 Muñoz Goyanes, 115.

126 AGS, Sección Contaduría del Sueldo, segunda serie, legajo 32, Madrid, July 1, 1678, asiento with Lorenzo de la Bastide; referenced in Goodman, 140.

explorations in the strategically situated Sierra de Segura (Andalusia) in 1738, which included the sources of both the Guadaliquivir River, which flowed to the Atlantic, and the Segura River, which flowed to the Mediterranean. The forest reconnaissance missions of the early modern era thus helped state authorities gain information about unfamiliar areas under its nominal control, extend bureaucratic authority over new resources, and expand its internal transportation infrastructure. The missions were an essential aspect of Spanish forest territorialization in the sixteenth, seventeenth, and, as we shall see, the eighteenth century.

Throughout the peaks and valleys of Spanish fortunes in the seventeenth century, the successive rulers of the monarchy maintained not only an appreciation of the value of forests for state interests, but also active involvement in conserving forests and ensuring that new plantations could serve future generations. The Spanish government developed a system of forest conservation a century before the French and English and continued to revise the ways it convinced local populations to adhere to this system. Rather than lose interest in forest conditions during a period of naval decline, the Habsburgs oversaw improved administrative methods, enhanced royal authority through forest legislation, and commissioned explorations of potential timber sources within its own borders.

The Bourbons relied on the Habsburg state forestry system for Felipe V's first quarter century of rule, and many naval experts who served under the Habsburgs continued to serve under the Bourbons. Historians seldom question the decline of the Spanish navy in the later Habsburg period, and they rarely connect the Bourbon naval renewal of the early eighteenth century to Habsburg precedents. However, the bureaucracy and forest legislation of the seventeenth century perhaps helped to lay the groundwork for the Bourbon naval renewal by creating and maintaining state access to strategic forests. The deforestation of the sixteenth and early seventeenth centuries, caused mostly by demographic and military expansion, has led some historians to conclude that the decline in shipbuilding from the middle of the seventeenth century until about 1730 meant that Spanish forests experienced a period of ecological reprieve. However, legislation that gave the crown primary access to shipbuilding timber in the later seventeenth century ensured that there was no corresponding political reprieve. The crown supported and enhanced its system of forest management near shipbuilding centers even during a period of shipbuilding decline.

On November 1, 1700 Carlos II died without an heir and war broke out over the Spanish succession. Carlos had named Philippe of Anjou, Louis XIV's grandson, as his successor, but England, Austria, the United Provinces, and other states feared a unification of Spanish and French power in Western Europe. Spain relied on the French navy during the war, but participation of Spanish mariners who had also served the Habsburgs would be crucial to the successful transition to Bourbon power in Spain, including power over the management of forests under the new regime.

CHAPTER 4

Bottoming Out and Revival under the First Bourbon, 1700–1746

Expansion of State Forestry in the Reign of Felipe V (1700–1746)[1]

Juan Valdés y Castro was a Spanish naval commander by profession, but his most noteworthy achievements did not occur on any vessel at sea. From 1737 to 1739 he carried out reconnaissance missions in the forested sierras of the Spanish interior along two of Spain's largest river systems, the Guadalquivir and Ebro, where he inspected the suitability of timber, hemp, pitch, and tar for naval shipbuilding. During the same period numerous other naval officers oversaw similar inspections that were collectively part of the first systematic assessment of all accessible forests suitable for the Spanish navy, and proved to be influential in the creation of Spain's first national forestry codes in 1748. The systematic survey of the country's forests occurred under the direction of Zenón de Somodevilla, the Marqués de la Ensenada, who was at the time Secretary of the Admiralty Board (*Junta de Almirantazgo*). From 1743 to 1754, as Secretary of Finance, War, the Navy, and the Indies, Ensenada oversaw a broad program of reforms meant to enhance state power.[2] Revitalization of the Spanish navy in the second half of the eighteenth century owed much to the reform program of Ensenada, but more work is required to investigate the relationship between naval expansion and improved management of domestic and colonial timber supplies. For a full understanding of eighteenth-century Spanish state-building and naval power, though, an examination of the interaction of naval administration and local communities brought on by the forest reconnaissance missions of 1737–9 is essential. Valdés's inspections covered more ground than most, but all inspections aimed to expand the range of naval oversight of forests deeper into Spain's river networks, from which timber

---

1 Felipe V briefly abdicated in 1724. His son and successor, Luis I, died shortly after and Felipe reascended the throne that same year.

2 On Ensenada, see Felipe Abad León, *El Marqués de la Ensenada, su vida y su obra* (Madrid, 1985); José Luis Gómez Urdáñez, *El proyecto reformista de Ensenada* (Lleida, 1996). The forest surveys could be seen as precursors to the famous *Catastro de Ensenada*, a large-scale survey of Spain's population and resources carried out between 1750 and 1754 as preparation for a single tax based on wealth, the *Única Contribución*.

 | DOI 10.1163/9789004261372_006

could be sent to shipyards at sea-level. The reports that Valdés produced were based on his own first-hand experiences, which required several months of reconnaissance in each region and addressed rather specific concerns of the crown. Wanting to know where it could find accessible and adequate timber for its program of naval renewal, the crown ordered Valdés and other officers to identify precise forest locations, enumerate the specific tree species there, remark on their quality, assess the status of roads nearby, and identify the distance to the nearest navigable river. Other helpful information included depictions of the local property regimes, descriptions of soil quality, potential for new plantations, and cost estimates for improving roads or clearing boulders from rivers. To fulfill their commands, Valdés and others combined expert knowledge of forest management, geography, hydrology, shipbuilding, and economics to indicate to the crown the potential value certain forests could have for the monarchy.

Although Valdés carried out his inspections within the Iberian Peninsula and not in some distant colonial possession, the forests of the regions being explored were nonetheless unfamiliar to the crown. The missions, then, also served as domestic trips of exploration that generated new insights about the monarchy's own shipbuilding materials and enhanced the crown's territorial governance. Spanish forest inspectors sometimes generated maps to aid in the communication of their findings, but the forest reconnaissance reports always combined qualitative and quantitative information in textual surveys. Such methods were generally more efficient than mapmaking, and had a long history in the collection of geographic information within the Spanish monarchy.[3] The collection of forest information into centralized "databases" benefited the operation of other early modern governments as well, and while scholars have emphasized the role of maps in generating geographic knowledge and enhancing territorial state power in the early modern era, many historians have recently emphasized the simultaneous expansion and employment of textual reports to facilitate such processes.[4]

This chapter analyzes the movements, aims, and reports of several naval officers as part of a broader process of internal territorialization by the Spanish

---

3 Richard Hunter and Andrew Sluyter, "How incipient colonies create territory: the textual surveys of New Spain, 1520s–1620s," *Journal of Historical Geography* 37 no. 3 (Jul 2011), 288–299.

4 Chandra Mukerji, "The Great Forest Survey of 1669–1671: The Use of Archives for Political Reform," *Social Studies of Science* 37 no. 2 (April 2007), 227–253. Also see Ricardo Padrón, *The Spacious Word: Cartography, Literature, and Empire in Early Modern Spain* (Chicago, 2004) and Antonio Barrera-Osorio, *Experiencing Nature: The Spanish American Empire and the Early Scientific Revolution* (Austin, 2006).

crown, especially in the years 1737 to 1739. While Valdés was sending back to the crown information from Andalusia, Aragón, and Catalonia, others were reporting on forest conditions and potential new sources of timber in the same years from Galicia, Asturias, Cantabria, the Basque Country, Navarre, Murcia, and Valencia. The reconnaissance missions expanded the crown's 'actionable' geographic intelligence, increased the royal presence in forest communities, and facilitated greater reliability of timber resources for Spain's naval shipbuilding program. Valdés's efforts led to long-term reforms in naval forest administration that sometimes ignited conflicts with forest communities.[5] Although many people resisted being told by the navy how to manage the forest, it had become much more difficult to ignore or evade forest directives from the crown by 1750. Naval administrators brought about major disturbances, requiring the cutting of many old trees as well as planting and managing many new ones. A noteworthy result of Valdés's travels was the eventual designation of the Sierra de Segura at the headwaters of the Guadalquivir and Segura Rivers as a special naval province in the 1748 ordinances. This region today is nearly coterminous with the largest protected natural park in Spain, partially as a legacy of territorialization efforts in this period, but its role as a major supplier of naval timber from the 1730s meant that the forest reconnaissance missions helped reshape the wider early modern world as Spain built up the world's second largest navy by the 1780s.[6] The long-term reforms that led to

5 Resistance in this period came in many forms and especially occurred in regions that had little experience with the crown's forest superintendents and royal forest laws before 1748. Royal representatives were at times ignored, harassed, or even killed by people who opposed the new regulations. Punishments for noncompliance or resistance also ranged widely, including small fines, imprisonment, or service in the presidios of North Africa. Abusive royal forest guards could also face punishment from the crown. Instances can be found at the Archivo General de Simancas (Hereafter, AGS), Secretería del Despacho de la Marina (Hereafter, Marina), legajo (Hereafter, leg.) 555, 557, 558, 559, and 571.

6 There were, of course, other matters that were crucial to naval resurgence besides timber supplies, including the acquisition of other materials such as sails, rigging, pulleys, and anchors; employing adequate numbers of skilled seamen; and operating an efficient financial system. For broader treatments of the Spanish navy in the eighteenth century, see José Merino Navarro, *La armada española en el siglo XVIII* (Madrid: Fundación Universitaria Española, 1981); John D. Harbron, *Trafalgar and the Spanish Navy* (Annapolis, 1988); Agustín Ramón Rodríguez González, *Trafalgar y el conflict naval anglo-español del siglo XVIII* (Madrid, 2005); Juan José Sánchez Baena, Celia Chaín Navarro, and Lorena Martínez Solís, eds., *Estudios de historia naval: Actitudes y medios en la Real Armada del siglo XVIII* (Madrid, 2011); Manuel-Reyes García Hurtado, ed., *La Armada española en el siglo XVIII: Ciencia, hombres y barcos* (Madrid, 2012); Manuel Díaz Ordóñez, *Amarrados al Negocio: Reformismo Borbónico y suministro de Jarcia para la Armada Real (1675–1751)* (Madrid, 2009). Towards the end of the

this resurgence began only after the near-total destruction of Spain's navy during the War of Succession (1701–1714).[7] The following addresses first the background that led to the well-orchestrated and wide-ranging naval forest reconnaissance efforts in the 1730s. Then, the chapter focuses on the activities of naval forest inspectors in each naval department as well as in the colonies.

Voices for Reform from the North Coast

During the War of Spanish Succession, Spain had no significant naval forces of its own, and depended on the support of the French. The Spanish Bourbons did not want to be dependent on the French for long, however, and aimed to develop Spain's ability to defend its empire and to restore its reputation on Europe's political stage. Not long after the war, top officials in the Spanish navy, such as Nicolás Manrique and Antonio Gaztañeta, informed Madrid of worsening conditions in the forests along the north coast, due mostly to a lack of administrative oversight. They argued that the condition of Spain's forests reflected the condition of Spain itself in the sense that without healthy forests, there would be no national security, no trans-Atlantic commerce, and continued dependence on foreign nations for support.

Spain's north coast remained the primary shipbuilding region in the eighteenth century. Local conditions in the forests of the region had influenced royal policy many times in the past, most notably in 1656 when Felipe IV adopted Toribio Pérez de Bustamante's Instruction for Cuatro Villas as a nationwide policy. That same year, one of Spain's most notable seamen, Antonio Gaztañeta, was born in Motrico, a seaside town in Guipúzcoa. Gaztañeta would go on to be a distinguished mariner and ship builder for the Habsburgs as well as the Bourbons; he commanded fleets under both dynasties. In 1702, he became a superintendent and inspector of forests in Cantabria. In 1717, he became the director of the new shipyard in Guarnizo (Santander). While there, he played an important role in revising regulations for naval recruitment.[8] The British

1700s, the navy employed 65,000 men and operated 64 ships of the line, 40 frigates, and 100 lesser vessels. See John Lynch, *Bourbon Spain, 1700–1808* (Oxford: Basil Blackwell, 1989) and Carla Rahn Phillips, "'The Life Blood of the Navy': Recruiting Sailors in Eighteenth Century Spain," *The Mariner's Mirror* 87, no. 4 (November 2001), 420–445.

7 For an overview of the eighteenth century in Spain, see John Lynch, *Bourbon Spain, 1700–1808* (Oxford: 1989). On the reign of Felipe V, see Henry Kamen, *The War of Succession in Spain 1700–1715* (London and Bloomington, 1969), Henry Kamen, *Philip V of Spain: The King who Reigned Twice* (New Haven, 2001).

8 Phillips, "Recruiting Sailors," 425.

captured him at the Battle of Cape Passaro in 1718, but he returned to Spain and achieved the rank of *Teniente General* (Lieutenant General) in 1720. Based on a lifetime's experience, he wrote multiple treatises that revised accepted notions of ideal ship configurations.[9] A *real cédula* from 1721 made his rules for ship construction the official rules throughout the monarchy. By the early 1720s he knew the conditions of Spain's most active shipbuilding region better than anyone else.[10] In addition to all he had to say about ship construction, naval recruitment, and piloting, Gaztañeta expressed ideas about forest conservation in Spain. In a letter sent to Spain's prime minister, José de Grimaldo, on July 18, 1723, he gave an account of continuous illicit cuttings made in the forests near the Basque town of Aramayona. He warned that if no one had a prompt remedy, the king would find himself left with no suitable trees. Gaztañeta feared that such illicit activities required the effective enforcement of extraordinary regulations to be applied within one year to all of the monarchy's maritime provinces.[11]

Other prominent administrative officials from northern Spain shared and expressed similar concerns. The Marqués de Villareal, the Superintendent of Forests in the Basque *Señorío* of Vizcaya, reported that failure on the part of the *señorío*'s *audiencia* to address complaints of abuse by local inhabitants had led to continued injustices in the forests.[12] Villareal proposed stronger regulations and laws, along with harsher penalties for abusers in the region in order to stop the destruction. He said that annual forest inspections would provide valuable information about the rate of deforestation over time, which would help the authorities assess the problem and reduce the abuse.[13] Vicente de Velasco, a forest inspector (*visitador*) in Cuatro Villas and Asturias, wrote about resistance and complaints from several communities in Asturias against crown

9 Carla Rahn Phillips, *The Treasure of the San José: Death at Sea in the War of the Spanish Succession* (Baltimore: Johns Hopkins University Press, 2007), 19.

10 See Merino Navarro, *La armada española*; and Didier Ozanam, "La política exterior de España en tiempo de Felipe V y de Fernando VI," tr. José Luis López Muñoz in Menéndez Pidal, Ramón, Dir. *Historia de España, tomo XXIX: La época de los Borbones – La nueva monarquía y su posición en Europa (1700–1759)* (Madrid: Espasa-Calpe. (1963-) 1985), 441–699.

11 AGS, Marina, leg. 552, July 18, 1723, Gaztañeta to Grimaldo. "...en las Provincias Marítimas de estos Reinos, que año tomarse alguna eficaz y extraordinaria providencia succederá infaliblemente, lo que recela don Antonio Gastañeta."

12 *Audiencias* were courts of appeal. The highest courts of appeal were the *Chancillería* courts in Valladolid and Granada, but other regions in Spain and possessions outside Spain had subsidiary *audiencias*.

13 AGS, Marina, leg. 552, De oficio, Madrid, August 27, 1723, Consejo de Guerra.

regulations on planting quotas and cutting restrictions which, they argued, abused their traditional liberties and exemptions.[14] In August of 1723, a treasurer named Nicolás Manrique de Lara advised the government to rectify excessive abuses committed in the forests of Navarre and near the border with France.[15]

While these officials from Asturias to Guipúzcoa expressed different ideas regarding the root causes of the problem, they all believed that the crown needed to curtail abuses in the navy's most valuable region of forest resources. Gaztañeta thought that a combination of unruly locals and bad weather had contributed to the abuses he observed. Villareal laid the blame on an ineffective *audiencia*. Manrique blamed a lazy and incompetent class of royal bureaucrats, who had become used to inheriting their offices. All proposed that Spain needed greater observation of forest activity and stronger enforcement of regulations, but only one of them spelled out in detail what he thought would work. Based on this flurry of letters from concerned ministers, the treasurer Nicolás Manrique composed a document of seventy-seven points addressing the principal aspects of forest and plantation conservation, forest visitations, and proper cutting methods in the manner best suited for building naval vessels. He sent his document to Grimaldo and the king, including a copy of Pérez de Bustamante's instruction from 1656.

Manrique divided his seventy-seven points into eight sections. The first twelve points introduced the concerns of Gaztañeta and the other ministers, and described the poor conditions of the kingdom's forests. Manrique argued that no one was doing anything to stop the usurpation of forests by the local population. He said that all of this is damaging to "Your Majesty and the public interest," and if they "do not attack the disorders of the region promptly, … before you know it, Your Majesty will find himself without one oak tree in the entire peninsula."[16] Manrique had harsh words for the ministers who had been allowing such abuses. He named Vicente de Velasco, the visitador of valleys in the Cuatro Villas region as someone who superficially imposed fines but did not ensure that villages fulfilled planting quotas. Vizcaya had an able minister in the Marqués de Villareal, but he had been too busy serving the king in his other position as *teniente* (lieutenant) to stop abuses. In Guipúzcoa, Gaztañeta had cited problems in Aramayona, and Manrique claimed similar abuses were occurring in other places in the province. Galicia, by contrast, seemed to be in good shape under the veteran José Bermúdez de Castro. Álava, Navarre, and

14 AGS, Marina, leg. 552, De oficio, Madrid, August 27, 1723, Consejo de Guerra.

15 AGS, Marina, leg. 552, De oficio, Madrid, August 27, 1723, Consejo de Guerra.

16 AGS, Marina, leg. 552, Madrid, August 27, 1723, Don Nicolás Manrique, article 1.

Catalonia contained vast forests suitable for shipbuilding, but they had been without a superintendent, leading to unchecked abuses in the region.[17]

Although Manrique's alarmist approach was rather common for early modern Europeans concerned with wood shortages, few documents at this time go into such detail regarding conservation methods and the benefits of forest conservation for the entire monarchy. He claimed that only immediate enforcement of forest regulations, inspections, and fines could help the king. He also argued that the construction of ships for war and commerce depended on these forests, which protected the monarchy from its enemies. In addition, forests and their resources were "venerated by everyone [for their] pasture and protection of livestock, [their fuel for] factories of iron and carbon, [their building materials for] temples, houses, bridges, mills, and other necessary materials that establish a happy Kingdom."[18] Overall, he said – in perhaps his most important point – the regeneration of these forests would enable the Treasury to stop buying costly foreign ships. The king could find all he needed within his own kingdom, and he could choose the best trees at a low cost. To Manrique, this goal required the inclusion of all forests, not just royal forests. He said that although it might appear contradictory to the liberty and nature of the area to prohibit people from doing whatever they wished in the forest, it would not be unjust to ensure the conservation of the forests and to apply severe penalties to those who destroyed valuable trees. He reemphasized that this approach was in the public interest; the king's uses of the forests were for the benefit of everyone.[19] In order to improve conditions, Manrique wanted to establish forest visitations to conserve and foment plantations, prevent further cuttings and extractions, facilitate the transportation and sale of wood, and regulate the cuttings.

The remaining sixty-five points explained how these goals should be accomplished. In the section, "Methods to prevent damages and assert regulations," Manrique repeated Gaztañeta's recommendation that a judge and two foremen should accompany the *visitador* of each province to inspect each forest's "fertility, distance to the sea or navigable river, abundance of, and quality of oaks, beech, ash, pine, walnut, alder, and other [trees] useful for vessels."[20] To make the timber extraction more affordable in the long run, Manrique called for a commitment to improve transportation to the shipyards from the mountains.[21]

17 AGS, Marina, leg. 552, Madrid, August 27, 1723, Don Nicolás Manrique, articles 3–7.
18 AGS, Marina, leg. 552, Madrid, August 27, 1723, Don Nicolás Manrique, article 9.
19 AGS, Marina, leg. 552, Madrid, August 27, 1723, Don Nicolás Manrique, articles 10–12.
20 AGS, Marina, leg. 552, Madrid, August 27, 1723, Don Nicolás Manrique, article 13.
21 AGS, Marina, leg. 552, Madrid, August 27, 1723, Don Nicolás Manrique, article 14.

Several points dealt with protecting new plantations, which were the most vulnerable to disturbances.[22] Judges needed to ensure that laborers planted new trees intended for the navy in areas conducive to healthy growth, for example, not exposed to heavy winds, but still near rivers and ports, while separately designating plantations that ironworkers would eventually use for fuel. Judges should provide oversight through annual inspections and appoint two or more guards to protect the new plantations on a day-to-day basis.

The third section of Manrique's report included ideas regarding the "Mode and Form of Provincial Visits."[23] An exemplary proposal for state forest territorialization, the section identified territory, outlined monitoring techniques, described methods of categorizing and quantifying resources, and recommended rules for controlling behavior at the local level. Primarily, every valley that fit the specific geographic criteria of accessibility had to be required to allow inspections by *visitadores*. Private forests and village commons that sloped to the coast or were near navigable waterways were not exempt. Provinces from Álava to Tortosa needed to install new *visitadores*, because the vast forests of the region had been without them for an extended period. A new *visitador* could not simultaneously serve as *teniente*, which would interfere with his duties, unless he received approval for a proper substitute. Manrique wanted the inspections to be brief, but thorough, so that the *visitador* could see every new forest plantation. The *visitador* would be required to record the planting obligations of each community, the number and quality of the trees, remark on any illicit cuttings or usurpations, and identify the culpable parties. *Visitadores* had to reprimand criminals promptly, but no longer could they penalize towns or communities in general. They had to name specific parties as those responsible for damages.

Manrique noted that valleys contained about twenty or thirty towns, and in many instances, the people were poor and needed to leave the valley to find seasonal work elsewhere. The annual flight of the population made planting quotas difficult to achieve for town officials such as *alcaldes* (mayors) and *regidores* (town councilors). Manrique declared that plantation quotas had to be a priority before people left the valley. If the villages did not meet requirements, the *alcaldes* and *regidores* had to answer to the *visitador*. Officials had to send visitation reports to the *Consejo de Guerra* and send the Secretary of the Navy reports on fines and appeals.

Manrique said he had stepped outside his specialty as a treasurer to report on matters perhaps better suited for a mariner or farmer, but he was certain he

22 AGS, Marina, leg. 552, Madrid, August 27, 1723, Don Nicolás Manrique, articles 15–21.

23 AGS, Marina, leg. 552, Madrid, August 27, 1723, Don Nicolás Manrique, articles 24–35.

was doing a service to the royal and public interest.[24] There is a good chance that Nicolás Manrique de Lara had some relation to the Manrique de Lara who was the Viceroy of Catalonia and who oversaw the 1589 inspection of forests in the borderlands between Valencia and Catalonia. In his conclusion, he offered

> just and suitable measures to achieve goals of the utmost importance and extinguish once and for all the damages and clearings experienced over the past two centuries, recognizable in the *consultas*, *instrucciones*, and *providencias* that have been given to stop them, especially the *pragmáticas* and laws from the Catholic Kings in 1480 and 1496, from Don Carlos in 1537 and 1548, and Don Felipe III in 1609. Given that such disorders and injustices against the public well-being continue, extraordinary remedies through penalties, fines, and condemnations are required, because crimes that are indulged persuade and encourage delinquents to continue.[25]

As part of his plea, Manrique pointed out that trees take a long time to grow, and during the long early stages they are very delicate. While people can use timber and fruit for various purposes after only a few years, it takes longer than fifty years of observation and management for trees to become suitable for naval uses. Therefore, he said, the trees that he and others benefited from in the present were there because of careful conservation in the past, echoing the words of Toribio Pérez de Bustamante from seventy-three years earlier. It was up to everyone in the present to care for the trees that would benefit future generations.[26]

In the past, communities in northern Spain responded in different ways to forest directives from the crown. Some resisted strongly, but most complied with the planting quotas, fines, and other rules communicated by superintendents.[27] In a period of recovery after the War of Succession, Spanish ministers made clear the distressing state of forestry in northern Spain. Manrique, Gaztañeta and others viewed the local populations as incapable of seeing the big picture the way they could. In some ways, however, they were sympathetic to what they considered the shortsighted focus of local residents, understanding the value of forests for their everyday lives. They also did not lay all the

---

24 AGS, Marina, leg. 552, Madrid, August 27, 1723, Don Nicolás Manrique.

25 AGS, Marina, leg. 552, Madrid, August 27, 1723, Don Nicolás Manrique, article 74.

26 AGS, Marina, leg. 552, Madrid, August 27, 1723, Don Nicolás Manrique, article 76.

27 David Goodman, *Spanish Naval Power, 1589–1665: Reconstruction and Defeat* (Cambridge: Cambridge University Press, 1997), 68–108.

blame on local villages. Manrique seemed to be more concerned about an ineffective administration than an unruly peasantry, accepting that locals would continue to destroy the forest if the crown did not prevent them. To Manrique, the cutting of trees for naval shipbuilding, by contrast, would not be damaging to the forests, since such activity would be selective, closely monitored, and would establish new plantations. He believed that the War of Succession had left northern Spain with a lack of effective oversight for its forests, perhaps with the exception of Galicia. The decade after the war had shown few signs of recovery of naval forest regulation. By 1726, however, forest ministers saw someone who held their interests in high esteem rise to power at the Bourbon court.

Patiño's Tenure and the Reestablishment of the *Almirantazgo*

José Patiño became the General Intendant (*Intendente General*) of the navy and President of the Tribunal of the House of Trade (*Casa de la Contratación*) on January 28, 1717. Cardinal Giulio Alberoni, Spain's acting prime minister of the time, bestowed numerous powers on Patiño in these posts, including the administration of plantations and forests. That year, Patiño placed Gaztañeta in charge of the Guarnizo shipyard, which would become Spain's most important shipyard for the early eighteenth century. More ships came out of Guarnizo than any other site in the peninsula between 1715 and 1759.[28] Patiño pursued a long list of naval reforms related to naval recruitment, arsenals, and primary resources, but Alberoni's foreign policy ambitions would not allow Patiño a chance to carry out all his plans. Patiño may have been an administrative reformer, but he was primarily a servant of the state who worked for the interests of the king and queen.[29] Queen Isabel Farnese exercised great influence over Spain's foreign policies. Alberoni, who had helped arrange her royal marriage, worked closely with her on devising plans to recapture the Italian possessions lost at the Treaty of Utrecht (1713) so that her sons could have a proper inheritance. First on her list were Sardinia and Sicily. Alberoni told Patiño to arrange a fleet for an expedition to the Italian islands in 1717. Patiño abandoned his administrative plans in order to buy foreign ships and prepare merchant vessels for war. He managed to assemble twenty-two navíos in 1717 and forty in 1718. Patiño placed Gaztañeta in charge of the fleet, which left from Barcelona. The activity stirred the British navy into action in the Mediterranean, leading

28 Ozanam, "La política exterior de España en tiempo de Felipe V y de Fernando VI," 489.

29 Lynch, *Bourbon Spain, 1700–1808*, 91.

to a surprise attack off Cape Passaro, Sicily, on August 18, 1718. Spain's poorly prepared fleet was nearly wiped out, while Gaztañeta was taken prisoner. Conditions worsened for Spain the following year, with a defeat at Messina and a combined French and English invasion of Spain itself. England invaded Galicia, and France crossed the border into the Spanish Basque region, destroying ship construction sites in Pasajes and in Santoña near Santander. The series of disasters and a new Quadruple Alliance of England, France, the Dutch Republic, and Austria against Spain led to Alberoni's ouster. Patiño retained his posts but did not yet move up in the bureaucratic hierarchy. The queen's ambitions did not subside after the War of the Quadruple Alliance, and she sought to make a deal with Austria for a favorable marriage arrangement for her sons and the acquisition of Tuscany, Piacenza, and Parma. Johann Wilhelm, known as Baron Riperdá, once a Dutch ambassador sent to the court in Madrid, rose in power by sticking close to the queen and devoting himself to Spanish government. The queen sent Riperdá to Vienna to negotiate a treaty with Austria in 1725. The agreement with the emperor proved unfavorable to Spain, granting Austria major commercial concessions in return for only vague promises of a marriage to one of the emperor's daughters and support to recapture Gibraltar. While the monarchs did not see the treaty as all bad, other powerful Spaniards called for Riperdá's ouster. The Baron sought shelter with the English in Madrid, but the Spanish arrested him and imprisoned him in Segovia.[30] The bold ambition of Alberoni and the foolishness of Riperdá failed to reestablish a favorable reputation for Spain in Europe. José Patiño aimed to build a solid foundation for naval power and renewed respect for Spain, and the king and queen gave him his chance in 1726, when he became a cabinet minister. He was named the Secretary of the Navy, the Indies, and Finance, and in 1731, he became the head of the Department of War.[31]

During Patiño's tenure, which lasted until his death in 1736, Spain acquired more than fifty warships, alarming the British ambassador Benjamin Keene. Spain held its own in a war with Britain in 1726–1728, took the strategic outpost of Oran in 1732, and invaded Naples and Sicily in 1734, securing for the queen's son, Charles, an Italian kingdom. Most importantly for the administration of

30 For events in this period, see Lynch, *Bourbon Spain, 1700–1808*, 73–89; Kamen, *Philip V*, 119–138; and Ozanam, "La política exterior de España en tiempo de Felipe V y de Fernando VI," 573–604.

31 For details on Patiño, see Lynch, *Bourbon Spain, 1700–1808*, 90–98; Ozanam, "La política exterior de España en tiempo de Felipe V y de Fernando VI," 459–460; Ildefonso Pulido Bueno, *José Patiño: El inicio del gobierno político-económico ilustrado en España* (Huelva: I.P. Bueno, 1998).

Spain's forests, Patiño created a new naval political geography to organize the administration of naval resources. To carry out his goals, he created three Departments, with headquarters in Cádiz, Ferrol, and Cartagena.[32] Patiño had plans to build technologically advanced arsenals in each to meet the demands of naval construction. A similar arsenal would be built in Havana, Cuba, a shipyard that would become the most productive in the entire monarchy during the first half of the century.[33]

Increasingly in the early eighteenth century, deforestation in regions accessible to the navy stalled shipbuilding operations or forced the government to purchase foreign timber.[34] In responding to the problems of increased deforestation, and in ensuring the continued expansion of naval construction, Patiño initiated new efforts to augment production in colonial shipyards and create new shipbuilding sites. This required assessments of forests surrounding potential shipyards and arsenals, such as in Coatzacoalcos, near Veracruz. These efforts had mixed results as colonial shipyards continued to be fraught with problems of cost, administrative ineptitude, and distance, despite the abundance of timber resources. While some locations continued to be productive, like Havana, the failure of others, like Coatzacoalcos by 1735, contributed to Spanish authorities' reinvigorated efforts within the Iberian Peninsula later in the decade.[35]

Patiño's death left a power vacuum, and his posts went to several different ministers. The king filled part of the void by reinstituting the *Almirantazgo*, or

32 Although, Iván Valdez-Bubnov pointed out that earlier in the eighteenth century "the administrative structure of the new Bourbon naval system was defined by Jean de Monségnur, a French advisor to Tinagero." Iván Valdez-Bubnov, "War, Trade and Technology: The Politics of Spanish Shipbuilding Legislation, 1607–1728," *International Journal of Maritime History* XXI, no. 2 (December 2009), 86–7.

33 Ozanam, "La política exterior de España en tiempo de Felipe V y de Fernando VI," 489.

34 Extensive commentary and information on the topic of timber shortages in Cádiz in 1738 from Ciprián Autrán and Intendente Francisco de Varas y Valdés can be found in AGS, Marina, leg. 553.

35 Gaspar de Aranda y Antón, "Las maderas de América en la arquitectura naval del siglo XVIII," *Revista de Historia Naval* 38 (1992), 7–31; José Manuel Serrano Álvarez, "Juan de Acosta y la construcción naval en La Habana (1717–1740)," *Revista de Historia Naval* 93 (2006), 7–31; Antonio Béthencourt Massieu, "El real astillero de Coatzacoalcos (1720–1735)," *Anuario de Estudios Americanos* 15 (1958), 371–428. In 1732, Patiño ordered Lieutenant General Rodrigo de Torres to inspect the Coatzacoalcos arsenal. One of Torres's criticisms was the failure of the overseer, Juan Pinto, to carry out a reconnaissance of the surrounding forests. In early 1733, Torres himself led a twenty-day expedition of surrounding forests to prove that lack of timber was not a problem.

Admiralty, on March 14, 1737.[36] Based partially on the English Admiralty, the king intended it to help improve naval recruitment and placed his son, Prince Felipe in charge as *Almirante*.[37] The secretary of the *Almirantazgo* was the Marqués de la Ensenada, who had worked his way through the bureaucracy under Patiño and gained his title of nobility after organizing the expedition to Naples in 1734.

By 1736, Spain had recaptured Naples and Sicily, territories lost in the War of Succession, and had a standing navy of thirty-four ships of the line, nine frigates, and sixteen lesser ships.[38] This demonstrated an impressive expansion, despite several years of war.[39] The British increased their complaints about Spaniards harassing their merchant shipping in the Caribbean. By April of 1738, the British ambassador in Madrid was demanding compensation for damage done to English shipping.[40] The memorable example of Captain Robert Jenkins exhibiting his severed ear in a jar to the House of Commons in 1738, sliced off seven years earlier by the Spanish as punishment for encroaching on Spain's claimed monopoly in the Caribbean, fanned the flames of war hysteria in England and led to the War of Jenkins' Ear (1739–48).[41] England's decision to fight a preemptive war with Spain in 1739 indicated the effectiveness of Patiño's policies to increase Spanish naval power and overseas commerce.

36 On the medieval *Almirantazgo* of Castille, see Francisco Javier García de Castro, *La marina de Guerra de Castilla en la Edad Media (1248–1474)* (Valladolid: Universidad de Valladolid, 2014), 35–66.

37 Phillips, "Recruiting Sailors," 425. The citation is from the Archivo General de la Marina (Museo Naval), Viso, Reales Órdenes, Cádiz, Legajo 6457, document dated March 14, 1737. The prince was not the first in line to succeed his father. Felipe was the younger brother of Carlos, the future Carlos III. Both brothers were sons of King Felipe V's second wife, Isabel Farnese.

38 Lynch, *Bourbon Spain, 1700–1808*, 129.

39 Spain had virtually no navy to speak of during the War of Succession, and had to rely on French support and vessels. Of 40 ships-of-the-line that entered into service between 1700 and 1726, the year Patiño became a cabinet minister, 21 were of foreign origin. Over the next decade under Patiño right up until the forest reconnaissance missions covered in this article, of the 39 ships-of-the-line that entered into service for the Spanish navy, 3 were of foreign origin. Over the rest of the century, about the same ratio of foreign ships-of-the-line was maintained (about 7.5%); 11 of 145 were of foreign origin. See Harbron, *Trafalgar and the Spanish Navy*. Of course, one must take into consideration that vessels built in Spain often were built with some amount of foreign timber. More research is required to get a sense of how much timber used in shipbuilding came from abroad.

40 Kamen, *Philip V*, 207.

41 Richard Harding, *The Emergence of Britain's Global Naval Supremacy: The War of 1739–1748* (Suffolk, UK: 2010). See in particular, "Chapter 1: The Route to War, 1738–1739," 9–28.

For Spain, however, this was not a good time for conflict with Britain. While the navy had come a long way, it was far from equal to the size of the British navy. The years 1734 to 1739 delivered the century's severest subsistence crises in the Iberian Peninsula, and in 1739, the crown declared bankruptcy and rescheduled its debts.[42] In short, the state of Spanish resources on the eve of war with Britain was dire. It was becoming more crucial by the day for the government to become self-sufficient with affordable, accessible war supplies, principally timber for shipbuilding from Spain's forests. To assess the available timber in Spain, the government would rely heavily on administrators in the three naval districts created by Patiño.

Forest Reconnaissance in the Naval Departments of Ferrol, Cartagena, and Cádiz

Ferrol

In March of 1738, Ensenada sent a letter to Juan de Revollar, the superintendent in Cantabria, stating that the *Almirante* needed to know about the forests along the coast in this very important region, "because the forests grown in this region have never failed the king" (in their use in shipbuilding).[43] Revollar sent a report in December and added some of his own ideas on how to stop abuses.[44] Scholars may be familiar with the role of the *Almirantazgo* in naval recruitment, but its role in gathering information on forests is not widely known. Within the first three years of the *Almirantazgo*, Prince Felipe and Ensenada sent several direct orders to superintendents and other forest inspectors to conduct reconnaissance missions in accessible forests for trees suitable for shipbuilding, and to report back to Madrid or to their department's *intendente*. The *Almirante* contacted next Juan de la Pedrueca, in charge of Cuatro Villas, for a report on the condition of forests in his region. He had to delay his response because he was away at Guarnizo.[45] He did answer, however, to an inquiry by the *Almirante* on the timber gathered at the port of La Chata.[46]

42 Kamen, *Philip V*, 209.

43 AGS, Marina, leg. 552, Madrid, March 12, 1738, from the Infante Almirante to Don Juan de Revollar in Cantabria. "Porque nunca falta al Rey madera criada en essos montes."

44 AGS, Marina, leg. 552, Avilés, December 27, 1738, from Don Juan de Revollar, "Acompaña relación del estado en que se hallan algunos montes de barios consejos que a reconocido en aquél principado proponiendo algunos abusos dignos de remedio."

45 AGS, Marina, leg. 552, Suances, March 2, 1738, Don Juan de la Pedrueca.

46 AGS, Marina, leg. 552, Guarnizo, April 17, 1738, Don Juan de la Pedrueca.

A few years earlier, he had written a ten-page report on fires and other damages that occurred in the forests of the region, an issue that also affected the nearby royal iron foundries of Liérganes and La Cavada (Cantabria).[47] That same month of March, 1738, the *Almirante* asked for a report on the forests of Galicia and on the timber sitting in the arsenal of La Graña, the precursor to the state-of-the-art arsenal in El Ferrol, from Superintendent Bernardo Freyre.[48] Freyre responded with a report on 524 pieces of timber gathered in the arsenal in May, but did not present a report on forest inspection.[49] Galicia had been an exception in Nicolás Manrique's report on forest abuses, thanks to the service of José Bermúdez de Castro. Bermúdez had died on November 30, 1735, but he had left several reports on the condition of Galicia's forests.[50] Perhaps Freyre referred the *Almirante* to these reports, but there is no proof of this. Vizcaya was also part of the Department of Ferrol. The Marqués de Villareal had issued reports on forest visits there in 1733 and 1734.[51] In 1738, Felipe Uriarte submitted a report on the forests of Navarre that were accessible by river.[52]

Considerable traffic grew between Guarnizo and La Graña in these years.[53] They were part of one department, and as El Ferrol became a de facto capital,

47 AGS, Marina, leg. 552, Guarnizo, October 13, 1735, Don Juan de la Pedrueca, "Representa las talas e incendios hechos en aquellos montes (10 pages)"; For a good study of the iron foundries, see José Alcalá-Zamora y Queipo de Llano, *Historia de una empresa siderúrgica española: Los altos hornos de Liérganes y La Cavada, 1622–1834* (Santander: Centro de Estudios Montañeses, 1974).

48 AGS, Marina, leg. 552, Madrid, March 12, 1738, Don Bern.no Freyre, A Galicia.

49 AGS, Marina, leg. 552, Aranjuez, May 13, 1738, A Don Bernardino Freyre, "Que remita una relación prompta de los árboles del norte que existen en aquel arsenal para arboladura de los navíos"; AGS, Marina, leg. 552, Graña, May 22 1738, Don Bernardino Freyre, "Con estada de los palos de pino para arboladura existentes en aquel arsenal."

50 AGS, Marina, leg. 552, Graña, December 6, 1735, Don Bernardo Freyre, "Que murió don Joseph Bermúdez Juez Conservador de Montes." After a reconnaissance of the forests on Galicia in 1719 (AGS, Marina, leg. 552, May 1, 1719, Reconocimiento de Montes de Galicia, in French), Bermúdez carried out his own inspections in 1724, 1729, 1730, and 1733. See Ofelia Rey Castelao, *Montes y política forestal en la Galicia del Antiguo Régimen* (Santiago: Universidad de Santiago de Compostela Servicio de Publicacións e Intercambio Científico, 1995). In 1737, Don Benito Salgado inspected the forests of Tuy in southern Galicia. An inspection in 1749 compared changes in forest cover.

51 AGS, Marina, leg. 577, "Visitas en 1733 y 1734 dos leguas de la mar y rios navigables."

52 AGS, Marina, leg. 577, San Sebastián, April 21, 1738, Don Phelipe Uriarte, "Remite relación individual de los montes que hay en aquella provincia y reino de Navarra, sus distancias a la mar, villas, y lugares a que pertenecen, de donde se pueden conducir los materiales, y a que rios."

53 AGS, Marina, leg. 553, for information on Guarnizo supplying La Graña in 1739.

the personnel and materials of Guarnizo transferred there. The relocation of smaller shipyards related to the disasters of the War of the Quadruple Alliance (1718–20), when England invaded Galicia, and France crossed the border into the Spanish Basque region. Traditional sites in northern Spain, such as Pasajes and Santoña, had been too exposed to enemy attack. El Ferrol sat near a well-protected harbor, but the government selected the city only after a lengthy debate over other possible locations.[54]

The *Infante Almirante* and Ensenada had shown great interest in the forests of the Department of El Ferrol almost immediately. They received quick and informative responses from the department's top officials and, apparently, they gained knowledge about the timber available in important shipyards as well as on the general conditions of nearby forests. The *Almirantazgo* understood that the north coast was no longer going to be sufficient to supply Spain's shipbuilding timber, so they turned their interest to the other departments at the same time.

Cartagena

In early August of 1738, Juan de León, a military officer, observed a large fire in the mountains outside the city of Tortosa in the royal forest. He immediately ordered the support of the royal forest guards and he recruited a large number of citizens to help fight the blaze. The forests of Tortosa had long been valuable for the king and as recently as three months earlier, Ensenada received a review of the royal forests from Antonio Sartine, the *Intendente* of Catalonia. Sartine said the forests were of very good quality, suitable for shipbuilding. He attributed the good quality to a series of royal orders, most recently from 1720, that prohibited any kind of cuttings without royal permission.[55] When the fire began, the people of Tortosa did not hesitate to risk their lives to put it out even though it was not an immediate threat to the city, but was only visible at quite a distance. The fire was unusual in its size and intensity, obviously dangerous, and required a great number of men to put it out. Sartine reported that over two hundred men died trying to extinguish the fire, which also damaged a space of five leagues of forest cover with its "extreme voracity."[56] Sartine sent to Madrid a six-page list with the names of every person who helped fight the fire to save what they could of the royal forest.[57] The *Almirante* wrote back in October with an order to pay ten *reales de vellón*, or

---

54 Ozanam, "La política exterior de España en tiempo de Felipe V y de Fernando VI," 491–494.

55 AGS, Marina, leg. 552, Barcelona, May 24, 1738, Don Antonio de Sartine.

56 AGS, Marina, leg. 552, Barcelona, August 30, 1738, Don Antonio Sartine.

57 AGS, Marina, leg. 552, Barcelona, August 30, 1738, Don Antonio Sartine.

about twice a journeyman's wage, to the people of the city of Tortosa who fought the fire.[58] The disaster meant that the royal forests of Tortosa would not be able to supply the arsenals of Barcelona or Cartagena for a long time, devastating news at a critical juncture for the Spanish navy.

Ever since the expedition to Oran in 1732, the Spanish outpost there had relied on the Department of Cartagena for its supplies, including timber for construction and weaponry.[59] In addition, plans for the arsenal at Cartagena meant the region would be producing and repairing ships in the near future, which would place a further demand on the forests of Valencia. On May 20, 1738, the Count of Clavijo notified the *Intendente* Alexo Gutiérrez de Ruvalcava of forests in Valencia that could be useful for the navy. Juan Valdés, the *Capitán de Fragatas* introduced at the start of this chapter, had been discussing timber with Clavijo, who recommended the forests of the Valley of Cofrentes, near the Júcar River, which contained many good pine trees. He also said "good pine trees of good proportions" grew in the Marquesado de Moya, along the "Valencia River," today the Turia River.[60] Extracting trees from either location would not be costly, he said, since they are rather accessible. Gutiérrez de Ruvalcava must have welcomed the news, because ten days later he confirmed that he had a skilled and intelligent man ready to carry out the reconnaissance.[61]

During the summer of 1738, from July 13 to August 24, José Maltés, Head of the Administration of Masts and Yards (*Ministerio Mayor de Arboladura*) of the Department of Cartagena, gave a daily account of his trip into the mountains near the sources of the Turia and Júcar Rivers to inspect the forests of Moya and Cofrentes.[62] His daily log confirmed that a good number of trees suitable

58 AGS, Marina, leg. 552, Barcelona, October 11, 1738, Don Antonio de Sartine. Lynch, *Bourbon Spain*, 237.

59 AGS, Marina, leg. 552, Cartagena, June 11, 1738, Don Alexo Ruvalcava.

60 AGS, Marina, leg. 552, Cartagena, May 20, 1738, el Conde de Clavijo.

61 AGS, Marina, leg. 552, Aranjuez, 30 May 1738, "Acuerdo con don Alexo de Rubalcava, se nombre sujeto practico y inteligente que passe a hacer el reconocimiento."

62 The following text is based on: AGS, Marina, leg. 552, Cartagena, September 10, 1738, Don Alexo Gutiérrez de Rubalcava, "Con la relación que formo el Ministro Mayor de Arboladura Joseph Maltés del reconocimiento que hizo en los montes del Valle de Confrentes y Marquesado de Moya, con expressan de la calidad y numero de árboles, que se hallan en ellos fines a que pueden applicarse, y los parages, por donde pueden conducirse a la costa. Que se dara quenta y avisara lo que S.A. resuelva dicho en 26 del mes."; and AGS, Marina, leg. 552, August 24, 1738, 12 pages, "Relación diaria de lo que por mi Joseph Maltés Ministerio mayor de Arboladura de este Departamento de Marina, sea executado y reconocido en el viage que de orden del Señor don Alexo Gutiérrez Rubalcava Intendente del mismo Departamento, he hecho al Reino de Valencia, y Marquesado de Moya, y Valle de

for shipbuilding grew in the region along his route – however, not in the Valley of Cofrentes. Moya, by contrast, had pine trees suitable for *arboladura* for *navíos* of sixty to seventy guns. Its forests were accessible by road to the river, which eventually flowed past Valencia. He inspected more than just those two forest regions in his survey, commenting on what he saw along the way. Maltés visited individual households, noted the types and quantity of timber in their possession, and entered the forests themselves. He also noted when access to a forest required a new road. For example, a forested mountain to the southwest of Albocacer needed a road to connect to another, which would lead to the port city of Vinaroz. He commented on the conditions of the river for transporting the logs. He traveled by river from Santa Cruz and encountered several locations that would require the clearing of large rocks on the way to Valencia. Along the Júcar near Cofrentes, a multitude of rocks impeded transport. With the region's predominantly hot and dry climate, the wetter and cooler mountains of Moya in northern Valencia would become the prime source of timber for the department capital at Cartagena, but would require much labor to clear the river of obstacles. The department's arsenal became busy enough by the 1740s for the government to officially shut down the arsenal of Barcelona and transport the entire crew to Cartagena.[63]

Cádiz

The naval shipwright Ciprián Autrán had experience in the shipyards of the north before he was promoted to Captain of the Arsenal (*Capitán de Maestranza y Constructor*) at the docks of La Carraca, in the Bay of Cádiz.[64] Cádiz had gained prestige when the House of Trade moved there in 1714. In 1717, Patiño fixed his residence there, granting it an honorific preeminence above Ferrol and Cartagena after 1726. Cádiz, however, was far from the main forests traditionally used by the navy. Reports in 1738 showed the docks would be short on timber for the repairs it needed to carry out in the coming year.[65] La Carraca was used mainly as a place to repair ships, and it often imported timber from the north coast or from abroad.[66]

Confrentes, y del Rio Júcar, y del Rio de Valencia, al fin de reconocer las maderas, que se encuentran en diferentes montes de estos reinos, y pueden servir para las navíos de S.M. y es en la forma siguiente."

63 Ozanam, "La política exterior de España en tiempo de Felipe V y de Fernando VI," 490.

64 Merino Navarro, *La armada española*, 48.

65 AGS, Marina, leg. 553, La Carraca, September 13, 1738, Don Ciprián Autrán.

66 See José Quintero González, *El arsenal de La Carraca, 1717–1736* (Madrid, 2000); and José Quintero González, *Jarcias y lonas: El renacimiento de la armada en la bahía de Cádiz* (Cádiz, 2003).

During Autrán's time in Cádiz, he became familiar with the wood from the forests of Soto de Roma in Granada and the forests near Málaga. In a letter from La Carraca in February 1738 to the *Almirante*, Autrán claimed that for the repairs needed at La Carraca, these forests were as good as the forests he had known in Asturias, Guipúzcoa, Vizcaya, Álava, and Burgos. Autrán did indeed have a lengthy career in both northern and southern Spain.[67] He advised the *Almirante* to start exploiting the forests closer to Cádiz to save money.

The Forest Inspections of Juan Valdés y Castro in Segura and Catalonia

Segura

Autrán knew of another source that had great potential for the *Almirante*. In 1733, Sebastián Caballero, the Superintendent of the Royal Tobacco Factory (*Fábrica de Tabacos*) in Seville, had run out of wood for the Factory's new building, which today houses the University of Seville, and was running dangerously low on funds to buy imported timber. He had known that, in the past, wood came to Seville from the mountains near the source of the Guadalquivir. His construction engineer (*Ingeniero de Obra*), Diego Bordic, inspected the wood used in houses and churches in Seville made with timber from the Sierra de Segura. Bordic saw that the quality was very good, and he and Caballero sent some trustworthy colleagues into the Sierra with the instructions to inspect the wood in the forest, assess their potential utility in constructing the Factory, and see how much trouble it would be to reestablish the old traffic down the river to Seville. The group soon came back, certain of the high quality of wood and optimistic about the reestablishment of an economical Segura-Seville river transport system. Caballero and Bordic notified the Secretary of State, José Patiño, of the possibilities for tapping into the forests of Segura for finishing the Royal Tobacco Factory. Patiño approved sending annual shipments of timber down the Guadalimar and Guadalquivir at the end of winter, beginning in 1734.[68]

The navy soon took notice of the timber arriving in Seville from Segura.[69] In 1734, Vicente de Acero, Architect and Master of Works in Seville, notified the

67 Merino Navarro, La armada española, 56.

68 *Expediente sobre el regimen y administración de los montes de Segura de la Sierra y de su provincia* (Madrid: Imprenta de Miguel de Burgos, 1825, originally 1811), 29. Held at the Archivo General de la Marina (Museo Naval), Madrid, Biblioteca Museo Naval, 9550.

69 *Expediente sobre el regimen y administración de los montes de Segura de la Sierra y de su provincia*, 32.

Intendente of the navy in Cádiz, Francisco de Varas y Valdés, of the potential of these forests for shipbuilding. That year, Varas ordered a local master carpenter, Francisco Gener, to make a reconnaissance trip with the experienced ship builder Autrán to examine the forests of Segura for naval interests. They went in December of 1734 and returned in January of 1735. Like the trip by Caballero and Bordic for the Royal Tobacco Factory, Gener and Autrán's inspection was cursory, and they proposed to Varas a more thorough inspection in the near future. Gener had assured Varas they would find an abundance of valuable timber.[70] Eventually, as with all important matters of the day, the call for an inspection of the forests of Segura wound up on Patiño's desk.

Patiño died before the inspections happened. By the time they did, word of the forests of Segura reached the *Intendente* of the Department of Cartagena, Alexo Gutiérrez de Rubalcava. He wanted the inspection to include the Segura River since it emptied into the Mediterranean not far from his city. Finally, in February of 1738, the *Señor Infante Almirante* commissioned the naval commander Juan Valdés y Castro of La Carraca and the previously mentioned carpenter Francisco Gener to inspect the montes of Segura.[71] They were to leave Cádiz on the first day of April to inspect the quality of trees in the sierra, the potential for their use in shipbuilding, and the potential for sending all of it downriver. A master carpenter and another skilled individual left Cartagena the same day to examine the entire course of the Segura River. It appears that Valdés and Gener were responsible only for inspecting forests along the Guadalimar and Guadalquivir Rivers, but their report mentioned that they inspected three leagues of the Segura River as well. They finished on April 26, 1738.[72]

The accounts of Valdés's reconnaissance are deserving of close analysis, because his work would leave a lasting legacy in the region and because historians have not given his inspections much attention. In early May, the *Infante Almirante* called Valdés to Madrid to report in person to the king's ministers. The written reports stated that the largest trees could serve the navy well for twenty years, but Valdés stressed the obligation for the local justices to keep an eye on things and oversee plantations. Valdés lamented the continuous burnings and clearings made for agriculture or livestock pasture that were destroying

70 *Expediente sobre el regimen y administración de los montes de Segura de la Sierra y de su provincia*, 32.

71 *Expediente sobre el regimen y administración de los montes de Segura de la Sierra y de su provincia*, 32.

72 *Expediente sobre el regimen y administración de los montes de Segura de la Sierra y de su provincia*, 32.

the forest. Roads were better near the Guadalimar and Guadalquivir, but three leagues down from the source of the Segura, where the river becomes wide enough for transport, some roadwork would be required.

Valdés and the foreman (*Capataz de Arboladura*) Juan Pinzón sent a more robust report from La Carraca to the *Intendente* Francisco de Varas in nearby Cádiz a few months later, on August 6, 1738.[73] This report provided an overview of the system that would send trees from Segura down the Guadalimar and Guadalquivir Rivers to Seville and the arsenal of Cádiz at La Carraca. Valdés and Pinzón argued for the most efficient methods of cutting and transporting the timber, the timetable that each transport needed to follow, the number of laborers required for its safe arrival, and the general costs for the workers' wages and the labor for making the cuts. Such a report would indicate to Varas how much timber to expect in Seville or La Carraca and the time of year such a delivery would arrive.

In the opening paragraphs, the authors referenced their experiences from the past year to demonstrate their credibility. They noted that they had inspected the forests themselves and worked with Francisco Gener, who by this point had the most experience, having first visited the forests of the Sierra two and a half years earlier. They referenced the recent use of timber from Segura at the Royal Tobacco Factory, and suggested the suitability of this timber for the arsenal of La Carraca. The rest of the document outlined how a successful transport of two hundred trees' worth of timber from a place called Arroyo de Peña Rubia to La Carraca might occur.

Valdés and Pinzón claimed much of the land nearby belonged to the crown, making access theoretically easier and more straightforward. Opening a road from Arroyo de Peña Rubia to Rio Badera and then to the Guadalimar, about half a league, required little work for the transportation of two hundred trees. They stated that tree felling should occur during a waning moon in August and September or a waxing moon in December and January, explaining that such timing is believed to be more useful for the conservation of the wood. They went on to emphasize that cuttings made in August and September would be better than in December or January in that venue, because snowfall in the winter prevents access to the forests. Since the rivers at the end of the summer were always too low to carry the wood, February and March were optimal times for transportation to begin, when melting snow raised the level of the water.

The authors calculated the cost of cutting and transporting two hundred trees from the cutting site to the Guadalimar River. To carry the timber into and down the river, Valdés and Pinzón recommended employing twenty men at a fair daily

73 AGS, Marina, leg. 553, August 6, 1738, "Prevenciones sobre la utilidad…," 7p.

wage. They determined that the first leg of the trip, down the Guadalimar to where it joined the Guadalquivir, is about twenty-five leagues of time-consuming turns and should take two months. Because this river was so narrow, the authors argued, the group would have to keep the timber in single file. It would take three more months to travel the fifty to sixty leagues down the Guadalquivir to Seville. However, this river was wider and the wood did not need to be in single file, except when going under bridges and through dams. Likely imagining the dexterity and vigilance required by the crew to carry out such a complicated, costly, and potentially dangerous operation, Valdés and Pinzón reminded the reader at this point of how indispensable those twenty men were. Employing fewer would be a risk to the entire operation. Transporting timber from Seville to Sanlúcar de Barrameda, following a much wider and calmer section of the Guadalquivir with the same crew, would take only fifteen days. In Sanlúcar, the wood could be loaded onto two small ships and taken to La Carraca. In total, the transportation would take about five and a half months, not counting the time needed to fell the trees. Half the total cost would go towards the felling and half towards the transportation. Valdés and Pinzón ended by estimating the costs of cutting wood into planks, beams, and other parts, but it is clear that the main goal of their report was to illustrate the feasibility of wood transport from the forests of the Sierra de Segura to the arsenal of La Carraca. In theory, the system appeared to work well, and it offered an affordable and less risky alternative to importing wood from abroad or even from northern Spain.

The report by Valdés and Pinzón described the geography of the Sierra de Segura and provided the crown with information on the methods it would need to exploit its resources using members of the local forest communities for their labor, materials, and time to transport wood to the arsenal. These few years of reconnaissance initiated significant changes for the navy and the region of the Sierra de Segura. Only ten years later, King Fernando VI (r. 1746–1759) signed the *Ordenanza para el aumento y conservación de montes y plantíos*, and designated Segura as a special province of the navy, the only territory mentioned in the ordinance that was not near a coast or previously under the jurisdiction of one of the three naval departments. The ordinance required superintendents of the navy to inspect the forests, make reports, and enforce planting quotas. The Departments of Cádiz and Cartagena both benefited from the strategically located forests and the rediscovered river transport system.

The work done by Valdés and his assistants ushered in a world of new information for the king's ministers about a forgotten region.[74] Valdés and Pinzón

74 Evidently, the region had been exploited for shipbuilding timber as late as the ninth century, when it was under Islamic rule. Emilio de la Cruz Aguilar, "La provincia marítima de Segura de la Sierra," *Boletín del Instituto de Estudios Giennenses* 107 (1981), 52.

demonstrated expert knowledge regarding tree felling, the Guadalimar and Guadalquivir river system, and the difficult task of transporting two hundred trees worth of timber from the sierra. By combining their knowledge of the needs of the crown with their experiences in the forests of the sierra, Valdés and Pinzón helped the Segura inspections of 1738 become one of the most successful – from the monarchy's point of view – of the many forest inspections carried out in the reign of Felipe V. However, the sudden interest the crown showed in the sierra caused considerable shock to the region, ignited some resistance that did not completely disappear during the eighteenth century, and imposed an additional jurisdictional body within an already complicated socio-legal patchwork.

The long-term effects of the navy's presence in the Sierra are ambiguous. Jesús de Cobo Guzmán y Lechuga has argued that social and economic conditions in the Sierra de Segura did not worsen because of the new naval administration.[75] He noted that conditions in the mountains had been difficult for a long time before the navy arrived, and that life was never easy in the sierra. People who could sell timber to the navy or work in the removal and transport stages could benefit from the presence of the navy. Many other scholars argue that life in the forest became more burdensome. Emilio de la Cruz Aguilar sees the involvement of the navy in the Sierra de Segura as ultimately causing deforestation and preventing adequate local access to forest resources. The jurisdiction of the navy only complicated the region's administration, he argues, because the navy did not displace the jurisdictions of the Military Orders, the Finance Ministry (*Hacienda*), or local municipalities that already existed.[76] Nevertheless, in terms of territorialization, the state had moved a long way toward integrating itself into the local milieu, and, largely due to Valdés's expedition of the Sierra de Segura, had identified untapped resources for naval interests.

Catalonia

Juan Valdés y Castro had about a month to rest before the *Intendente* Francisco de Varas, under orders from the *Infante Almirante General*, called on him to undertake an even more ambitious expedition. The instructions, written in October of 1738, told Valdés to execute a reconnaissance of the pine forests of Catalonia for wood, pitch, tar, and hemp to supply the Armada.[77] Catalonia at

75 Jesús Cobo de Guzmán y Lechuga, *Estudio sobre las Ordenanzas de Montes del año 1748 y del expediente sobre el régimen y administración de los montes de Segura de la Sierra y de su provincia marítima de 1811* (Madrid, 1994).

76 Cruz Aguilar, "La provincia marítima de Segura de la Sierra," 67–72.

77 AGS, Marina, leg. 553, October 13, 1738, Francisco de Varas y Valdés, "El Señor Infante Almirante General ha resuelto…."

this time had been experiencing major changes. The abolition of its *fueros* in 1716 removed its distinct Catalan government. Barcelona remained an important city for the navy after the War of Succession, launching the fleet that took Sardinia in 1717, but the waning of the galley era slowed production in the city's arsenal and factories for naval stores.[78] It had become commonplace to buy naval stores and masts abroad. The royal forest around Tortosa remained valuable, but a severe fire raged through it in the summer of 1738, mentioned above.[79] On top of all this, forest clearings and burnings accelerated with Spain's expanding population.

These themes were the motivations behind Varas's instructions. Although Valdés worked in the Department of Cádiz, he had demonstrated his competence during his inspection of Segura earlier in the year. The crown needed information on the condition of Catalonia's forests and its naval stores industries in order to increase efficiency and remove the need to rely on foreign materials. Three documents in particular tell the story of Valdés's trip, including his instructions, his daily log, and a final assessment in the form of a letter establishing a franchise to provision wood and naval stores to all three naval departments, called an *asiento de arboladura*.

The instructions described his intended itinerary, the time he needed to complete all of it, the types of trees to look for, and the elements he needed to include in his report.[80] The instructions named a construction supervisor to travel with Valdés and summoned both of them to the court in Madrid to discuss the mission with Ensenada and the *Almirante*. The planned itinerary took Valdés from Cádiz to Madrid to Calahorra, a town on the Ebro River in La Rioja. Valdés was supposed to inspect several of the important rivers that flowed into the Ebro, and report on any obstructions to the transport of wood from the mountains. These rivers included the Aragón, Gállego, Aguas, Martín, Guadalope, Nogueras, and the Segre. Along the way, Valdés was to write in his diary in detail about lands that were suitable for growing hemp and the condition of facilities used to make pitch. The instructions called for descriptions of the methods used and the quality of the pitch made in each of these locations. The navy wanted mixtures with no sand, dirt or rocks, like the kind it had been importing from foreigners. When Valdés finally reached Barcelona, he was to visit the *asentista* (contractor), José Basora, and inspect his rigging factory, based on a royal ordinance from December 1737 that described the proper way of making rigging and preserving the materials.

78 Henry Kamen, *Empire: How Spain Became a World Power, 1492–1763* (New York: Perennial, 2002), 453.

79 AGS, Marina, leg. 552, Barcelona, August 30, 1738.

80 AGS, Marina, leg. 553, October 30, 1738, "Ynstrucción que ha de observar…," 17 articles.

All of this was to be done in the winter, and was intended to make the most efficient use of Valdés's trek to Barcelona. For the second and more important leg of the trip, he had to wait until April because the snow in the mountains made it impossible to enter many of the forests. Once the spring arrived, he was to follow a precise itinerary, and record everything in his daily log. The instructions stipulated that he needed to record the type, size, and number of pine trees in each forest, and their distance to a road or navigable river. He was to record the condition of the roads in each area. Trees should be cut before winter, when the forest was accessible. The instructions named four types of pine trees and ranked them by quality. They also identified specific forests, including those near Sort and Riat and those northwest from Urgel, that were considered potentially valuable for the navy. Valdés was to enlist people who could help him, apparently meaning knowledgeable local experts, not just any local peasant.[81] From Barcelona, he would follow a route that took him across the mountains near the headwaters of the Segre, Noguera, Gállego, and Aragón rivers. When he was done, he needed to report in person again in Madrid to Ensenada and submit his report. Francisco de Varas repeatedly emphasized the importance of this reconnaissance to Valdés and allowed him to take about four months in the spring to complete the forest inspection.

Although the instructions ordered Valdés to cover a lot of ground, the navy did not require him to inspect all the forests of Catalonia, or even those that had been historically the most important for the region, such as those near Barcelona, the shipbuilding center Sitges, or the royal forests of Tortosa. Instead, the trip aimed to expand the range of naval forestry and, indeed, much of Valdés's trip occurred in Aragón and Navarre, outside Catalonia. His trip was multi-purpose, indicating the myriad ways the navy felt it could improve its network of supply, particularly in becoming more self-sufficient in shipbuilding timber, rigging, and pitch. This required more information on the hemp-growing industry and the pitch factories near Barcelona and Tortosa. Overall, the instructions outlined an ambitious mission covering about one thousand miles, along a dozen rivers and over difficult terrain, to be accomplished in eight or nine months.

It appears that Valdés did not accomplish everything listed in the instructions for the first leg of the trip from Madrid to Barcelona.[82] He did inspect the Cinca River from its source to the Ebro and various locations in Aragón and Catalonia suitable for growing hemp, but by January 1739, he still had not

81 AGS, Marina, leg. 553, October 30, 1738, article 17.

82 AGS, Marina, leg. 553, January 17, 1739, letter of Valdés's arrival in Barcelona.

inspected the other rivers. He made a favorable report on the Cinca River and the region's suitability for hemp growing, saying there were only two sections that might block wood transport on the river, and made an estimate of 260,000 *arrobas* worth of hemp that could be harvested in Catalonia and Aragón.[83] It is possible that winter conditions in the region prevented Valdés from accomplishing everything, or he might have been concerned about the time he needed in Barcelona to prepare for the spring.

In and around Barcelona, from January through March, Valdés inspected pitch production facilities and factories for rigging.[84] Repeatedly, the facilities failed inspection. The main rigging factory in Barcelona, run by the *asentista*, José Basora, lacked proper protection of materials from the elements, and Valdés recommended building a new roof. A pitch-making center outside of Barcelona consisted of only the remains of a factory. Another made pitch with too much sand. A lack of trees in the royal forest around Tortosa meant they would have to buy trees from privately owned forests, but Valdés was certain the Armada could be well served from this location in the future. In each place, Valdés recorded that he asked about the methods in use, and gave instructions for how to improve the process. In a letter written in March, Valdés sent a review of what he had seen, and notified the authorities that he was planning to leave Barcelona on April 6, a week or two ahead of schedule.[85]

The previous November, Francisco de Varas noted that he had appointed Valdés's former partner on the Segura inspection, Juan Pinzón, to go to Barcelona as a knowledgeable and experienced aid.[86] Perhaps Valdés would have been happy to reunite with Pinzón, but it is not clear whether Pinzón ever arrived in Barcelona or went with Valdés to inspect the forests. His name does not appear in later documents and in a letter Valdés sent upon his arrival in Barcelona he remarked that they had not seen each other yet.[87] Other people to join Valdés included the *Comisario de Guerra de Marina* and a Minister of the *Almirantazgo* in Barcelona, Antonio Gállego y Montemayor. They employed a local carpenter and a knowledgeable assistant for the duration of the trip.

83 An *arroba* was both a weight and volume measure equaling 25 Castilian pounds of 16 ounces; 8 *azumbres* or 32 pints; 11.5 kilograms. See Carla Rahn Phillips, *Six Galleons for the King of Spain: Imperial Defense in the Early Seventeenth Century* (Baltimore: Johns Hopkins University Press, 1986), 228.

84 AGS, Marina, leg. 553, "Diario, especulación y reconocimiento hecho por el Capitán de Fragata don Juan de Baldés y Castro...."

85 AGS, Marina, leg. 553, Barcelona, March 14, 1739.

86 AGS, Marina, leg. 553, Cádiz, November 25, 1738, from Varas.

87 AGS, Marina, leg. 553, January 17, 1739, letter of Valdés's arrival in Barcelona.

In April of 1739, almost exactly one year after his expedition to the Sierra de Segura, Valdés set out with his crew to begin his forest inspection of Catalonia, Aragón, and Navarre.[88] In his daily log, he followed his instructions closely. In addition, he calculated and recorded the estimated costs of repairing or improving the roads and rivers in order to facilitate access. He also specified who owned the forests he inspected, whether a noble, a town, or the king himself. The rivers he visited included the Segre, Noguera Pallaresa, Noguera Sola de Aragón (today the Noguera Ribagorçana), Noguera de Tor, Gállego, Aragón, and Esca. He maintained an official tone throughout, but commented on his delight at finding a couple of particularly rich forests in the northwest corner of Catalonia, and his disappointment in the forests near the source of the Gállego River. In a letter from Jaca near the end of his trip in June, he disclosed that there should be enough wood for the armada for forty to fifty years, without requiring much cost to open the roads or clear the rivers. However, something had to be done about the fires that damaged the forests every year.[89] For the remainder of the trip, Valdés worked hard to make up for what he had not accomplished in the winter. After finishing his forest inspections, he returned to inspect hemp production near Zaragoza and along the Aguas, Martín, and Guadalope rivers, finishing in mid-July. By August, he had submitted his diary and presumably met with the Marqués de la Ensenada in Madrid that summer.

Although the official instructions given to Valdés failed to mention it, the *Intendente* Francisco de Varas made it clear in several letters that the navy was hoping this reconnaissance mission would result in establishing an *asiento de arboladura*, or contracting franchise that would be in charge of supplying the entire Spanish navy with wood, pitch, tar, and other naval stores.[90] By October, one year after Varas commissioned the trip, Valdés put together a report declaring the establishment of this *asiento*.[91] The *asiento* would begin the next year in 1740, and would be granted for twelve years. Its purpose was to provide naval stores to the three naval departments, but it did not name who would receive the *asiento*. Valdés specifically cited the forests in the northwest corner of Catalonia, called the *montes* of Señora de Caldés and San Nicolás, as the best and the ones that

88 AGS, Marina, leg. 553, "Diario, especulación y reconocimiento hecho por el Capitan de Fragata don Juan de Baldés y Castro…."

89 AGS, Marina, leg. 553, Jaca, June 17, 1739.

90 For example, Francisco de Varas's letter that came with the instructions mentions the *asiento*. AGS, Marina, leg. 553, Cádiz, November 4, 1738.

91 AGS, Marina, leg. 553, "Papeles concernientes al reconocimiento de los Montes de Cataluña, hecho por don Juan Valdés Capitán de Fragata de la Real Armada: Assiento de Arboladura, Tabloneria, y Vetunes," 30p., 23 articles.

should be exploited immediately. For other naval stores and hemp, he recommended areas around the *montes* of Beumor and Carén. To make pitch, Valdés preferred the royal facilities of Tortosa. He made clear the need to open roads and clear rivers of obstacles. The work required employing village carpenters, laborers, river specialists, and their carts and oxen to cut and transport the wood. All day wages were supposed to be paid promptly at a fair rate.

The establishment of the *asiento* after Valdés's forest inspection may have had significant long-lasting results, but more research is needed to draw this conclusion. The 1748 forest ordinance does not mention it, but an *asiento de arboladura* still existed in the 1770s.[92] If it was established successfully, it was due in no small part to Valdés's travels and reports. At the very least, Valdés's expedition into the forests of Catalonia, Aragón, and Navarre demonstrated the difficult situation the Spanish navy faced in this period and the interest of the crown's highest officials in improving Spain's domestic supplies of naval stores. The inspections of Valdés added new information to that collected from the inspections of the Ebro River network in 1639 and 1677.

Valdés's reports provided a lot of new information for the Bourbon government concerning the poor state of the naval stores industry, the positive outlook for hemp supplies, and the accessibility of high quality pine forests. It becomes clear in the documents that the navy required the active support of knowledgeable locals and their resources such as pack animals, carts, boats, and tools. The forest ordinance of 1748 did not refer to most of the territory Valdés inspected in Catalonia and none that he visited in Aragón and Navarre. Perhaps these regions were too distant or the infrastructure was too underdeveloped for the navy to incorporate a new Province of the Navy, as in Segura. In the end, the forests of Tortosa and Barcelona retained their primacy in naval matters in eastern Spain, but Valdés's reconnaissance mission reveals that the crown was at least well informed about the possible uses of other venues.

In 1740, one year after Valdés completed his reconnaissance of forests in the Ebro River network, Juan José Navarro, the first Marqués de la Victoria (1687–1772), a naval officer who would later become Captain General, published a map to illustrate the forests, mountains, rivers, and towns covered in Valdés's mission. The map is impressive for its artistry and for the wealth of information presented, demonstrating cartographically how the navy had begun taking greater territorial control over its forest resources. Accompanying it are extensive descriptions of forests, hemp plantations, and tar production

92 Archivo General de Indias (Hereafter, AGI), Correos 457B, 1772, "Correspondencia y expedientes de la Dirección General de Renta."

facilities inspected by Valdés. The text covers half of the entire document and even includes information from José Maltés's inspection of the forests of the Marquesado de Moya from 1738. It conclusively states that the king of Spain "is found to be the richest prince in the world (with the Indies) in wood for the construction of *navíos*." Such a map would not have been possible before Valdés completed his mission. The map was published in 1756 with Victoria's masterful collection of other illustrations depicting aspects of the shipbuilding industry (Figure 6).[93]

Colonial Forest Territoriality in the Early Eighteenth Century

The reconnaissance of forests also continued to occur in colonial contexts during the eighteenth century so as to revive and expand naval shipbuilding. Spanish naval operations had been performed largely by French allies before and during the war. Even before the War of Succession had ended, officials in Spain, led by Bernardo Tinajero de Escalera argued for a greater and more permanent naval presence in the Americas, particularly in the Caribbean and Gulf of Mexico. Tinajero claimed that "if the *navío* is built in Vizcaya, it will no doubt last ten years, but from the Indies, it would last thirty."[94] Tinajero was the secretary of the Council of the Indies when Agustín de Arriola, a Cuban mayor, visited the court near the end of the war to promote the establishment of a better shipyard in Havana. At the time, the Spanish court was under the heavy influence of French ministers, including Jean Orry, who urged Spain to focus its attention on the Mediterranean rather than the Atlantic. In 1712, a royal commission appointed by Felipe V charged Tinajero with the task of defining Spain's naval needs. Tinajero's strategy focused not on the Mediterranean, but on the Atlantic and Caribbean. His recommendations included the establishment of a new shipyard in the region for building and servicing large vessels; the use of American timber from Mexico, Venezuela, and Cuba; and the adoption of Antonio de Gaztañeta's naval construction guidelines. While Havana would retain its preeminence in the region due to its resources, personnel, and defenses, Tinajero inspected other locations for a permanent shipyard in Tlacotalpán, Campeche, Maracaibo, and again in Jagua. The most important

93 Archivo General de la Marina (Museo Naval), Madrid, BMN-14430, "Álbum del Marqués de la Victoria, Hoja 22ª, Mapa topográfico de todos los bosques que el Rey tiene en el Río Ebro…." Navarro did not draw the maps and other illustrations himself, but supervised the production of the album.

94 Merino Navarro, *La armada española*, 185.

FIGURE 6 *Juan Valdés's reconnaissance of the Ebro River system – This map was produced in 1740, one year after Valdés completed his reconnaissance of forests in the Ebro River network, and was published in the Marqués de la Victoria's* Dictionary of Naval Construction *in 1756. The map depicts the forests, mountains, rivers, and towns covered in Valdés's mission. Accompanying it are extensive descriptions of forests, hemp plantations, and tar production facilities inspected by Valdés. The text even included information from José Maltés's inspection of the forests along the Turia River from 1738.*
SOURCE: ESPAÑA. MINISTERIO DE DEFENSA. ARCHIVO DEL MUSEO NAVAL. DICCIONARIO DE CONSTRUCCIÓN NAVAL DEL MARQUÉS DE LA VICTORIA. AMN. PL 235–6, HOJA 22. MAPA TOPOGRÁFICO DE TODOS LOS BOSQUES QUE EL REY TIENE EN EL RÍO EBRO, LA NOGUERA, EL CINCA, EL SEGRE Y ESCA DE NAVARRA, LUGARES QUE TIENEN CÁÑAMO Y LOS QUE TIENEN HORNOS DE ALQUITRÁN Y BREA, ETC.

factor for any new site was its proximity to quality forests, but the failure of all of these locations proved to be some combination of high costs and lack of defenses or other resources.

The site of Coatzacoalcos continued to be discussed as well. In 1706, a proposition to establish a shipyard at Coatzacoalcos was made in Veracruz after a reconnaissance of the Coatzacoalcos River and nearby forests.[95] The following year, two maps were created depicting the coastline, rivers, mountains, and water depths of the region (Figure 7).[96] In 1720, the governor of Guatemala, Francisco Güiral, sent a report to the crown endorsing the creation of a new shipyard at Coatzacoalcos, saying "this river has on either bank copious forests of cedar, pinzapo, haví, and quiebrahachas woods, so regular for building *navíos*, larger and more convenient amounts will not be found, one can navigate 36 leagues upriver in any boat." And while there is a sandbar at the mouth of the river, an 800-*tonelada* vessel could pass over it. In 1726, *teniente general* Antonio Serrano asserted to Patiño that the forests of the Tehuantepec Isthmus possessed an abundance and great variety of woods.[97] Inspections also occurred at San Marcos de Apalache, near Pensacola, where there was an abundance of large trees, but also an abundance of armed Indians who resisted the Spanish.[98] In 1731, *comisario de marina* Juan Pinto and *constructor* Pedro de Torres arrived. After many delays, however, Patiño requested a full report from Pinto, who claimed that timber and labor shortages, difficulties communicating with authorities in Veracruz, and the unhealthy environment of the site were the principal causes for the delays. Ultimately, the shipyard produced only one vessel at an astronomical cost. In 1732, Patiño ordered Vice Admiral Rodrigo de Torres to inspect the Coatzacoalcos arsenal. One of Torres's criticisms was the failure of the overseer, Juan Pinto, to carry out a reconnaissance of the surrounding forests. In early 1733, Torres himself led a twenty-day expedition of surrounding forests to prove that lack of timber was not a problem, because, he said, "the forests are like the mines in that only if they are worked can one come to know what they produce."[99] Indeed, even after the shipyard had folded, the region continued to supply mast timber to the Havana shipyards.[100]

---

95 AGI, Contratación, 637; made before General Diego Fernández de Santillán.

96 AGI, MP-Mexico 573; AGI, MP-Mexico 574.

97 Béthencourt, 14–15.

98 Béthencourt, 20.

99 Béthencourt, 35.

100 Béthencourt, 401–2; Marina Alfonso Mola and Carlos Martínez Shaw, "Los Astilleros de la América Colonial," in *Historia General de la América Latina*, v. 3, n. 1, Alfredo Castillero Calvo and Allan J. Kuethe, eds. (Madrid: Trotta, 2000), 297–298.

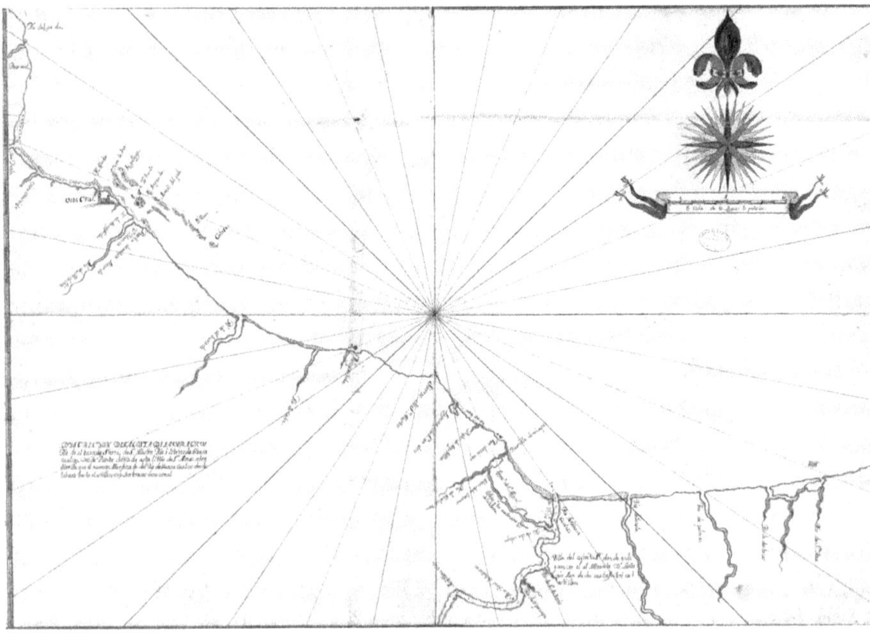

FIGURE 7 *Coatzacoalcos, México – Potential shipbuilding sites that could rival Havana continued to be sought by naval personnel in the Caribbean region into the eighteenth century. Reconnaissance of the forests and rivers near Coatzacoalcos led to the establishment of a shipyard there in 1731, but it had failed by 1735 due to high costs. This map from 1707 depicts part of the Gulf of México coast, major rivers, mountains, and water depths along the Coatzacoalcos River. The city of Veracruz is visible on the upper left corner of the map.*
SOURCE: ESPAÑA. MINISTERIO DE EDUCACIÓN, CULTURA Y DEPORTE. ARCHIVO GENERAL DE INDIAS. AGI, MP-MÉXICO, 573 – DESCRIPCIÓN DE LA COSTA DE LA VERACRUZ.

An Atlantic-oriented royal naval policy and construction of new facilities led to a revival of shipbuilding in Havana in the first half of the eighteenth century. A *real orden* in 1713 obligated Manuel López Pintado to build ten *navíos* in Havana. Construction on a new shipyard began in 1722. Of the 70 *navíos* built for the Spanish navy in the first fifty years of the century, most were built in Havana and Guarnizo and most were built after 1723.[101] Until the shipyard was deactivated in 1747 and superseded by a new arsenal from 1748, Havana produced twenty-two ships-of-the-line, six frigates, and three packet boats. The *asentista* Juan de Acosta was responsible for much of the construction from 1717 to 1740, and the Havana Company built vessels for the crown

101 Merino Navarro, *La armada española*, 182.

from 1740 to 1749. From 1748 to the British occupation of 1762, Havana produced eight more ships-of-the-line, two frigates, three bergantines, one packet boat, and one Goleta. Havana remained an active shipbuilding center for the rest of the century, and its output by 1800 totaled seventy-four ships-of-the-line, thirty-two frigates and jabeques, and another forty-two smaller ships.[102]

In the first half of the century, the navy continued to assert its interests ahead of sugar producers and herdsmen in the forests around Havana. In 1719, the shipyard was granted authority to remove trees from anyone's land, while sugar planters had to apply for licenses. The cuttings were carried out in fourteen designated cutting zones, mostly in the forests near Havana.[103] In 1721, cedar, which was desired by both the navy and the sugar planters, was off limits for house construction.[104] Spain's forest ordinances of 1748 gave the navy control of woodlands twenty-five leagues from the coast and navigable rivers. In theory, it applied to the colonies as well, but Ensenada indicated a more locally-appropriate law for Cuba. In 1748, Ensenada approved an edict issued by the captain general Francisco Cagigal y de la Vega that reserved and developed the best wood for the navy.[105] In many ways, this was an update of the 1622 legislation. It increased the area reserved for the navy and the number of species that required a license to cut down, including sabicú, chicharrón, yaba, ocuje, guaiacum, and three species of oak. The jurisdiction of the new regulations included

> all the coast on both seas for forty leagues windward and leeward of this port [Havana] for a distance of six leagues inland from the north coast and twenty from the southern one beginning at the anchorage of Batabanó, with no distance indicated, because in terms of distance what is important is to conserve all the [wood] that exists in the narrow width of the island.[106]

In lands cleared for cane or crops, trees reserved for the navy would be left standing or a person could be fined 100 ducats. For each tree cut down, they

102 Alfonso Mola and Martínez Shaw, 299.

103 John Robert McNeill, *Atlantic Empires of France and Spain: Louisbourg and Havana, 1700–1763* (Chapel Hill: The University of North Carolina Press, 1985), 132.

104 Miguel Jordan Reyes, "La deforestación de la isla de Cuba durante la dominación española, 1492–1898" (PhD diss., Universidad Politécnica de Madrid, 2006), 53–54.

105 Reinaldo Funes Monzote, *From Rainforest to Canefield in Cuba: An Environmental History Since 1492* (Chapel Hill: University of North Carolina Press, 2008), 23.

106 Funes Monzote, 23.

were to plant four more in the same area using cuttings from the same tree. In forests reserved for the king, no trees could be cut down, including palm trees because the fruit was suitable for feeding pigs and the branches were useful for many things. When forest reserves proved inadequate for naval needs, the crown signed contracts with private individuals, especially in the 1760s.[107]

The Havana shipyard drew its resources from Cuban forests, of course, but also from a wider region, including Florida, Mexico, and Venezuela. At the same time, Cuba exported timbers from acaná, ocuge, guayacán, sabicú, and mahogany trees to Spain for various purposes, including shipbuilding and general construction. The Royal Palace in Madrid had to be rebuilt after a fire on Christmas Eve 1734, and much of the wood came from America. Havana relied heavily on Cádiz and its arsenal at La Carraca for hardware, iron, rigging, sails, arms and other supplies. Ciprian Autrán signed an instruction in 1745 detailing the cutting, refining, and transport of cedar from Cuba to Spain.[108] Autrán recommended exploiting forests some distance from the city of Havana to avoid contraband wood sales, including sites near Santiago and the Bay of Jagua. Overall, though, the wood trade to Spain never operated in a systematic way.

Guayaquil's shipyards remained in operation in the eighteenth century, and private shipbuilders complained about state monopolies of palo maría in the forests of Bulubulu.[109] By the end of the century, additional Pacific coast shipyards operated in San Blas, Realejo, and Callao. In addition, Asunción and Buenos Aires built vessels on the South Atlantic side. However, none of these would rival Havana. In the Philippines, Cavite remained the chief shipyard, but construction continued to occurr in several places. Improvements were made to the port and warehouses at Cavite in the first decades of the century.[110] Some of the largest vessels in the world were built in the shipyards of the Philippnes in the late seventeenth and early eighteenth-centuries.[111] This was not perceived as a good thing in Madrid. For example, in 1704, an 800-tonelada galleon built in Bagatao was

---

107 Funes Monzote, 24.

108 José Quintero González, "La Carraca y La Habana: Aproximación al intercambio de pertrechos entre arsenales," *Revista de Historia Naval* no. 93 (2006), 62.

109 Alfonso Mola and Martínez Shaw, 299.

110 AGI, Filipinas, 332, L.10, F.189r–189v; AGI, Filipinas, 332, L.11, F.70v–72v.

111 Greg Bankoff, "'Tree as the Enemy of Man': Changing Attitudes of the Forests of the Philippines, 1565–1898," *Philippine Studies* 52, no. 3 (September 2004), 324.

FIGURE 8 *Manila, Philippines – This map from 1715 was based on reconnaissance of forests near Subic Bay (visible in the upper right-hand corner), carried out by* sargento mayor *José Cortés Monrroy in 1698. According to the map, the forests of Subic Bay could provide resources for many galleons and was connected to Manila Bay and Cavite by a royal road. One galleon was already under construction there.*
SOURCE: ESPAÑA. MINISTERIO DE EDUCACIÓN, CULTURA Y DEPORTE. ARCHIVO GENERAL DE INDIAS. AGI, MP-FILIPINAS, 146 – MAPA DE LA ENSENADA DE SUBIG Y COSTA DE CAVITE Y MANILA.

criticized for being too large.[112] New inspections of forests brought about recommendations for exploiting new areas for the navy. Explorations of the forests near the Bay of Subic by *sargento mayor* José Cortés Monrroy in 1698 led to the creation of a map in 1715 (Figure 8). The report and map indicated the potential value of the forests near Subic, which was connected by a royal road to Manila Bay.[113] The Cagayán River valley in northern Luzon was also mapped in 1719, just as the Spanish were facing a major rebellion there (Figure 9).[114]

112 AGI, Filipinas, 332, L.11, F.97r–98v.

113 AGI, MP-Filipinas, 46.

114 AGI, MP-Filipinas, 22; Rebellions occurred in Pangasinán and the Cagayán valley between 1717 and 1719 and again in Pangasinán in 1762 and 1764.

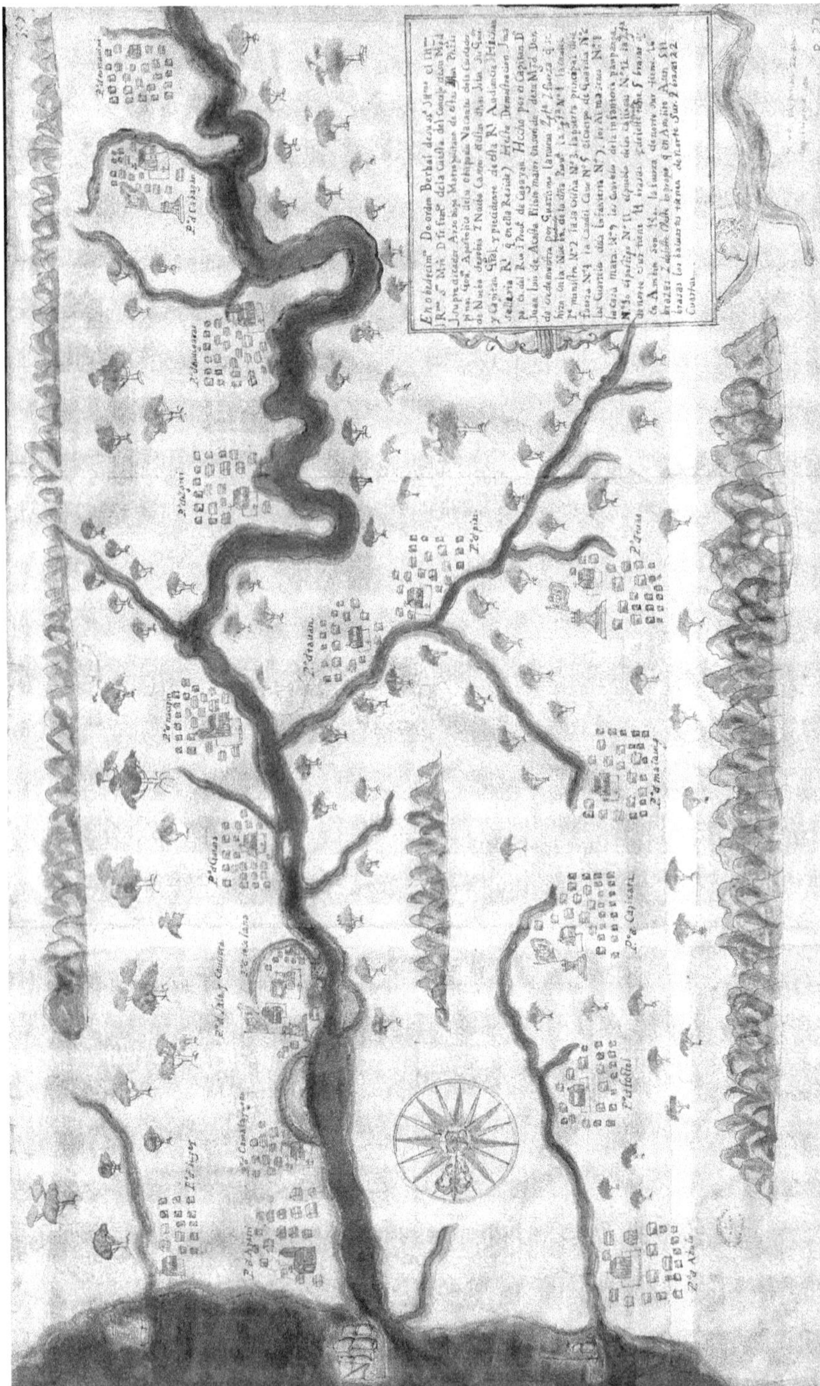

FIGURE 9 *Cagayán, Philippines – The Spanish produced several vessels in the Cagayán River valley, depicted in this map from 1719 by the capitán Don Juan Luis de Acosta. The map shows physical features, towns, as well as the Spanish religious and military installations. The Spanish were facing a major rebellion in the region around the time this map was made.*
SOURCE: ESPAÑA. MINISTERIO DE EDUCACIÓN, CULTURA Y DEPORTE. ARCHIVO GENERAL DE INDIAS. AGI, MP-FILIPINAS, 22BIS, RECTO – MAPA DEL RÍO Y PROVINCIA DE CAGAYÁN.

Accomplishments of Felipe V's Reign

In many ways, the Spanish Bourbons differed very little from the Habsburgs in their approach to forest exploitation and stewardship. The methods of monitoring, extracting, and transporting timber in Spain had their origins in the Habsburg period. For example, the Bourbons continued to rely on the bureaucracy to monitor forest conditions and make reports. They sought the same materials; oak trees were still valuable for the hulls and pine trees for various other components, particularly masts. The same technologies existed for transporting wood, typically by floating logs downriver, carrying them on barges, or hauling them overland with oxen. As a result, the navy continued to seek materials in locations that lent themselves to existing technologies for transport. In other words, agents of the crown continued to look for trees near the coasts, on mountains that sloped to the water, or along navigable rivers. This meant that important forests for the Habsburgs remained important for the Bourbons, particularly along the northern coast of Spain and forests near colonial shipyards. Under the Bourbons, however, ministers were eager to find more locations that fit these criteria, resulting in an expansion of state forestry sites.

After the administrative reforms under José Patiño, the forest reconnaissance missions of the 1730s under the direction of the *Almirantazgo* resulted in the production of new information for the crown and an increased royal presence in forest communities. Centralization of forest management increased under the first Bourbon, and forest inspectors expanded their work to new regions in the monarchy. The experiences of Ciprián Autrán and Juan Valdés y Castro show us that the three naval departments did not operate in isolation from one another. Autrán's experiences in the north became useful in his role as an official in Cádiz, and Valdés's successes in Segura led to his appointment to inspect the forests of Catalonia. The reports that the forest *visitadores* produced illustrate a combination of expertise in forest management, geography, hydrology, shipbuilding, and economics to indicate to the crown the potential value such forests could have for the monarchy. Similar explorations occurred in the colonies, including Coatzacoalcos, San Marcos de Apalache, Jagua, Subic, and the Cagayán valley. In some cases, however, the crown must have been disappointed in what they found. Maltés's journey in Valencia proved that the forests of Cofrentes were not as good as one nobleman had claimed, and the Júcar River was too rocky to use for transporting logs. Valdés had expressed his disappointment at the condition of forests along the Gállego River in Aragón. Valdés also revealed the poor conditions of factories for rigging and pitch in and around Barcelona. Good news or bad, such efforts generated a new phase of dynastic state territoriality by describing new forests, outlining management procedures, expressing ways to categorize and record forest resources, and defining regulations for local communities that were intended for all of the forests of Spain. The crown's geographic knowledge of its realm was enhanced by these reports, and would eventually be able to make more reliable estimates of available timber at any given time in order to manage more efficiently its shipbuilding efforts.

As in the Habsburg era, Bourbon inspectors who carried out exploratory missions in Valencia, Andalusia, and Catalonia, relied on the participation of local residents. While differing state and local interests in forest use led to numerous conflicts, the crown and local communities frequently relied on each other to achieve their goals. Bourbon legislation and bureaucratic reports reveal the continued need for participation from knowledgeable local residents for forestry expertise, labor, and materials. The limitations of early modern bureaucratic government meant that the state could not possibly maintain strict regulations of forest use or continuously observe forest conditions. The territoriality of Spanish state forestry, then, operated through a series of compromises between state goals and the realities of accessibility and control at the local level. The collection of such a vast amount of information about its forests in a brief period revealed how much the crown still had to learn about its own geography. By 1740, officials and locals alike were monitoring the forests near Spain's coasts to a greater degree.[115] The information gathered from forest reconnaissance missions in the 1730s led to major changes in forest management by 1748.

115 Officials focused much more on preventing the export of timber to foreigners after 1738. For example, AGS, Marina, leg. 553, Madrid, January 14, 1739, Don Joseph de la Torre Escovedo, "Preveniendole la orden que se comunica a don Juan de Rebollar prohibiendo absolutamente la corta, y extracción a reinos estraños de las maderas que puedan aprovecharse en los construcciones y carenas de los navíos del Rey…"; AGS, Marina, leg. 553, Avilés, January 31, 1739, Don Juan del Revollar, "Acusa el recivo de la orden de 14 del corr. te y contexta el desorden que se ha tolerado en la corta, y quema de mucha parte de maderas de los montes de aquel Principado, y el abuso de extraerlas a dominios extrangeros. Que absolutamente no permita se vaguen del Principado maderas algunas de las que directa o indirectamente puedan servir para construcción y carena sin expressa licencia de S.A.," (response from 18 feb); AGS, Marina, leg. 553, Madrid, February 18, 1739, A Don Juan de Revollar, "Contextando VM con fecha de 31 del passado el desorden que seá tolerado en la corta y quema de mucha parte de maderas de esos montes y el abuso de extraerlas a dominios extrangeros; me manda el don Infante Almirante General repetir a VM que sin expresa licencia suya, no permita, y antes si prohiba, y defienda la corta, y saca de maderas que se reputen utiles para la construcción y carena de los navios del Rey, en inteligencia de que al actual viceregente de essa Audiencia se le tiene comunicada igual orden y que no se duda la observará inviolablemente aún en la parte de recoger las licensias que aya concedidas por otro que por S.A. passandoso p. VM el oficio conveniente a este fin."

CHAPTER 5

The Triumph of State Forestry, 1748–1754

Ensenada's Push for Peace and Naval Revival, 1743–1748

From 1737 to 1739, the Marqués de la Ensenada oversaw a program of forest reconnaissance in nearly all forests within accessible range of Spanish naval shipyards to determine their suitability for shipbuilding. The scale and urgency of the forest reports were unprecedented and appeared to signal a new commitment from the crown to fulfill Ensenada's desire to strengthen Spanish naval power. However, during the War of Jenkin's Ear and the War of Austrian Succession (1739–1748), the institution charged with overseeing the Spanish navy's forest inspections, the *Almirantazgo* (Admiralty), effectively dissolved, and Ensenada left for Italy in 1742 with the former *Almirante* (Admiral), Prince Felipe. Ensenada returned to Spain after the death of the prime minister José del Campillo in 1743, and was soon appointed secretary of Finance, War, Navy, and the Indies, becoming Spain's new prime minister in practice. Ensenada found it difficult to achieve anything with a king entrenched in a war that was draining military resources in Italy.

The death of King Felipe V (1700–1746) came two years before peace and Ensenada wasted no time in presenting the new king, Fernando VI (1746–1759), with his proposals for reform and his pleas to put an end to the war. Ensenada addressed foreign relations, the state of the royal treasury, and Spanish government and power. Primarily, Ensenada wanted a single tax based on income, and a stronger state bolstered by an enhanced navy.[1] At the end of Felipe V's reign, Ensenada was able to increase the traditional range of coastal forest inspections from two leagues to fourteen leagues from the sea or a navigable river,[2] and in 1745, he ordered a general report on forests from all three departments, the first single representation of forest conditions near all the coasts of

1 John Lynch, *Bourbon Spain, 1700–1808* (Oxford: Basil Blackwell, 1989), 164.

2 Archivo General de Simancas (Hereafter, AGS), Secretaría del Despacho de Marina (Hereafter, Marina), legajo (Hereafter, leg.) 571, Madrid, July 17, 1742. Some towns disputed the royal order. Town officials in Osuna claimed that their forests were more than fourteen leagues from the sea. The crown responded by saying that while the town was seventeen leagues from the sea, it was fourteen leagues from Seville, which was a port, and therefore within the navy's forest jurisdiction. AGS, Marina, leg. 571, April 24, 1747.

 | DOI 10.1163/9789004261372_007

Spain,[3] but forests continued to show signs of rapid depletion.[4] To persuade Fernando VI that Spain required peace to generate the finances to build up its navy, he wrote numerous well-reasoned memoranda. For example, in 1747 he wrote, "There is no power in the world that is in greater need of maritime forces than Spain; it is a peninsula, and has to protect its vast American possessions."[5] Having a large navy, Ensenada argued, would make Spain a pivotal player in European politics once again.

On January 31, 1748, King Fernando VI issued one of Ensenada's most ambitious measures, the *Ordenanza para el aumento y conservación de montes y plantíos* (Ordinance for the increase and conservation of forests and plantations). Forest conservation was part of a larger body of reforms designed to increase state control over Spain's resources, but it is fair to say that it was among Ensenada's top priorities, because it was among his earliest legislative achievements and it related directly to his mission of naval resurgence. This chapter places the January 1748 forest conservation ordinance in historical context and argues that while it emerged from Bourbon reformism, most of its goals and methods retained a close connection to Habsburg legislative and methodological precedents. Many of the articles in the January 1748 ordinance differed little from their seventeenth century counterparts. However, the degree of territoriality represented by the ordinance was new, with important implications for the future. Due to its link with the goals and methods of the past, the implementation of the forest ordinance under Ensenada can be viewed as a triumph for state forestry, which had been designed principally to accommodate the immediate interests of naval supply by maximizing accessibility to suitable shipbuilding resources for the long term. Then, the chapter examines other pieces of forest legislation issued by the crown between 1748 and 1751. In this time, the crown extended the principles of naval forestry to the interior, no longer limiting the ordinances to twenty-five leagues from the coast or navigable rivers. Some resistance to the new legislation, examined in the next section, occurred from all over Spain, particularly where the rules of naval forestry were completely new. Naval administrators produced numerous forest reports from both the interior provinces and the coastal naval

3 AGS, Marina, leg. 571, "Estado por mayor de los árboles se han encontrado existentes en los montes de los tres departamentos."

4 AGS, Marina, leg. 554, Guarnizo, May 28, 1744, Don Juan de Rebollar, "Sobre talas hechas en las montes de la Montaña: y la necesidad de remediar este daños."

5 Didier Ozanam, "La política exterior de España en tiempo de Felipe V y de Fernando VI," tr. by José Luis López Muñoz, in Menéndez Pidal, Ramón, Dir. *Historia de España, tomo XXIX: La época de los Borbones – La nueva monarquía y su posición en Europa (1700–1759)* (Madrid: Espasa-Calpe, 1963–1985), 465.

departments to indicate conditions of forests, which included numbers and ages of different tree species in each province, numbers of new plantations, and new plantation quotas for each year. Beyond Spain, the direct impact of the 1748 ordinances in Guayaquil and the Phillipines is unclear. Where the crown was most active in naval forest territorialization, Havana, royal legislation catered to local conditions. Only in the 1770s and 1780s did a new systematic round of colonial forest inspections take place in Guayaquil, Cuba, and elsewhere in the Caribbean.[6]

Ensenada understood that Spain would not be able to match the power of the English navy, but he believed Spain could be powerful enough to tip the balance in the escalating rivalry between the French and English. Ensenada argued that once at peace Spain could build a force in eight years that would deserve the respect of both rival countries.[7] At the time, however, it was not clear which side Spain preferred to support. Both France and England had advocates at court. In addition to Ensenada, José de Carvajal also rose in the government of Fernando VI, and attained the posts of secretary of state, governor of the Council of the Indies, and president of the *Junta* of Commerce in 1746. Carvajal frequently disagreed on Spanish policy with Ensenada, and argued for supporting British interests. Ensenada thought that England was the greater threat and favored an alliance with France. Generally, however, the men agreed to pursue Spanish national interests by limiting foreign influence at court.[8] Their contentious co-administration may have served as more of a "medium of consensus" than as "a house divided," and would continue in that mode until Carvajal's death in 1754.[9] Despite disagreements with Carvajal over foreign policy and economic matters, Ensenada continued to pursue his

6 Archivo General de Indias (Hereafter, AGI), Santo Domingo, 1599A, "El Capitán de Fragata Dn Jph Villena, Avisa haben elegido…," Havana, May 3, 1788; AGI, Santo Domingo, 1599A, "El comandante general de marina, Acompaña un informe," Havana, July 25, 1787; AGI, Santo Domingo, 1599A, "El capitán de fragata Dn Jph Manuel de Villena, Propone nuevos…," Havana, July 8, 1787; AGI, Santo Domingo, 1599A, "Transferido a esta corte…Reconocimiento de la parte oriental de Cuba," Madrid, February 11, 1790; AGI, Santo Domingo, 1599A, "Dn Nicolas de Villa, Presenta un papel sobre los motivos…," Madrid, December 11, 1784; AGI, Santo Domingo, 1599A, "Papel: En que pone a la vista de VM Dn Nicolas de Villa, los motivos," December 11, 1784; AGI, Quito, 239, No. 141, Guayaquil, October 1, 1778, "Fomento y conservación de montes."

7 Lynch, *Bourbon Spain, 1700–1808*, 166.

8 Lynch, *Bourbon Spain, 1700–1808*, 161. For example, Carvajal and Ensenada agreed to appoint a Spanish Jesuit to the position of the king's confessor in 1747 to replace a French incumbent.

9 Lynch, *Bourbon Spain, 1700–1808*, 163.

agenda. Early reforms, including the forest conservation ordinances of 1748, served to improve the underlying infrastructure that would support the growth of naval power.

Spain's Naval Forest Conservation Ordinance of January 31, 1748

The Bourbons brought French administrative innovations to Madrid, such as the regional *intendentes* and the government secretaries that largely replaced the councils used by the Habsburg kings, and Ensenada looked abroad for expertise in some naval matters. The French forest ordinance of 1669 may very well have been a model for Ensenada, for example,[10] but the articles of the January 1748 ordinance mostly incorporated earlier Spanish forest regulations. The ordinance itself had no organizational subheadings, only seventy-nine articles. The material is divided below into five different thematic categories, including inspections, plantings, fines and licenses, benefits, and territoriality.[11]

The ordinance began with a statement on the poor conditions of naval forests. In the voice of the king, the document recognized "the decadent state" of Spanish forests, "especially those near the sea."[12] The king attributed their poor condition to "the frequent unduly cuttings, damages and burnings" committed without proper employment of plantings and visitations, "as the laws of these kingdoms and various resolutions from kings before me prescribed." From the beginning, he invoked a lineage of Spanish kings who had attempted this sort of thing before. The preamble ended by stating that forest conservation also provided a "useful service to my vassals," referring to the benefits they derived from forest resources. This had been a common way to introduce forest legislation in early modern Spain. In 1518, Carlos I's and Juana's *pragmática*, which was reissued several times in the sixteenth century, began with the recognition

10 On the influence of Jean-Baptiste Colbert's *Ordonnance des eaux et forêts* of 1669, see Luis Urteaga, *La tierra esquilmada: Las ideas sobre la conservación de la naturaleza en la cultura española del siglo XVIII* (Madrid: CSIC, 1987), 128; Gaspar de Aranda y Antón, *Los bosques flotantes: Historia de un roble del siglo XVIII* (Madrid: ICONA, 1992), 55 and table, "Modelo de los aprovechamientos forestales de la marina"; Jesús Cobo de Guzmán y Lechuga, *Estudio sobre los Ordenanzas de Montes del año 1748 y del expediente sobre el régimen y administración de los montes de Segura de la Sierra y de su provincia marítima de 1811* (Caja de Jaén: Obra Socio Cultural, DL, 1994), 211.

11 *Novíssima recopilación de las leyes de España* (Madrid, 1804–1829), tomo III, libro VII, titulo XXIV: "De los montes y plantíos, su conservación y aumento," ley XXII, 532–543.

12 The forests nearest to the sea had been the greatest concern of the crown since naval forestry took shape in the middle of the sixteenth century due to issues of accessibility.

of damage and destruction to their forests, and they lamented how few new trees were being planted.[13] Toribio Pérez de Bustamante had expressed the idea that forest conservation would be good for vassals as well as for the king in 1650, as a kind of enticement to increase plantations. In 1675, Queen-regent Mariana's *real cédula* began in a similar manner by stating that people were destroying the forests of Galicia, and that conservation techniques were not working.[14] In the 1720s, several bureaucrats similarly warned that forests all along the north coast were disappearing, and methods to stop deforestation were not effective.[15] In fact, the crown had always recognized the dependence its vassals had on forests, from building material to pig fodder. All of these administrators sought solutions to the problems of deforestation and the challenges of effective enforcement. Forest legislation and state forestry policies since the sixteenth century aimed to address the crisis of increased competition over increasingly scarce resources in the coastal forests of Spain for naval shipbuilding interests.

Inspections[16]

The first article of the 1748 ordinance stated that *intendentes* of the navy from each of the three naval departments would continue to be in charge of the coastal forests. Their local representatives (*subdelegados*) would help to carry out policy. *Intendentes* reported to the crown on private forests, common woodlands, and other various public lands in the areas that required visitations. *Visitadores* had to make annual visits to forests, record the latitude of the location, its distance from the coast, the potential for road construction to the forest, and the exact quantity of trees, dividing them into three categories of "new," "growing," and "old." The department *intendente* could request more frequent inspections if he felt it was necessary. Not all trees were of interest to the navy, so the ordinance specified eleven tree species the *visitadores* needed to identify. These included common oak, holm oak, cork oak, elm, white poplar, ash, alder, walnut, beech, chestnut, and pine. The *visitadores* had to specify which tree was most prevalent in each location in order to know which species grew best. This helped the inspectors decide which trees to plant in new plantations.

Justices from each location had the duty of providing the *visitador* with the local plantation testimonies, or the records of cuttings and plantings. Each

---

13 *Novíssima recopilación*, tomo III, libro VII, título XXIV, ley II.

14 AGS, Marina, leg. 571, February 10, 1675.

15 AGS, Marina, leg. 552, Madrid, August 27, 1723, De officio.

16 The articles of the ordinance I consider part of this category include numbers 1–6, 15, 23, 33, 35–41, and 44–46.

year, the inspector visited the town and went out to inspect the forests with a local bailiff and scribe. Each night after a full day of forest inspections, the scribe had to enter information regarding each forest's size and general quality in foliated registers under proper headings. The register had to include information on all the trees planted during the year, at least three for each citizen, and the number of all the trees cut down. For each tree cut down, the register had to say who did it, for what reason, and which type of license they received to do it. They made copies of these registers for the provincial and local naval ministers, the *intendente* of the department, and the *Secretario del Despacho de la Marina* (Secretary of the Naval Office). Salaries for the bailiffs and scribes came from fines collected from the justices and citizens. In towns that complied fully with all the regulations, the salaries came from the extra fines from other towns. Forest inspectors could not accept additional salaries or rewards, nor claim anything other than a village house in which to live while performing their duties. The provision meant to eliminate the problem of absenteeist bureaucrats, who had inspired complaints in the 1720s.

Each villager had to plant three trees of the type specified by the forest inspector. The ordinance made clear that no private forest owner or town could claim a traditional right, or *fuero*, to excuse themselves from the duty of planting. The only people who could be excused were disabled citizens and "poor widows" ("viudas pobres") who did not have any children over eighteen years old living with them. Justices had the responsibility to oversee the distribution of the work and the proper planting procedures. In some cases, naval ministers or forest inspectors called shipbuilding specialists from the coast into the forests to examine the quality of plantations. The specialists could judge whether the young trees would grow to produce suitable shipbuilding timber. The forest inspectors had to mark the trees they claimed for naval use, and to order replacement plantings. Inspectors also involved themselves in forestry matters not directly connected to shipbuilding. For example, they had the duty of arbitrating price disputes between firewood sellers and *asentistas*, who were under contract to supply timber for the navy.

It should be clear that forest inspection procedures described in the 1748 ordinance had much in common with earlier practices. José Patiño created the departmental administrative organization in 1726, placing the *intendentes* at the head of each department. They and their *subdelegados* and forest inspectors had evolved from the position of forest superintendent used in the sixteenth and seventeenth centuries. It had been the superintendent's duties to make annual visits, consult the annual planting testimonies, and bring along local bailiffs and scribes. Mariana of Austria had requested the expertise of shipbuilders to assess the suitability of timber in Galicia's plantations. She had also asked for more kinds of geographic categories in the forest inspections to help understand the accessibility of forests and their potential productivity. The innovative use of three descriptive categories, "new," "growing," and "old," in the 1748 ordinance may be seen as an extension and continuation of such practices. The 1748 ordinance gave the local justices greater responsibility, especially with regard to assigning and distributing the planting obligations to the citizens. Earlier, this had been the duty of the forest superintendent. Perhaps, reports of overextension by Barros, Pérez de Bustamante, and others in earlier times, had led the crown to realize the limits of the government officials and recognize the benefits of delegating some of their responsibilities to the local citizenry.[17]

Planting[18]

Rules for planting in 1748 also had much in common with earlier practices. New plantations needed to be in places that received plenty of sunlight but were sheltered from heavy winds. People were to plant seeds from the largest and strongest oak trees. The winter was the best time to plant, especially January. The ordinance stated that one should "open tiny grooves" in the soil and place the seeds inside by hand, then close the grooves with the same dirt, being careful not to trample on the seeds. People should leave the rest up to the weather. Nevertheless, they could not leave things completely alone once they planted the seeds, because some care was necessary to help the plantation grow successfully. Every year, people had to add manure and root out harmful weeds. Grasses and other harmless vegetation should be allowed to remain, because they retained the moisture and dew of the summer. At the three-year mark, and each year after that, people had to prune branches delicately and leave only the main trunk and the strongest branches. This could help struggling trees grow better. Pruning happened between November and February during waning moons. During these early stages, people had to be particularly vigilant against the entrance of livestock and other animals that could gnaw away at and damage the young trees.

When the trees were of a certain size, 3½ to 4 *pulgadas* in circumference and 3½ to 4 *varas* in height,[19] the people had to transplant them from the nursery to more protected areas of the forests. The times to do this were between the

17 See David Goodman, *Spanish Naval Power, 1589-1665: Reconstruction and Defeat* (Cambridge: Cambridge University Press, 1997), 73.

18 For this category, I include articles 7–16.

19 A *pulgada* is about 1 inch or 2.54 centimeters. A *vara* is about 33 inches or 83.5 centimeters.

middle of December and the middle of February during a waxing moon. The ordinance stipulated that the distance between trees depended on the terrain, with poorer terrain requiring trees that were farther apart. Generally, people planted according to what they had seen in the nearby forests or according to what knowledgeable townspeople said. The hole had to be about one *vara* deep and wide to provide room for the roots. The dirt should be soft and should fully cover the roots once the tree was in the hole. The dirt should cover enough of the base of the tree so it stayed strong against the forces of winds and cattle. If the tree was still weak, people could tie a sturdy stake against it to keep it upright. Once successfully transplanted, people had to protect the trees by observing the same rules of the nurseries.

These provisions in the 1748 ordinance also echo earlier precedents. Among the practices described in Barros's letter to Riva Herrera in 1594, many of which might have been based on long-standing practices developed at the local level, he described the need for a hole of similar size to ensure the roots took to the soil and to collect moisture for the early stages of growth. He knew people had to plant trees farther apart in poorer soils. It had also been standard practice to plant in the winter, since it was easier to deal with trees without their leaves and with a larger labor supply than in the summer. In 1650, Pérez de Bustamante clearly stated that the period between the middle of December and the middle of February was the best time to make cuts and plantings. Some of the Habsburg forestry documents even included concerns not addressed by the 1748 ordinance, including the use of thornbushes to protect plantations against livestock, and the need for vigilance not just against animals, but also against the ever-present danger of forest fires.[20]

Fines and Licenses[21]

The idea of requiring licenses to cut down trees in order to curb deforestation dated at the latest to the Catholic Kings and their *Pragmática sobre cortas* of 1496.[22] Over time, the crown made such regulations more forceful, and installed more royal officials to oversee behavior in the forests to help enforce the law, but enforcement remained a serious challenge throughout the early modern era. The 1748 ordinance made it clear that people still required licenses to cut down trees. The removal of some branches from a tree during pruning, or from the ground did not require a license, because these practices would not harm

---

20 See Cristóbal de Barros's letter to Hernando de la Riva Herrera in AGS, Guerra Antigua, leg. 403, folio 102.

21 For this category, I include articles 17–21, 31–37, 41–43, and 49.

22 *Novíssima recopilación*, tomo III, libro VII, título XXIV, ley I.

the trees. Article 21 stated that people could not reduce a tree to a state where it could not produce new branches, however. A set of drawings from 1773 helps to illustrate which kinds of cuts would or would not do long-term damage[23] (Figure 10). The top row shows cuts that would allow new branches to sprout. Several examples in the bottom row show trees that had been cut in ways that would not allow for regeneration; these types of cuts required licenses. An interesting part of this *demostración* is that the author stated that the pruning techniques were typical of Spain and had been described in the 1650 Instruction of Toribio Pérez de Bustamante.

The main condition that came with licenses was that people had to plant three trees for each one removed. As stated earlier, the town had to keep a record of all its licenses and cuttings. The crown officials ordered fines based on the testimonies written in these books. Towns could incur fines for not making nurseries, not planting, not taking the correct precautionary measures against damages, not preventing fires, not complying with the visitations, cutting trees without licenses, or not adhering to any of the other regulations described in the ordinance. When ordering fines, the crown official had to state which provision had been broken, because article 42 stated that fines had to be justified. The forest inspectors took confessions from the lawbreakers and, with the assistance of the town's *asesor*, issued sentences. Those charged could make an appeal before the *intendente*, whose sentence could be appealed to the *Secretario del Despacho de la Marina*. The local justices were responsible for collecting and submitting the fines. If certain regions felt they needed some extra help managing their forests, they could request to have a naval minister appoint a guard, who would have authority to apprehend criminals. Heavy fines could result from sending wood to foreign domains without expressed approval from the king himself. Here, too, the 1748 ordinances echoed earlier practice. Spanish kings at various times since the Middle Ages had outlawed the export of timber, since it could easily be used as a war material by their enemies.

Benefits[24]

The 1748 ordinance described ways that the local populations benefited from observing the laws of forest conservation, much like other forest legislation from the past. Pérez de Bustamante emphasized the material benefits people enjoyed from their trees, even if the navy laid claim to them, and described the financial benefits that they could enjoy from years of caring for their

23 AGS, Mapas, Planos, y Dibujos, 57-013.

24 For this category, I include articles 22, 24–30, and 32.

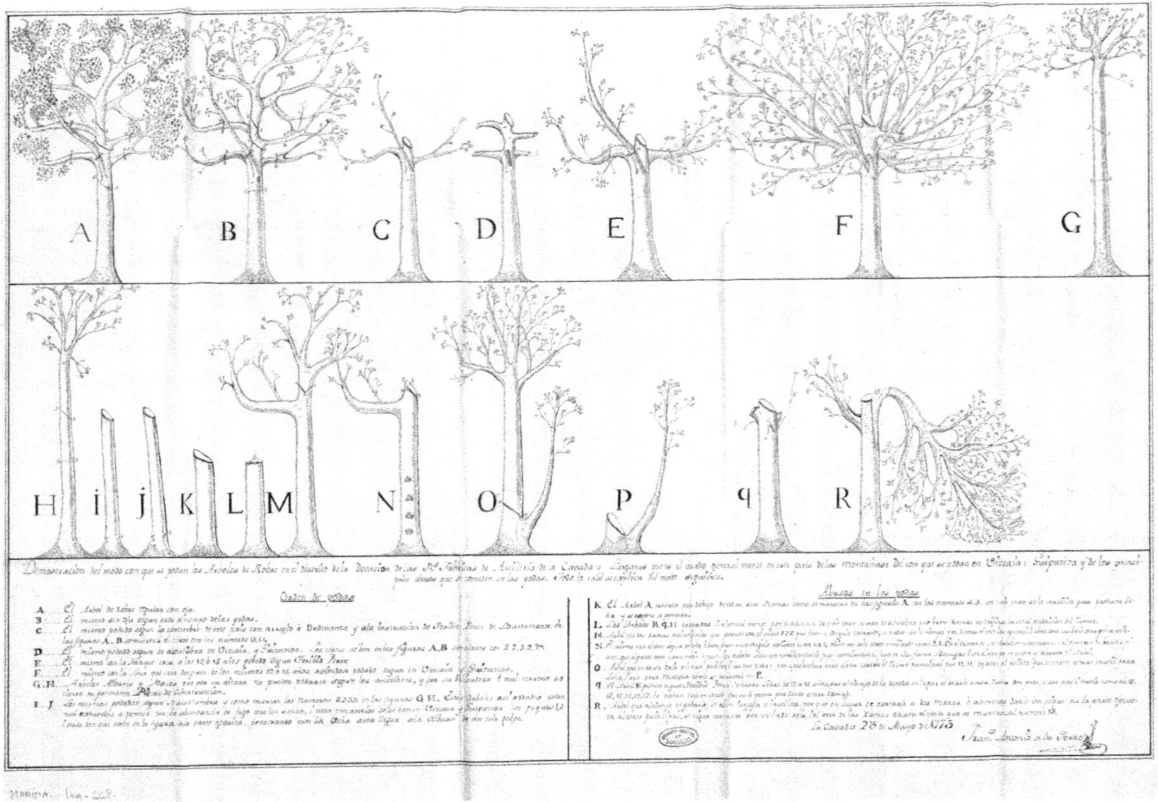

FIGURE 10 *Illustration of cutting and pruning methods – These drawings indicated ways of pruning oak trees near the royal artillery factories of La Cavada and Liérganes (Cantabria) as well as in the Basque provinces of Vizcaya and Guipúzcoa, following customary techniques and those stipulated in Toribio Pérez de Bustamante's 1650 instructions. Letters A through J describe proper methods, while letters K through R show pruning abuses. The drawings were made by Francisco Antonio de la Torres in La Cavada (Cádiz) in 1773*

SOURCE: ESPAÑA. MINISTERIO DE EDUCACIÓN, CULTURA Y DEPORTE. ARCHIVO GENERAL DE SIMANCAS. MPD,57,13.

plantations. Ensenada's ordinance similarly emphasized material and financial benefits. The ordinance noted that towns could make a substantial amount of money by selling firewood surplus to towns that were located far from forests, which was a practice common since the Middle Ages.[25] Another incentive for towns to plant more trees was the offer of one *real* for each cubic *codo* of wood cut from their forests for building and repairing vessels in the royal armada.[26] In short, towns could turn their plantations into an investment that would help them pay taxes, tribute, fines, and other municipal obligations. They could put their profits toward public works and new plantations, but the towns had to report any wealth they generated from wood sales to the naval *intendentes*.

The ordinance attempted to balance the interests of those working for the crown with local needs for wood. Although, while *asentistas* paid the same prices for timber as any townsperson, the regulations of the ordinance gave them a certain privileged position regarding access. *Asentistas* acquiring wood for artillery and ironworks, as well as those supplying any industry that served royal interests, had the opportunity to buy firewood surpluses before ordinary people who needed timber. However, *asentistas* working in the forests had to leave the branches and wood leftover from cuts for local domestic use.

Territoriality[27]

The most innovative and transformative aspect of the 1748 ordinance for state forestry in Spain was the territorial hierarchy it imposed over all coastal forests. No other Spanish forest legislation for the conservation of naval timber included so many forests. Organizationally, José Patiño's naval administrative changes from 1726 served as the basis, but criteria for identifying the most important spaces and methods for controlling behavior within those spaces resulted from years of accumulated experience and geographic knowledge. The naval forest administration no longer operated with one forest superintendent for each province, as the Habsburgs had done. Rather, the navy had an *intendente* at the head of each naval department, ministers for each province within the department, and several more forest guards and inspectors within each province.

Thirty-one articles of the 1748 ordinance identified the territories in which its rules applied. Each of the three naval departments contained smaller

25 María del Carmen Carlé, "El bosque en la Edad Media (Asturias-Leon-Castilla)," *Cuadernos de historia de España* 59–60 (1976): 346.

26 Article 27 made clear that the measuring of cubic *codos* would be made after the wood had been cut, smoothed off, and prepared for river transport.

27 For this category, I include articles 3, 13, 24, and 52-79.

provinces. For each province, such as Ferrol or Tuy, both in Galicia, the ordinance listed several smaller jurisdictions, such as Villanueva de Redondela and Vigo, both in Tuy. One region, the Sierra de Segura, received special attention. While most locations appeared in list form with several nearby towns, Segura alone received five articles to describe its reestablished river transport system for supplying pine timber to the arsenals at La Carraca and Cartagena, and to identify its territorial divisions. The side of the sierra that fed the Guadalquivir and Guadalimar Rivers that flowed westward fell within the Department of Cádiz, and the side that fed the Segura River served the Department of Cartagena. Article 71 mentioned that the sierra was sparsely populated, and surmised that it would be difficult to encounter or identify delinquents who caused damage to the forests there. The typical laws of forest observation and oversight depended on a great deal of communal cooperation. The Sierra de Segura lacked the population density of coastal forest villages, and would therefore have trouble enforcing the regulations of the ordinance. As a result, the article charged the *intendente* of Cádiz with the task of informing the crown of suitable measures and rules to overcome that problem.

Even within the smallest jurisdictions listed in the ordinance, a spatial hierarchy existed. In any forest, for example, towns and other forest owners were required to reserve the most productive terrain that was nearest a port or coastline for growing oak trees of the highest quality. It listed several regions in Asturias specifically, because they grew oak trees of superior quality faster than other places. In these areas, people were told to cut down and remove old and useless trees (useless for naval purposes), and make room for new oak trees. These areas met certain highly valued geographic criteria for naval interests, and the navy made its strongest presence felt in these areas by carefully controlling the behavior of people in local forest communities.

The realities of early modern society and the limitations of early modern government and technology meant that the crown could not control all forests with the same intensity. Too many people in the king's realms relied on timber resources for the crown to prevent all domestic use of coastal forests. If the crown attempted to prevent citizens from gaining access to forest resources entirely, the crown would either face the impossible task of enforcing the measure or the undesirable social unrest that would come from it. Article 30 expressly recognized this point by stating,

> Because the absolute prohibition of cutting wood and trees could be prejudicial to my vassals, [who are] short of the necessary materials for building and repairing houses, mills, and other things that require the use

> of wood, whose lack I desire they do not experience, the *intendentes* will order their *subdelegados* to allow citizens to cut trees when they need to.

There had been signs of tension between how much control the crown desired and how much it could effectively employ since the sixteenth century. When Cristóbal de Barros was sent to oversee forest conservation in northern Spain by Felipe II, he was told to use no more than persuasion to induce the local populations to plant more trees. The naval forest territoriality of the 1748 ordinance grew out of a long history of state forestry, and clearly relied on forest inspections, plantations, prohibitions and incentives developed for the political and ecological realities of Spain over the previous two centuries.

Other Forest Legislation, 1748–1751

Ensenada and the Spanish government were far from finished with forestry matters after January 31, 1748. The peace of Aix-la-Chapelle ended the War of Austrian Succession on October 18, 1748, and put the crown in a strong position to implement more of Ensenada's reforms. The crown recognized the effectiveness of these comprehensive rules in naval forests, and later that year, it took steps to employ them in Spain's interior forests near the court. Over the next few years, the Spanish government issued several more pieces of forest legislation, one of which placed forests twenty leagues from the court under direct crown supervision.[28] This ordinance, issued on December 7, 1748, aimed to address the quickly dwindling supplies of timber in the sierras near Madrid and the royal palaces. During Felipe V's reign, the court had carried out major construction and reconstruction projects of royal palaces. The Royal Palace in Madrid was built after a fire destroyed the old palace, the Alcázar, in 1734. Felipe and his queen, Isabel Farnese, made improvements at the palace of Aranjuez, south of Madrid, after 1730, and a fire there soon after required further renovations. Improvements were also made during the reign at El Pardo,

28 *Novíssima recopilación*, tomo III, libro VII, título XXIV, ley XIV. The original ordinance set the limit at thirty leagues around the court. This must have changed at some point between December, 1748, and the following year, because early forest inspection reports included territory only twenty leagues out. In 1755, the limit changed again when the reports included territory twenty-five leagues from the court.

north of Madrid.[29] Such massive building projects during the reign of Felipe V may have caused severe deforestation in the area, prompting the ordinance of December 1748.

The December 1748 ordinance placed the *corregidores* in charge of interior forests instead of the naval *intendentes*. As in the naval forests, the *corregidores* gathered testimonies about plantings, collected fines, and oversaw annual inspections with local experts. Cutting trees down required licenses, and damages of any kind otherwise resulted in fines that increased for multiple offenders. The forests along the banks of the Manzanares River in Madrid nearest the court received particular attention. The same species of trees valued by the navy remained important for the crown, since these species were common, and were used for most domestic needs and building projects.

A set of undated revisions to an early draft of this forest ordinance exists in the archive of Simancas. The editor was an official named Alonso Pérez Delgado.[30] A copy of the draft is not extant; we only have the suggested revisions. They provide an interesting window into how the government considered extending state forestry practices from coastal areas to the interior of Spain. The editor took the navy's earlier ordinance as the model. He began by saying the introduction on the draft "is very [read: too] erudite. Perhaps it would be better to adopt the introduction from the Forest Ordinance made for the navy: it refers to the damage that is occurring…and the necessity to apply remedies." The editor said the draft was "full of defects in its explanation" and it had "too much insubstantial material." The full draft apparently required a lot of trimming. It also required slight revisions in its policies. For example, the draft evidently expressed an obligation for everyone to plant ten trees each year, which the editor said was excessive, and he suggested lowering it by half. We know that at least some of his suggestions were incorporated into the final version, because article eleven of the December 1748 ordinance required each citizen to plant five trees per year, not ten.

Finally, the suggested revisions to the draft provide some insight into the crown's future goals for state forestry. It seems that the crown had considered having one minister in charge of all of Spain's forests. The editor wrote, "It is true that to achieve this end, it is very advantageous that everything proceeds under only one hand." However, legislation issued on December 12, 1748, reveals that at this point one minister would have been unable to oversee all of

29 Henry Kamen, *Philip V of Spain: The King who Reigned Twice* (New Haven: Yale University Press, 2001), 199.

30 AGS, Marina, leg. 571, "Borrador, sin fecha, de reparos a la ordenanza de montes de lo interior del Reino. Es de don Alonso Pérez Delgado."

Spain's forests.[31] The ordinance chose two ministers from the king's *consejo* to be in charge of conserving his forests in the interior. One of them would be in charge of forests twenty leagues from the court, and the other would be in charge of the provinces and towns that fell outside both this interior zone and the navy's coastal zone. The duties of the forest minister in the middle zone between the navy and the court are unclear, but they were probably not too different from the other zones. The head minister had to submit an annual report of plantations, just like his counterpart in charge of forests closer to Madrid. In short, by the end of 1748, the Spanish state claimed regular administrative authority over all of Spain's forests, a dramatic extension of state forest territorialization (Map 5).

By the summer of 1749, the crown already faced problems of enforcement. The king responded to tacit resistance in Guipúzcoa with an ordinance written especially for the province.[32] In one sense, the creation of a separate piece of legislation for the Basque province might dispel the notion that Spain had a uniform national forestry code. However, evidence mentioned above indicated that the code was far from uniform at its inception. Some areas already had special distinction, whether as a naval province in the high sierras near Segura, in a particularly suitable area for growing oak trees in Asturias, or in the colonial forests of Cuba, all of which required separate rules and regulations. Also, this was not the first time the crown faced resistance from Guipúzcoa over forest policy. The two sides had a long relationship of back-and-forth disputes over legitimate practices that usually ended with some compromise.

This instance was no different. The crown granted the provincial government greater authority in overseeing the rules and regulations of the 1748 ordinance. However, at the same time the crown extended a stronger form of state territoriality through the legal system. The state granted Guipúzcoans the right to challenge articles in the ordinance, but it maintained ultimate authority over the forests by settling the disputes within naval courts, in effect extending the jurisdiction of Secretary of the Naval Office. If the local government opposed some of the regulations, they had recourse to the navy's provincial minister, who would hear the complaint. If he found the complaint justified, he would remit the testimonies to the provincial congress (*Diputación de la Provincia*). If the settlement reached there did not satisfy the parties involved, then the account would be sent to the king's Secretary of the Naval Office. In short, the ordinance set up a system that would allow Guipúzcoans to maintain some of their traditional practices, if found agreeable by a royal judge.

31 *Novíssima recopilación*, tomo III, libro VII, título XXIV, ley XVI.

32 *Novíssima recopilación*, tomo III, libro VII, título XXIV, ley XXV.

MAP 5 *1748 forest ordinances – The January ordinance set rules for naval forest regulation in the three naval departments of Ferrol, Cartagena, and Cádiz. The December ordinance set the rules for forest regulation 30 leagues from the court, indicated by the circular shape in the center, as well as the territory that fell between this circle and the naval departments.*

In 1751, the crown issued more legislation regarding the naval forests to give the local justices even more authority. They now could grant forest cutting licenses and oversee the behavior of forest guards appointed by the naval minister, in addition to presenting books of testimony to the forest inspectors.[33] The special legislation freed royal bureaucrats from several burdensome tasks, which were delegated to the local justices, and it settled a potentially fruitless and time-consuming dispute with local authorities. The years 1748–1751, therefore, extended naval forestry regulations to all of Spain. The major question remained whether these forest ordinances would produce results from which the navy (and the royal court) could derive benefit.

Resistance to the New Legislation

The resistance from Guipúzcoa had been peaceful, characterized by noncompliance. Reactions to the forest ordinances of 1748 were sometimes much more violent. In 1751, not far from the royal palace of Aranjuez in a forest that belonged to a town called Don Fadrique, two men were caught cutting down some trees. The two men were from a neighboring town that had customarily been able to enter and use the forest. The mayor of the town, Manuel de Mendoza, enforced the regulations of the ordinance that pertained to forests twenty leagues from the court, and arrested the two men for cutting without a license. This legislation was still new to the area and the mayor had only been carrying out his recently specified duties, but the two men responded to their apprehension by killing Mendoza.[34] In response, early in 1752 Blas Jover, the crown-appointed forest superintendent for the area twenty leagues from the court, visited the town in the company of twelve armed soldiers and a lawyer to verify the crime, listen to testimony, and provide a verdict and sentence. José Moñino was the crown-appointed lawyer, and within twenty days he found the two men guilty of murder and of disobeying the forest ordinance. At some point, one of the men fled and became a fugitive. The other man, however, was incarcerated in a royal prison.

In the district of Jeréz (Andalusia) in 1748, a trigger-happy forest guard named Juan Muriel killed a citizen of the nearby town of Alcalá de Guadaíra named Juan Jarillo, who, it turned out, had only been setting rabbit traps in the forest. The Duquesa de Medinaceli wrote to the *intendente* of Cádiz about

33 *Novíssima recopilación*, tomo III, libro VII, título XXIV, ley XXIII, articles 4 and 12 respectively.

34 AGS, Marina, leg. 558, March 10, 1752, Don Blas Jover.

the matter, who imprisoned the guard.[35] In 1749, near Alcalá de Henares, located within twenty leagues of the court, Blas Jover examined a case involving shepherds who destroyed the town's forest plantations. Rather than punish the shepherds, however, Jover reprimanded the *corregidor* and the *regidores* of Alcalá for allowing such abuses to unfold right under their noses. He even sent the town's scribe to prison in Toledo.[36]

Several disputes involved questions of exemption. The Marqués de la Candia wrote to Francisco de Varas, the *intendente* of Cádiz, to ask if the military orders had exemption from the rules of the ordinance. The reply was predictable, "Que no hay exento nadie," no one was exempt.[37] The Duquesa del Infantado wrote to the *intendente* of Ferrol relating that people in Potes, a small town in Cantabria near the border with Asturias, managed their forests according to ordinances approved by the crown from 1622 and 1678. She said people had access to timber for building material and for fuel without problems. The town had continued to abide by these older regulations for two years after the king issued the 1748 ordinance, until a forest inspector put a stop to unlicensed cuttings. The Duquesa asked for the town to be exempt from the ordinances, because the regulations from the last century had been working fine. That, too, was unacceptable to the navy, and the response was for the town to observe the new regulations.

Most people expressed resistance in other non-violent ways directed toward the forest inspectors. There is evidence of non-compliance in Gibraltar in December of 1749, about which a minister of the navy reported to the Department head, Francisco de Varas, who commanded the minister to act immediately and punish those culpable.[38] In 1749, people from the town of la Puebla, near Seville, complained that the forest guards were "public criminals" ("ladrones públicos"). Ensenada ordered the *intendente* of Cádiz to examine the case. If he found the town's complaint justified, he would punish the guards. If not, the town had to pay the guards' salaries.[39] Such cases were common and verdicts could fall either way.[40]

Some instances of non-violent resistance resulted in severe penalties. One case involved the area around the town of Pliego, which was not far from

35 AGS, Marina, leg. 571, Cádiz, March 26, 1748, Don Francisco de Varas.
36 AGS, Marina, leg. 555, December 22, 1749, Don Blas Jover.
37 AGS, Marina, leg. 571, August 27, 1748, El Marqués de la Candia.
38 AGS, Marina, leg. 555, Gibraltar, December 23, 1749.
39 AGS, Marina, leg. 555, La Villa de la Puebla, 1749.
40 Erich Bauer Mandersheid, *Los montes de España en la historia*, 3rd ed. (Madrid: Fundación Conde de Valle de Salazar, 2003), 470.

Murcia. The Pliego River flows into the Segura River, so the area came under the jurisdiction of the Department of Cartagena, according to the 1748 forest conservation ordinance. While the *subdelegado de la Marina*, Diego Antonio de Molina, carried out his visitation of the area, a large crowd composed of local mayors and townspeople bombarded him with insults. Some people received fines, but the principal aggressors were sent to presidios in North Africa. One mayor named Gonzálo Martínez was sent to Oran for four years.[41] In 1751, several mayors near Huete, which was located within twenty leagues of the court, complained that a man named Juan Serrano repeatedly cut down and damaged trees in their jurisdictions without license. Blas Jover examined the case, found Serrano guilty, but said the town of Huete had no secure jail. He decided that it would be too costly to send Serrano to royal prison, so he sentenced him to four years in a presidio.[42]

The crown was capable of revising sentences and fines if provided with persuasive evidence. Some locations wrote to the naval *intendentes* about how the unsuitability of their terrain prevented successful plantations. The crown typically responded by telling the town they had failed to follow the ordinance's rules for planting, and they should try again.[43] However, a record of fines collected from Valencian towns in the district of Alzira in 1754 indicates that in some cases the land really was unsuitable for plantations.[44] In Alcazer, the *intendente* of Cartagena revoked a fine given for not starting a plantation because of the "poverty of the land" ("estrechez de la tierra para el plantío"). A forest inspector also absolved the town of Yátova of the obligation to have a nursery. In some instances, the navy lessened the original fine for various reasons. For example, although a forest inspector initially fined the town of Almutajes for not submitting its testimonies, the *intendente* revoked the fine after receiving them from local justices. This document also reveals how towns committed offenses such as failing to prune properly or clear out weeds, which typically amounted to about two *pesos* for each year for each plantation.[45] Towns received fines for having no nursery or no book of testimonies, which normally came to four *pesos* each. One town made a nursery two years too late and had to pay fines of four *pesos* for each of those years. Annual fines

41 AGS, Marina, leg. 555, October 4, 1749, Don Francisco Barrero.

42 AGS, Marina, leg. 557, Orcajada, August 25, 1751.

43 For one example, see AGS, Marina, leg. 557, La Villa de Elche, February 23, 1751.

44 AGS, Marina, leg. 559, August 28, 1754, "Resoluciones."

45 A *peso*, also *peso de a ocho reales*, equaled 272 *maravedís*, and for purposes of conversion it equaled 20 *reales de vellón*. Lynch, *Bourbon Spain, 1700–1808*, xi.

collected by the state could be quite substantial.[46] The total amount collected from the interior forests twenty leagues from the court in 1750 came to 604,400 1/3 *maravedís*, or 17,776.16 1/3 *reales*. For some comparison, in Madrid at this time, daily wages were about 15 *reales* for master weavers, 5 *reales* for journeymen, and 4 *reales* for shepherds.[47]

Forest Inspection Reports after 1748

Despite causing many problems, the ordinances undoubtedly led to the planting and pruning of millions of trees, and the collection of more detailed reports on forest conditions from all over Spain. In the years after 1748, the crown received numerous "*estados de montes*," which set a base of forest information on which it could measure future forest growth. Compared to the forest information-gathering projects commissioned by the *Almirantazgo* during 1737–39, these reports were more numerous and more thorough, aided by guidelines set out in the 1748 ordinances.

For the navy, these reports consisted of similar, but variously organized tables and texts that characterized forests, their accessibility, and their suitability for shipbuilding. Department *intendentes*, naval ministers, forest inspectors, carpenters, scribes, and thousands of citizens contributed to these reports. Local populations provided testimonies on the quality of the land and forest property divisions. Carpenters used their sylvan expertise to inspect the species, ages, and quality of trees in each district. Scribes collected and recorded the forest numbers, the socio-geographic data, and local testimonies. Naval ministers oversaw the operation and prepared the final reports for the department heads.

For the rest of Spain, the two royal forest ministers appointed by the king in 1748 produced and submitted annual reports from each forest administration zone. Blas Jover produced reports on forests twenty leagues from the court, and José Bermúdez reported on forests in the middle zone of Spain, outside the jurisdictions of the court and the navy. The information desired by the crown, as described in the ordinances of 1748, included specific species of trees, ages of trees broken down into three categories, and basic accessibility conditions, such as distances to the nearest port, and options for transporting timber to nearby rivers. Some reports included the information required, some

---

46 AGS, Marina, leg. 556, Don Blas Jover, "Razón de lo que ha pertenecido a S.M. por las Quartass partes de penas impuestas…."

47 Lynch, *Bourbon Spain, 1700–1808*, 238.

provided less information, and some provided even more information that the ordinance required, depending on the naval minister leading the visitations.

Interior Forest Reports

Jover produced his first report on forests twenty leagues around the court on July 28, 1749. It included data on the numbers of towns and people in each of the fourteen *partidos* that fell within twenty leagues of Madrid.[48] Other information included numbers of trees planted, sown fields created, and numbers of inspected forests, all according to testimonies submitted by the respective *corregidores* and *alcaldes mayores*. However, Jover's notes stated that the testimonies did not conform to the guidelines of the December 7, 1748 ordinance. Many places did not send testimonies, and the ones he did receive suffered from the "incompetence, negligence, or poor intelligence of the executors." Due to these problems, Jover stated, this report stood only as "an example in form for subsequent [reports]." Clearly, he was disappointed in the first attempt, despite recording that 254,528 trees had been planted.

The following year, Jover issued a more robust report, including a list of fines collected from each *partido*.[49] After thirty-seven pages of data on each town, Jover provided a summary table similar to the previous year's. The report stated that the population of 51,502 (and a half)[50] planted 244,678 trees since the last collection of testimonies, indicating that people came close to meeting the ordinance's obligation to plant five trees per person, with an average of 4.75. They came up short by 12,835 trees.

Totals from Jover's 1751 report included 73,914 people and 303,670 trees planted, down to an average of 4.1 trees per person. It also reported that 2,859,481 trees had been pruned and 21,560 *fanegas* of land had been sown.[51] The next report, from 1753, indicated that 168,198 trees had been planted by

48 AGS, Marina, leg. 555, Madrid, July 28, 1749, "Plano en que se demuestran los catorze Partidos que se comprehenden en las veinte leguas de la circunferencia de esta Corte." The fourteen *partidos* included: Madrid, Toledo, Cuenca, Segovia, Ávila, Guadalajara, Ocaña, Alcalá, Huete, Talavera, Almonacid, Uclés, Yllescas, and Manzanares.

49 AGS, Marina, leg. 556, Don Blas Jover, "Relación de los lugares que comprehende la comisión de plantíos de las veinte leguas de la circunferencia de esta corte" and "Razón de lo que ha pertenecido a S.M. por las Quartas partes de penas impuestas…."

50 The report showed several towns counted ½ of a person, but did not explain the criteria for doing so. Probably there was some categorization by age and ability.

51 AGS, Marina, leg. 557, Madrid, September 2, 1751, Don Blas Jover, "Remite estado anual de los montes de las 20 leguas de esta corte." When a unit of area, a *fanega* is the amount of land required to grow one *fanega* (amount) of the dry good, about 3.5 square km or about 2.2 square miles.

56,552 people, down again to an average of 2.97 trees per person.[52] Perhaps due to the decreasing average of trees planted by each person, in 1755 the crown extended the jurisdiction to include towns 25 leagues from Madrid. However, the average continued to drop to 2.33 trees per person.[53] Pedro Vidal replaced Jover, who died in 1754. He oversaw the 1756 inspection, and restructured the data, replacing the category "trees planted" with "trees planted and in existence," obscuring the number of trees planted per person that year.[54] The decline in the rate of trees planted per person is not surprising, considering that suitable land had already been used in many *partidos*, limiting the space for new healthy growth. While the crown was unable to put the ordinance into full practice, it managed to get a population of about 60,000 people living twenty leagues around Madrid to plant 863,018 trees between 1748 and 1755 (Table 1).

Meanwhile, José Bermúdez collected testimonies from scribes who visited forests in 87 *corregimientos* between the zones of the court and the navy. His

TABLE 1 *Interior forest reports, 1748–55 (14 partidos)*

Year	Population	Trees planted	Trees per person
1749	77,478	254,528	3.29
1750	51,502	244,678	4.75
1751	73,914	303,670	4.1
1752	–	–	–
1753	56,552	168,198	2.97
1754	–	–	–
1755	62,653	146,472	2.34
Total	64,420	1,117,546	3.49

SOURCES: "ESTADOS DE MONTES," AGS, MARINA, LEG. 555, 556, 557, 558, AND 559.

52 AGS, Marina, leg. 558, Madrid, November 8, 1753, Don Blas Jover, "Estado general de las ciudades, villas, y lugares que comprehenden las veinte leguas de la circunferencia de esta corte."

53 AGS, Marina, leg. 559, Madrid, December 16, 1755, "Estado general de las ciudades, villas, y lugares que comprehenden las 25 leguas de la circunferencia de esta corte."

54 AGS, Marina, leg. 560, Madrid, November 30, 1757, Pedro Vidal, "Estado general de los partidos y provincias que comprehende la comisión de montes y plantíos de las 25 leguas del contorno de la corte."

report from 1753 provided totals on the previous four years. The year 1749 included the smallest total of trees planted, and had been only an important first step to enforce the ordinance, much like Jover's first report. Mirroring Jover's reports in another way, the totals then increased, but at slower rates each year from 1750 to 1753. The total number of trees planted and pruned came to 7,561,036.[55] During these years, the state made its presence felt in unprecedented ways by mandating laborious tree plantings and management practices throughout the interior of Spain, exacting fines on those who failed to meet the requirements of the ordinance, and generating a centralized storehouse of forest and population data (Table 2).

Naval Forest Reports

As for the areas under naval jurisdiction, the first completed *visita de montes de la Marina* came from the unlikely source of Mallorca.[56] From May 7 to July 9, 1748, the naval minister Pedro de Hordeñana inspected and catalogued the island's forests with the help of a scribe, a bailiff, a carpenter, and four horses. The bottom line, according to Hordeñana, was that Mallorca contained enough

TABLE 2 *Middle zone forest reports, 1749–1753 (between court and coast)*

Year	Trees planted
1749	225,741
1750	2,393,418
1751	1,797,433
1752	1,541,817
1753	1,602,627
Total	7,561,036

SOURCES: "ESTADOS DE MONTES," AGS, MARINA, LEG. 558.

55 AGS, Marina, leg. 558, Madrid, July 26, 1753. One major drawback regarding these data is the lack of information we have on the rates of tree removals at this time. We now know how many trees were planted, but we cannot have a clear understanding of overall forest regeneration until we know how many trees were lost during the same period. To find these data, historians can search for indications through fuel consumption, changes in timber prices, or perhaps in forest testimonies held in local archives.

56 AGS, Marina, leg. 554, Palma, July 11, 1748, Don Pedro de Hordeñana.

trees for about four frigates of fifty guns. Hordeñana's crew visited the forests of 37 cities and towns. For each town, they recorded information on the age and suitability for naval shipbuilding of nine different tree species. They gave the distance from the towns to the nearest ports, and commented on the ease or difficulty of river transport. Most forests were privately owned. Hordeñana recommended that Mallorcans could increase the number of poplar, walnut, and ash trees that grew in the meadows and dry lands by planting more nearer the towns, which generally had land that was more fertile and better irrigated. He said chestnut trees did not grow well in Mallorca, and there were few oak trees, but they could grow more holm oaks with better assistance. The meticulousness of Hordeñana's report let the crown know exactly how few chestnut trees grew on Mallorca (2) and how many pine trees (4,708,447).

While other naval ministers in Spain struggled to elicit forest testimonies during the first year after the ordinance's issue, Hordeñana provided an exemplary report on Mallorca, even adding a report on two other Balearic Islands, Ibiza and Formentera, in October.[57] The crown approved and granted the 2,276 *reales* and 16 *maravedis* in expenses for the visitations, with extra gratitude. Pedro Antonio de Hordeñana received special praise from Ensenada and the king the following May for accomplishing an impressive task so soon after the issue of the ordinance.[58] His was the only major naval *visita* submitted in 1748.[59]

Soon, other reports arrived at the departmental capital of Cartagena. In September of 1749, Francisco Barrero submitted an *estado* of forests from 74 towns in Valencia, 30 towns in Murcia, and 25 towns in Granada.[60] The Valencian towns were accessible by the Guadalaviar and Júcar Rivers, while the other locations were accessible by the Segura River. In Valencia, as in Mallorca, pines were the most common species, and totaled 2,907,427. From the population figures of each Valencian town he visited, Barrero calculated the next year's planting quotas. He calculated that the 36,381 people he counted in the province of Valencia had to plant 109,143 new trees by the next inspection; the 28,404 people in Murcia had to plant 85,212 new trees; and the 13,943 people in Granada had to plant 41,829 new trees.

Early the following year, Gabriel López added a report on 14 more towns in Valencia, 9 in Murcia and 2 in Granada to Barrero's earlier work.[61]

57 AGS, Marina, leg. 554, Palma, October 9, 1748, Don Pedro Hordeñana.

58 AGS, Marina, leg. 555, May 12, 1749.

59 On November 5, 1748, the crown grew concerned, and demanded the completion of more reports. AGS, Marina, leg. 554, Madrid, November 5, 1748.

60 AGS, Marina, leg. 572, September 20, 1749, "El Barrero, sobre visitas de montes."

61 AGS, Marina, leg. 556, Cartagena, February 25, 1750, Don Francisco Barrero.

The Valencian inspections took place in and around the northern town of Morella, not far from Antonio de Alçatte's inspection of 1589, and were eight to ten leagues from the nearest port, Vinaroz. López's report also provided information on the number of new trees to be planted in each town, based on their populations.

In 1751, Juan Francisco de la Torre submitted the most ambitious *estado de montes* to date.[62] He inspected the forests of 84 towns in Murcia and Granada. In addition to providing the population size and expected new plantings from each town, the report included detailed geographic information on forest quality and accessibility. Where possible, he divided each of ten tree species into the categories of "new," "growing," and "old." He even added a fourth category, "marked," to indicate specific trees that were marked as suitable for naval shipbuilding. De la Torre counted 601,806,209 trees, of which pines amounted to 513,313,072. His brief descriptions of each of the 84 locations provided more information. Some examples include:

> *Pliego* – [Its size is] one-half league, north-to-south and one-quarter league, east-to-west. [It is] 14 leagues to [the nearest port of] Cartagena... This town has two water sources that irrigate its huerta (irrigated plain) Its terrain is not the most advantageous, but pines in the forest will be of some utility. Pruning techniques continue to be beneficial.
>
> *Almansa* – holm oak is the dominant species. 4 leagues north-south, 5½ leagues east-west. 12 leagues from Alicante. This territory promises the growth of trees because of the high quality of the land. One can find hemp growing there. The small forests contain many decrepit pine trees. There is no river of consequence, nor more water than an abundant spring and an artificial reservoir, but the flat land allows for easy wood transportation to the indicated port.
>
> *Albacete* – holm oak is the dominant species. 6 leagues north-south, 4 leagues east-west. 22 leagues from Alicante. The natural humidity of this area is good for growing poplars. It does not have mountain forests (*montes*), but its plains produce pines and holm oaks. The Júcar River irrigates part of the land and can only be useful for transport if it is made navigable.
>
> *Murcia* – holm oak and walnut are the dominant species. 7 leagues north-south, 7 leagues east-west. 9 leagues from Cartagena. The Segura River fertilizes the expansive huerta of this city with its abundant waters

62 AGS, Marina, leg. 572, Cádiz, March 31, 1751, El Intendente de Cartagena, "Yncluye estado de los montes de Murcia y Granada."

> to produce all types of trees, and they can grow without damage to the hemp harvests. In its few forests (*montes*), there are pines but they are not currently usable.
>
> *Segura de la Sierra* – holm oak is the dominant species. 12 leagues north-south, 10 leagues east-west. 2 leagues to the Segura River. The Segura River has its source in this region, on which one can transport the white pines that grow with superior quality and size. Roads need to be built, but at small cost.
>
> *Siles* – oak is the dominant species. 1½ leagues north-south, 1 league east-west. 38 leagues to Cartagena. It is a short distance from this town to the Guadalimar River, which irrigates its huerta... The terrain of its forests is suitable for pines and oaks, which will be serviceable in a short time.
>
> *Cartagena* – oak, holm oak, and walnut are the dominant species. Poplars grow around six small springs nearby. The land nearest the city is salty. The few pines that are in its forests are useless at present, but people there have sown many seeds, which are already growing. They have planted 40,000 small trees so that in the future they will not be lacking.

These descriptions reveal that most locations had both positive and negative aspects for the navy. While many forests had some trees that were immediately useful, the inspections indicated that towns were taking steps to increase the potential timber supplies where they were lacking. Also in 1751, Gabriel López Peña and José Marco y Espejo inspected more towns near Morella and Peñíscola.[63] Their report included information on the population of each town, expected new plantings, distance to the port of Vinaroz, and numbers of seven different tree species. Some 304,530 trees grew in the area, and the *visitadores* calculated that the citizens needed to plant 118,733 more by the following year.

Finally, there was a forest inspection report from the *corregimientos* of Catalonia, including Barcelona, Mataró, Vigué, Manresa, Comprendon, Gerona, Talarn, and Tortosa.[64] The report did not divide the trees into categories of "new," "growing," and "old," but included ten different tree species. The 810,021 pines amounted to 82% of the trees counted (Tables 3 and 4).

63 AGS, Marina, leg. 572, Cartagena, February 24, 1751, "Estado en que manifiesta el número de árboles...en diferentes pueblos de las governaciones de Morella y Peñíscola del Reino de Valencia."

64 Bauer, *Montes*, 139–141.

TABLE 3 *Department of Cartagena forest reports, 1748–1751*

Province (districts inspected)	Year	Trees counted	Dominant species	Number of dominant species
Mallorca (37)	1748	7,186,710	Pine	4,708,447
Ibiza and Formentera	1749	2,419,227	Pine	2,418,503
Valencia (74), Murcia (30), Granada (25)	1749		Pine (Valencia)	2,907,427
Valencia (14), Murcia (9), Granada (2)	1750	1,121,577		
Murcia and Granada (84)	1751	601,806,209	Pine	513,313,072
Morella and Peñíscola	1751	304,530		
Catalonia (8)	1751	993,416	Pine	810,021

SOURCES: "ESTADOS DE MONTES," AGS, MARINA, LEG. 554, 556, AND 572. ERICH BAUER MANDERSCHEID, *LOS MONTES DE ESPAÑA EN LA HISTORIA* (MADRID: FUNDACIÓN CONDE DE VALLE DE SALAZAR, 2003), 141–142.

TABLE 4 *Department of Cartagena forest report, 1749*

Province	Population	Planting quota
Valencia	36,381	109,143
Murcia	28,404	85,212
Granada	13,943	41,829
Total	78,728	236,184

SOURCES: "ESTADOS DE MONTES," AGS, MARINA, LEG. 572.

The Department of Ferrol generated its first reports in 1749. José Colossia inspected forests in 39 towns in the western region of Asturias.[65] In this region, oaks were most abundant, making up about 63% of the 1,619,602 trees counted,

65 AGS, Marina, leg. 572, January 30, 1749, Don Joseph Colossia, "Estado del en que se hallaba la visita de montes de Asturias."

with no pines mentioned. Francisco Xavier García y Sarmiento, *capitán* and *subdelegado* of the navy, led a visitation of Tuy, in southwestern Galicia at the Portuguese border.[66] He brought Juan Francisco González as his scribe and Francisco Rodríguez as his carpenter. They collected data on town populations; on royal, communal, and private property holdings; on oak, chestnut and pine trees; and on different shapes of timber necessary for shipbuilding. They counted 94,541 trees, of which oaks made up over 94%. After gathering all of the data, they made a comparison to a forest inspection of the same area from 1737 by Benito Salgado. They found that every species had increased in number over the past decade, oaks by about 3%, pines by about 43%, and chestnuts by about 57%.

José de las Cuevas inspected forests in the Valley of Mena, in the mountains between Bilbao and Santander, during September and October of 1751.[67] Holm oaks were most abundant, with 164,438 of the 288,078 trees counted, or about 57%. Other oak trees made up 27% and were broken down into categories of "new," "growing," "serviceable," and "old." Of the 76,132 oak trees, de las Cuevas deemed only 1,547 serviceable, about 2% (Table 5).

Forest inspectors also produced several visitation reports based on the 1748 ordinance for the Department of Cádiz. In 1749, Tomás de Guerra counted trees in twenty towns from the province of Huelva.[68] He counted 3,871,201 trees, 76% of which were pine trees. Manuel Samaniego and others

TABLE 5 *Department of Ferrol forest reports, 1749–1751*

Province (districts inspected)	Year	Trees counted	Dominant species	Number of dominant species
Asturias (39)	1749	1,619,602	Oak	1,020,349
Tuy, Galicia	1749	94,541	Oak	88,869
Valle de Mena, Cantabria	1751	288,078	Holm oak	164,438

SOURCES: "ESTADOS DE MONTES," AGS, MARINA, LEG. 572. VALLE DE MENA: ERICH BAUER MANDERSCHEID, *LOS MONTES DE ESPAÑA EN LA HISTORIA* (MADRID: FUNDACIÓN CONDE DE VALLE DE SALAZAR, 2003), 145.

66 AGS, Marina, leg. 572, March 17, 1749, Don Francisco Xavier García y Sarmiento, "Provincia de Tuy – Demarcaciones, extensión, y distancias de sus montes y dehesas."

67 Bauer, *Montes*, 145.

68 Bauer, *Montes*, 128–129.

gathered forest inspections from nearly a dozen towns in the province of Tarifa.[69] These inspections reported on population, forest sizes, trees categorized by three ages and nine species, road and river transport conditions, and the distances to the nearest port. Out of 7,575,116 trees, 60% were cork oaks. In a report on 39 locations near Seville from 1750, Luís Milhau inspected a diverse group of 21 tree species, gave distances to Seville or other nearby ports, and counted 3,369,446 trees.[70]

In early 1751, José Gutiérrez de Rubalcava completed an inspection of lands held by the house of Medina Sidonia, known as the Coto de Doñana.[71] For each of twenty locations visited, Gutiérrez de Rubalcava reported the ages of trees and included a category of "construcción" for those trees suitable for naval shipbuilding. He recognized the strategic location of the Coto de Doñana, which was on the coast and within a short distance of the royal arsenal of La Carraca in the Bay of Cádiz. He specifically cited article number 67 from the naval forest ordinance of 1748 to promote planting pines and poplars. The article stated,

> In the plain of the Kingdom of Seville there are no other trees useful to the service of the navy than pines and poplars, whose growth, conservation, and augmentation should be guarded and promoted along the banks of the Guadalquivir. I order that the justices and hacienda owners in places along the river from Villanueva de Úbeda to Sanlúcar de Barrameda plant as many pines and poplars as the land can handle.

Gutiérrez de Rubalcava also visited the forests of the Sierra de Segura later that year, one of Spain's most well-forested regions.[72] He counted 22,264,991 new, growing, and old trees of eight different species. By the mid 1750s, some regions could start reporting on the effects the ordinance had on local forests. From the province of Jeréz de la Frontera, Francisco de Sandoval submitted totals from inspections in 1751 and 1754, revealing increases in the "new," "growing," and "old" tree age categories.[73]

---

69 AGS, Marina, leg. 572, Partido de Tarifa, no date.

70 AGS, Marina, leg. 556, Cazalla de la Sierra, July 10, 1750, Milhau, "Con un estado de arboles reconocidas en varios lugares, partido de Sevilla."

71 AGS, Marina, leg. 557, Almonte, February 13, 1751, Don Joseph Gutiérrez de Rubalcava.

72 AGS, Marina, leg. 557, August 14, 1751, Don Joseph Rubalcava.

73 AGS, Marina, leg. 559, "Estado de los Montes y Árboles de la Provincia de Xerez de la Frontera, resultánte por las dos visitas que practicó Don Francisco Fernández de Sandoval."

Perhaps the most impressive document of all the forest inspections made during this period was an assemblage of information from major provinces of the Department of Cádiz from 1753.[74] Álvaro Bermúdez authored the final set of tables and submitted it to the king on Christmas Day. José Gutiérrez de Rubalcava submitted the information from the Sierra de Segura. He reported the region as having 33 towns, 136 *fanegas* of plantations, 481 forests, 8 rivers, and 2,121,051 usable trees. Similar information came from other naval ministers working in Málaga, Tarifa, Jérez, Cádiz, Sanlúcar, Seville, and Ayamonte (Figure 11). In October of 1753, the crown had asked the Department of Cádiz to submit information regarding the employment of the 1748 forest conservation ordinance and on the 1748 ordinance regarding the matriculation of men suitable for naval service.[75] The ministers responsible for the report on matriculation submitted their report in November, placing greater pressure on the naval ministers to complete their report by the end of the year. Indeed, they submitted the final version without a full report from Sanlúcar and without an active head minister in Seville. Despite these shortcomings, there was no comparable report made during this period from the other two departments of Ferrol and Cartagena (Tables 6 and 7).

Other Reforms and Ensenada's Fall from Power, 1749–1754

As noted earlier, the forest legislation of 1748 was part of a broader program of government reform that included changes in the sectors of finance, naval technology, and administration. Like the forest legislation, these reforms enhanced state control over its resources. Ensenada's financial reforms included a proposal for a single tax based on income to replace the traditional taxes based on consumer goods and services. A series of royal orders in October of 1749 eliminated the *alcabalas*, *cientos*, and *millones* taxes; eliminated private collection by tax farmers; and initiated an ambitious project to collect information from every household in Spain, known as the *Catastro* (cadastral survey) of Ensenada.

This information was supposed to go towards determining rates of taxation, but by 1754 the project had failed. Such a revision of the fundamental tax structure of Spain would clearly have altered aspects of the social order and, as a result, the proposal met with strong resistance from the nobility and clergy. There is an interesting parallel with Felipe II's questionnaires of the 1570s; both

74 AGS, Marina, leg. 558, December 18, 1753, "Estado de Montes, Departamento de Cádiz."
75 AGS, Marina, leg. 558, December 25, 1753, Alvaro Bermúdez.

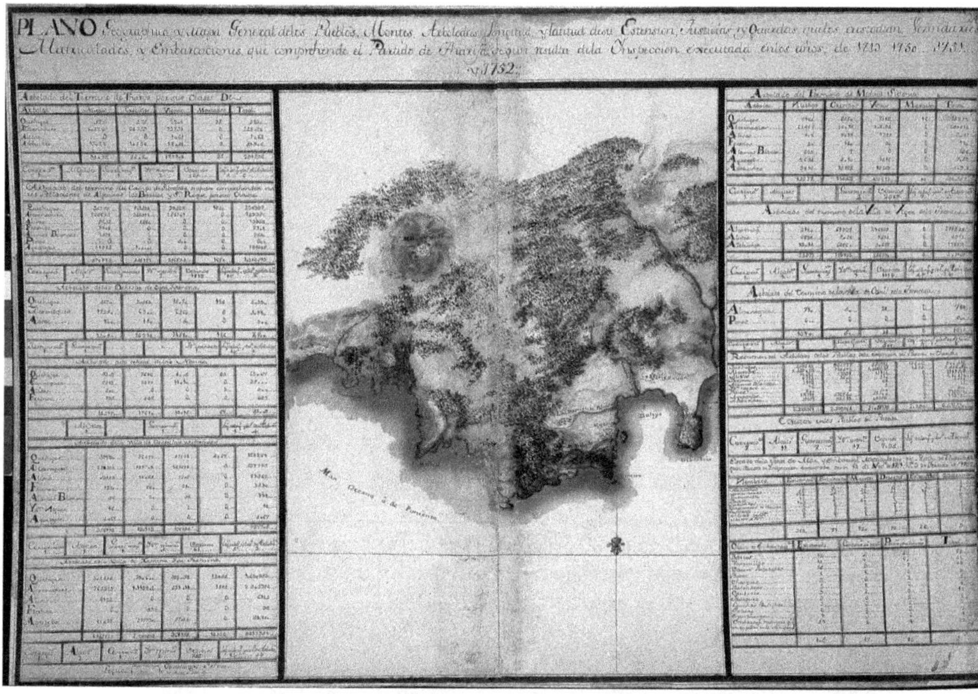

FIGURE 11 *Forest inspection of Tarifa, 1749–52 – The information gathered for this chart and published in the Marqués de la Victoria's* Dictionary of Naval Construction *came from inspections executed in 1749, 1750, 1751, and 1752 in the naval district of Tarifa, the southernmost part of the Iberian Peninsula. The map and tables identify the towns, forests, and groves in the region, the longitude and latitude of their extension, the numbers of justices and guards who protected the forests in each town, and the populations of each town*

SOURCE: ESPAÑA. MINISTERIO DE DEFENSA. ARCHIVO DEL MUSEO NAVAL. DICCIONARIO DE CONSTRUCCIÓN NAVAL DEL MARQUÉS DE LA VICTORIA. AMN. PL 235-1, HOJA 24. MAPA DE LOS MONTES Y ARBOLEDAS QUE TIENEN LOS TERRITORIES, DESDE CONIL Y JIMENA Y DESDE MEDINA SIDONIA HASTA TARIFA Y MONTES DE GIBRALTAR.

had a small impact on government, but they now provide the modern historian with a wealth of contemporary social, cultural, economic, and geographic information. Together, they reveal how little the government's needs had changed, because both projects aimed to gather similarly detailed information about the monarchy's resources for taxation and other purposes. They also show how difficult it was to achieve social and economic reform on a large scale at the time. Other financial reforms by Ensenada proved more effective, such as the *Giro Real* (1751), which was inspired by the Bank of England, and it

TABLE 6 *Department of Cádiz forest reports, 1749–1753*

Province (districts inspected)	Year	Trees counted	Dominant species	Number of dominant species
Huelva (20)	1749	3,871,201	Pine	2,942,113
Tarifa (~12)	?	7,575,116	Cork oak	4,545,070
Sevilla (39)	1750	3,369,446		
Coto de Donana (20)	1751			
Sierra de Segura	1751	22,264,991		
Jerez de la Frontera	1751, 1754			
Department-wide (5 provinces)	1753	79,484,465	"Usable mix"	2,161,511

SOURCES: "ESTADOS DE MONTES," AGS, MARINA, LEG. 556, 557, 558, 559, 572. HUELVA: ERICH BAUER MANDERSCHEID, *LOS MONTES DE ESPAÑA EN LA HISTORIA* (MADRID: FUNDACIÓN CONDE DE VALLE DE SALAZAR, 2003), 128–129.

TABLE 7 *Department of Cádiz, Department-wide forest report, 1753*

Province (districts inspected)	Trees counted
Segura (33)	13,264,991
Malaga (310)	27,793,607
Tarifa (10)	6,576,604
Jerez (9)	28,098,707
Cadiz (3)	3,750,556
Sevilla	0
Ayamonte	0
Total	79,484,465

SOURCES: "ESTADOS DE MONTES," AGS, MARINA, LEG. 558.

ensured that the state managed foreign exchange transactions, rather than leaving them in private hands.[76] Ensenada also enacted stricter regulation of American trade and cut down on fraudulent reports of treasure remittance in

76 Lynch, *Bourbon Spain, 1700–1808*, 170–172.

1749 and 1750. A concordat with Rome in 1753 increased royal power over the resources and finances of the Spanish Church as well.[77]

To improve the navy, the crown issued legislation on naval registration and arsenal construction, in addition to forest conservation. In 1749, Ensenada sent Jorge Juan to England and Antonio de Ulloa to France and other parts of continental Europe to spy on foreign dockyards to improve Spanish knowledge of the latest technology used in shipbuilding.[78] Juan and Ulloa had achieved fame by participating in a scientific voyage to South America from 1735 to 1743.[79] Both were knowledgeable and trustworthy servants of Spain, and after spending well over a year collecting shipbuilding information, they returned in 1750 with blueprints, secret information, and even foreign experts and technicians. Over the next several years, they employed this information and the foreign experts hired to help in the design and construction of state-of-the-art arsenals in the navy's departmental capitals of El Ferrol, Cádiz, and Cartagena. Reasons for changing state forestry administration in the eighteenth century came from a combination of technological and political developments that required new state-of-the-art shipbuilding facilities in well-protected harbors. Spain's older shipyards were not equipped to build and arm the latest types of vessels. With the technical help of Juan and Ulloa, these new vessels would be built in the new arsenals. In short, the Bourbons did not take over a decayed system of forest management left by the later Habsburgs. Rather, developments particular to the first half of the eighteenth century caused many of the old methods to change. In the 1750s, while the new shipbuilding centers were under construction and not yet operating at full strength, the arsenals of Guarnizo, near Santander in northern Spain, and Havana, Cuba, continued to be productive.

In 1754, Ensenada fell from power. Carvajal died in April that year, alarming a powerful pro-English cohort of the possibility that Ensenada's policies would move forward unchallenged. The navy by this time had already shown significant signs of growth, adding to the concerns of British supporters. Spain's fleet

---

77 Lynch, *Bourbon Spain, 1700–1808*, 171–173. On the crown's attempts to control the church in the eighteenth century, see Andrea J. Smidt, "Enlightened Absolutism and New Frontiers for Political Authority: Building Towards a State Religion in Eighteenth-Century Spain," in Tonio Andrade and William Reger, eds. *The Limits of Empire: European Imperial Formations in Early Modern World History, Essays in Honor of Geoffrey Parker* (Farnham, UK: Ashgate, 2012), 33–57.

78 Ozanam, "La política exterior de España en tiempo de Felipe V y de Fernando VI," 465–466.

79 Kamen, *Philip V*, 234; Larrie D. Ferreiro, *Measure of the Earth: The Enlightenment Expedition that Changed the World* (New York: Basic Books, 2011).

had grown from about a dozen vessels at the time of peace in 1748 to forty-five warships in 1754.[80] A well-orchestrated conspiracy led by the British ambassador Benjamin Keene, the duke of Huéscar, and Ricardo Wall led to Ensenada's removal by the summer of 1754. They also made sure that Ensenada's supporters were purged from the government, creating a new bureaucratic regime under Fernando VI. As a consequence, many of Ensenada's reforms were put on hold, and ship construction slowed for the next five years.[81]

Despite this setback, Ensenada had laid the foundation for naval revival between 1748 and 1754 by enabling greater access to primary materials, gathering technical experts, and employing the latest technical knowledge to build new arsenals for an eighteenth-century navy. Ensenada had also improved the state's finances so that by the end of Fernando VI's reign in 1759, Spain had a larger financial surplus than ever before. As a result, Carlos III (r. 1759–1788) inherited both stable finances and a relatively strong navy.

Accomplishments of Fernando VI's Reign (1746–1759)

Forest legislation had never been directed to specific provinces of the navy along all Spanish coastlines and under one unified system of management before the 1748 forest conservation ordinances. The Marqués de la Ensenada wanted to produce a forestry code that made the growth, conservation, and extraction of naval timber systematic, repeatable, and centrally controlled. Above all, Ensenada aimed to enhance Spanish power, and he looked to specifically Spanish methods to meet the demands for domestic timber. The crown and its forest communities developed forest regulations that fit into a variety of environmental, political, and social contexts. The ordinance defined its policies based on historic precedent, *in situ* information-gathering, and reports from local inspections all over Spain.

The great challenge for the crown and its bureaucracy in the decade after 1748 was the proper enforcement of the ordinances through forest inspections and full reporting. With the system in place, and with a clear legal framework backing up the activity of forest bureaucrats, the crown could expect very specific data on forest conditions from all over Spain at regular annual intervals. The forest reports created in the decade after 1748 made the inspections by

80 Lynch, *Bourbon Spain, 1700–1808*, 178.

81 Ozanam, "La política exterior de España en tiempo de Felipe V y de Fernando VI," 466. Here is a list of *buques* built (per year): 15 (1754), 13 (1755), 8 (1756), 3 (1757), 3 (1758), 2 (1759).

Valdés (1737–39) and the map by Navarro (1740) appear rudimentary, particularly in comparison to the extensive detail found in Juan Francisco de la Torre's work in the Department of Cartagena or Álvaro Bermúdez's compilation of information from the provinces in the Department of Cádiz. The numeric and descriptive reports de la Torre gathered from Murcia and Granada were more complete than anything the crown had accumulated on its forests up to that point. Overall, each report was slightly different, but each produced information that was more detailed and more thorough than forest inspections produced in the 1730s. Habsburg forest superintendents had produced similarly detailed reports in the early seventeenth century, but these covered much smaller areas.[82] The crown still had the most to learn from the departments of Cádiz and Cartagena, and that was exactly where most of the new information came from, because the forests of northern Spain in the department of El Ferrol were more familiar. As the culmination of forest territorialization in the early modern era, the improved inspection methods and guidelines supplied by the 1748 ordinance resulted in the state's firmer control of Spain's forests.

Fernando VI's reign tends to be overshadowed by the successful establishment of the Bourbon line during the lifetime of his father Felipe V and the enlightened reforms during the reign of his half-brother and successor, Carlos III. Both Felipe V and Carlos III had long and eventful reigns, punctuated by major foreign wars. By contrast, the reign of Fernando VI was generally peaceful, and historians of his kingship often focus either on the cultural life at court or on the king's personal battles with incapacitating depression. Nonetheless, the reign of Fernando VI marked an important period of transition in the crown's management of natural resources. Forest conservation proved to be a major part of state-building projects, spearheaded by the Marqués de la Ensenada. The patient years of administrative reform, and the laborious, repetitive tree-counting presented above contributed to a firmer foundation of information and material networks on which the naval expansion planned by Ensenada could eventually be built. Ensenada's primary legacy was his naval program, which depended on accessible supplies of shipbuilding timber.

Carlos III would build upon Ensenada's foundation during his reign. The new king came to Madrid in 1759 with over twenty-five years of governing experience in Italy as duke of Parma (1732–1734) and king of Naples and Sicily (1734–1759). He decided very early on to end Fernando VI's policy of Spanish neutrality in the Seven Years' War, then raging in Europe. A Bourbon family compact signed with France in 1761 led to Spain's entry into the war; for which

82 Goodman, 105–107.

the country was ill-prepared. Spain had forty-nine ships of the line (vessels with between 112 and 58 guns) and twenty-one frigates in 1761, but, by then the conflict had already turned decisively in England's favor. Spain went down in defeat, along with its French Bourbon ally.

Three years after the defeat in 1763, a series of economic reforms put forward by the Marqués de Esquilache set off riots in Madrid, known as the *Motín de Esquilache* (1766). While Ensenada was allowed to return to Madrid and work on financial reforms under Carlos III, he never achieved the level of power he had held under Fernando VI before 1754. In fact, he was exiled again in 1766 after several members of government expressed their suspicion of his role in the popular uprising in Madrid. Nonetheless, the naval reforms had taken on a life of their own. Continued government commitment to Spanish naval power after Ensenada's second exile led to a full recovery from the Seven Years' War by 1770. The navy performed well enough in the War of American Independence (1775–1783), fighting on the side of the American rebels, to recover earlier territorial losses, including Florida and Menorca from Great Britain. The enlarged navy reinforced the defense of Spain's American possessions, and even served as a tool of foreign policy by deterring enemy activity in the Americas.

It is no exaggeration to say that the territorialization of Spain's forests enabled the state to procure enough timber to create the second largest navy in Europe by 1783. That year, the navy had sixty-seven ships of the line and thirty-two frigates. Eighteenth-century Spanish naval strength would peak in about 1792, when it had eighty ships of the line and fourteen frigates.[83] Not only was the navy building more ships, it was building ships that were able to serve longer. Vessels made between 1714 and 1724 had an average lifespan of 12.6 years. Between 1725 and 1749, vessels were in use for an average of 14.7 years. Between 1750 and 1774, they were in service for an average of 31.6 years.[84]

There would be additions and revisions made to Spain's forest ordinance of January 1748 over the next several decades, but it would remain the official set of regulations guiding naval forestry through the eighteenth and into the nineteenth century. Forest inspections continued to take place in the later eighteenth century, and they covered more ground and gave more detailed reports than during Fernando VI's reign. Whereas one inspection of the Sierra de Segura yielded information on some 22 million trees in 1751, a more

83 Lynch, *Bourbon Spain, 1700–1808*, 315.

84 Ozanam, "La política exterior de España en tiempo de Felipe V y de Fernando VI," 497.

thorough inspection in 1785 counted over 260 million trees.[85] The 1748 ordinance was completely replaced only by the general forest ordinance of 1833. The nineteenth century regulations spelled the end of Spanish naval forestry, because the Department of the Interior would assume management of Spain's forests from then on. The navy that grew after 1750 became almost as impressive as Ensenada had hoped, and it owed a great deal to his commitment to creating an enforceable national forestry code. Although the ordinance of 1748 was based on historic precedent, in many ways it pushed beyond past efforts to reinforce the state's territorialization of the forests of Spain.

85 Bauer, *Montes*, 127. In comparison to a century earlier, the actual total number of trees in the Sierra de Segura had probably decreased due to the high rate of extraction for naval shipbuilding and the sparse population available to replant.

General Conclusion

The Argentine writers, Jorge Luís Borges and Adolfo Bioy Casares wrote a one-paragraph short story, titled *"Del Rigor en la Ciencia"* ("On Exactitude in Science"), about an empire that developed the science of cartography to the point where it eventually made a map that was exactly the size of the empire.

> ...In that Empire, the craft of Cartography attained such Perfection that the Map of a Single province covered the space of an entire City, and the Map of the Empire itself an entire Province. In the course of Time, these Extensive maps were found somehow wanting, and so the College of Cartographers evolved a Map of the Empire that was of the same Scale as the Empire and that coincided with it point for point. Less attentive to the Study of Cartography, succeeding Generations came to judge a map of such Magnitude cumbersome, and, not without Irreverence, they abandoned it to the Rigours of sun and Rain. In the western Deserts, tattered Fragments of the Map are still to be found, Sheltering an occasional Beast or beggar; in the whole Nation, no other relic is left of the Discipline of Geography.[1]

Borges and Bioy Casares wrote their spoof as a literary forgery, falsely crediting it as a quotation from "Suarez Miranda, Viajes de varones prudentes, Libro IV, Cap. XLV, Lérida, 1658." While creating a map the size of the empire is a fanciful idea, the first part of the story describes an empire's desire to know ultimately everything about the geography of its possessions. It suggests how state power and geographic knowledge reinforced one another. Indeed, during the early modern period, the Spanish crown explored forests in its territory, gathered information about them, mapped them, and described them in order to control them better. The crown's efforts to gain and maintain access to shipbuilding timber unfolded through a process of internal territorialization, which involved an accumulation of pertinent geographic knowledge and required the deployment of state agents who communicated and enforced royal policies through the power of law backed by the strength of arms. As a result, internal territorialization became essential to the process of establishing and performing state power.

---

1 Jorge Luís Borges and Adolfo Bioy Casares, "On Exactitude of Science," in Jorge Luís Borges, *A Universal History of Infamy*, tr. Norman Thomas di Giovanni (London: Penguin Books, 1975), 131. Thanks to Bradley Wilson for introducing me to this story.

 | DOI 10.1163/9789004261372_008

The second part of the story by Borges and Bioy suggests that complete territorial mastery is an impossible ideal even for an empire that could create a 1:1 scale map. The long-term viability of the endeavor proved futile and resulted in the decay of geography, the science of the state. The story that unfolded in early modern Spain contained similar difficulties, even if the empire persisted. The crown's efforts to control its forests faced socio-economic, technological, and geographic restrictions. From the crown's point of view, the ability to gain and maintain access to forest resources was conditioned by the topography, vegetation, and hydrology of Spain and its geo-political position *vis-à-vis* other states, continents, and oceans. Just as importantly, the crown had to maintain the social order and not drastically alter traditional forest regulations. State forest conservation in early modern Spain was an issue of balancing the resource needs of the local scale (Guernica oak) and the national or imperial scale (galleon hull). At times, such restrictions forced the crown to look to foreign sources of timber, particularly for masts, but the crown continued to prefer and promote domestic self-sufficiency, aided and supplemented by colonial forest resources. Naval timber had to be easily accessible given the technological limitations of the time, and the crown could not afford to transport wood much farther than a few leagues from the sea or a navigable river, because of the costs of overland transport. While recognizing some of its limitations, the crown remained primarily concerned with achieving its goal of ensuring long-term supplies of shipbuilding timber for its shipyards. Forest territorialization in Spain brought forest resources and communities under closer state regulation for the specific interests of the navy. Even with the crown's new demands on coastal forest communities, the general trend towards deforestation did not abate. Forest loss may have been a major concern of the crown's, but strategies of imperial defense required greater control over forest access, which could be achieved without necessarily stopping the retreat of forests. The extension of a forest-regulation bureaucracy strengthened naval resources, but forest territorialization alone did not turn Spain into a cohesive nation-state. The realities of the state's limitations prevented a uniform treatment of its territory. Imperial expansion motivated a response to the potential crisis of material shortages, while the identification and interpretation of the crisis contributed to the expansion of state power. Spanish forest regulation for primarily naval interests points to potential avenues of exploration into the process of resource identification and control in other states, and over other resources within the early modern world.

Spain's environment posed numerous challenges to the crown's ability to exploit forest resources. Formidable mountains and seasonal climate extremes limited the places and times of year people could extract and transport timber from the forests to the nearest marina. While abundant forests existed throughout Spain in ancient times, settled agriculture had deforested much of the lowlands by the Middle Ages. The Iberian Peninsula's rugged terrain made transportation in and out of the *montes* very difficult. The preferred mode of transportation was by river. However, Spain contains relatively few rivers large enough for dependable transport of timber. Some of the largest rivers, such as the Duero, Tajo and Guadiana, flow through Portugal before reaching the sea, and Portugal was an enemy of Spain for periods during the early modern era. Other important rivers included the Ebro, Turia, Júcar, and Segura, which all flow into the Mediterranean, and the Guadalquivir, which reaches the sea on Spain's southern Atlantic coast. The pattern of rainfall in much of Spain made many of its rivers only seasonably capable of transporting timber. During the hot months of summer, they would become too dry, and seasonal rains had the potential of causing flash flooding. The mountains of the Mediterranean endured harsh winters, and in parts of Spain, such as in the Pyrenees, periods of heavy snowfall made many forests in the *montes* inaccessible.[2] The crown managed forests for specifically naval interests, so certain tree species received special consideration. In general, vessels required sturdier timbers for the hull and more lightweight ones for the superstructure. Many forests in the colonies contained species that were unfamiliar, but the Spanish adapted the resources of these environments to fit the familiar European practices and designs of shipbuilding.

Felipe II's appointment of Cristóbal de Barros as the first superintendent of forests and plantations marked a turning point in the formation of long-term forest management administration, but the reign of Carlos I set important precedents. In the first half of the sixteenth century Carlos I issued the first legislation in Spain that aimed to conserve forests specifically intended for shipbuilding, an important step in the formation of state forestry. The idea of employing a forest superintendent to oversee the forests on the north coast emerged during Carlos I's reign, but it was only during his son's reign that one was appointed. It was also under Carlos I that the idea for systematic collection of socio-economic and geographic information in the empire took shape. Eventually, ministers during Felipe II's reign carried out the task. Together, the reigns of Carlos I and Felipe II witnessed the foundation of state forest territorialization in Spain. No longer was the crown an arbiter of intermunicipal disputes, but actively engaged in identifying specific areas where adequate and accessible timber existed by employing agents of the crown to collect this

2 Fernand Braudel, *The Mediterranean and the Mediterranean World in the Age of Philip II*, tr. Siân Reynolds, (New York: Harper Colophon Books, 1972), 27.

information. Other royal personnel were charged with communicating the interests of the crown through regular forest inspections and by disseminating new legislation. The forest superintendent guided the planting, managing, guarding, cutting, and transporting of timber for the navy. They were also charged with enforcing royal rules and regulations in the stipulated jurisdictions and possessed the authority to fine or punish violators. With closer observation and regular inspections of forests by royal bureaucrats and members of the citizenry, the search for and identification of valuable sources of timber enabled an extension of state power.

Inspection and planting techniques matured under Cristóbal de Barros's successors in the early seventeenth century, though not without conflict. Difficulties securing foreign timber spurred the crown to control its own forests more carefully and to deploy additional reconnaissance missions within Spain to find new stands of suitable timber. Such a system helped Spanish naval power recover after the failed armada against England in 1588 and helped Spain survive the demanding wars between 1618 and 1659. While a similar permanent bureaucracy of forest superintendents did not emerge in the colonies, the Habsburgs established new restrictions in the forests of Cuba and Ecuador to protect naval timber. Forest reconnaissance work occurred in colonial settings as it did in Spain, helping to extend the resource networks of major colonial shipyards.

Spanish state forestry depended on the labor, tools, animals, and land of local forest communities to help the crown locate new stands of timber, start and maintain plantations, prune trees, clear underbrush, defend against animals and fire, cut and shape the timber, load and transport the wood to the naval shipyard, and record all of these activities. Inducing full participation in naval forestry was a challenging task, and the crown employed multiple methods, ranging from persuasive rhetoric to imprisonment in Spain and forced labor drafts or slavery in the colonies. Forest communities did not oppose forest conservation, but they preferred it on their terms. Resistance also came in many forms, including noncompliance, physical violence against crown representatives, and outright rebellion.

For many reasons other than a failure of the timber supply, Spanish naval power waned in the middle of the seventeenth century. However, the crown's commitment to maintaining access to coastal forests and its desire to induce its people to plant more trees continued throughout the second half of the century. This underscores the conclusion that the significance of forest territorialization rests not so much on the sheer quantity of timber that wound up in the ships of the Spanish navy, but rather on the ability of the state to direct human behavior towards achieving state goals ahead of the goals of traditional local land use practices. Landmark forest legislation of this period included Toribio Pérez de Bustamante's Instruction of 1650 and Mariana of Austria's *real cédula* of 1675. Both documents included management techniques that would be used throughout the next century. After the Bourbons ascended to the throne and secured their position by winning the War of Succession, the crown commissioned forest inspections on a national scale for the first time in Spanish history. The most ambitious of which were the explorations of forests in the Sierra de Segura and the Ebro River valley by Juan Valdés y Castro. Such explorations extended Spanish state forestry to new sites and helped in the recovery of Spanish naval power after the War of Succession.

The considerable continuity from one reign to the next also indicates the fruitfulness of studying the Habsburg and Bourbon periods together, rather than ending or beginning a study at the conveniently round date of 1700. Bourbon era forestry had much in common with the state forestry of the late sixteenth and seventeenth centuries. Changes in naval forest administration after 1726 had more to do with new technological demands, such as the need for arsenals in El Ferrol, Cartagena, and La Carraca, and new international conditions requiring greater protection of Spanish coasts, than with the failure of forest management under the later Habsburgs. This continuity becomes clear by examining in detail the 1748 ordinances. At the same time, though, reforms carried out by the navy under the Bourbons spearheaded the technological advacements of the day, including ship designs, innovations in matters of navigation and oceanic sciences, use of mathematics and the systematization of measurements. By the middle of the century, the navy carried out forest inspections more systematically and with greater precision.

To defend and connect its vast empire Spain required broader strategic control from the center. However, the crown eagerly sought knowledge of local conditions and suggestions from its forest superintendents. The experience of people like Barros, Antonio de Alçatte, Agustín de Ojeda, Martín de Vallecilla, Toribio Pérez de Bustamante, and Juan Valdés y Castro indicate that Spanish forestry evolved through a familiarity with local conditions that developed over the entire early modern era. The crown took steps to conserve its forests for naval purposes with a permanent system of observation and inspection long before France and England. Forestry lawmakers, forest inspectors, and forest superintendents in Spain all worked in the interest of the navy, limiting the debilitating conflict of interests that formed between France's navy and treasury and England's navy and Parliament. State territoriality in Spain was unique in its scale and timing, and was remarkable for its continuity, particularly in the northern provinces.

Naval historians are well aware of the resurgence of Spain's presence at sea in the late eighteenth century under Carlos III (r.1759–1788) and into the reign of his successor, Carlos IV (r.1788–1808). The study of that resurgence would be incomplete, however, without recognition of the long process that enhanced state control over forests of interest to the navy. The monarchy's ability to have a predictable supply of timber in the major ports in any given year enhanced the state's ability to plan and organize naval power. That ability had developed during the course of nearly three centuries of territorialization over the forests of Spain.

Bibliography

Abad León, Felipe. *El Marqués de la Ensenada, su vida y su obra*. Madrid: Editorial Naval, 1985.

Aedo Pérez, Carlos. *El bosque en Cantabria*. Santander: Universidad de Cantabria, 1990.

Albion, Robert. *Forests and Sea Power: The Timber Problem of the Royal Navy, 1652–1862*. Cambridge, MA: Harvard University Press, 1926.

Alcalá-Zamora y Queipo de Llano, José. *Historia de una empresa siderúrgica española: Los altos hornos de Liérganes y La Cavada, 1622–1834*. Santander: Centro de Estudios Montañeses, 1974.

Alcalá-Zamora y Queipo de Llano, José. *Altos hornos y poder naval en la España de la edad moderna*. Madrid: Real Academia de la Historia, 1999.

Alfonso Mola, Marina, and Carlos Martínez Shaw. "Los Astilleros de la América Colonial." In *Historia General de la América Latina*, vol. 3, no. 1, edited by Alfredo Castillero Calvo and Allan J. Kuethe, 279–303. Madrid: Trotta, 2000.

Alfonso X, King of Castile and León. *Las Siete Partidas*, translated by S.P. Scott, edited by Robert Ignatius Burns. 5 vols. Philadelphia: University of Pennsylvania Press, 2001.

Allen, Paul. *Philip III and the Pax Hispanica, 1598–1621: The Failure of Grand Strategy*. New Haven: Yale University Press, 2000.

Anderson, Jennifer L. "Nature's Currency: The Atlantic Mahogany Trade and the Commodification of Nature in the Eighteenth Century." *Early American Studies* 2, no. 1 (Spring 2004): 47–80.

Anderson, Jennifer L. *Mahogany: The Costs of Luxury in Early America*. Cambridge, MA: Harvard University Press, 2012.

Appuhn, Karl. "Inventing Nature: Forests, Forestry, and State Power in Renaissance Venice." *The Journal of Modern History*, 72, no. 4 (Dec., 2000): 861–889.

Appuhn, Karl. *A Forest on the Sea: Environmental Expertise in Renaissance Venice*. Baltimore: The Johns Hopkins University Press, 2009.

Aragón Ruano, Álvaro. *El bosque guipuzcoano en la Edad Moderna: Aprovachamiento, ordenamiento legal y conflictividad*. Donostia: Sociedad de Ciencias Aranzadi, 2001.

Aranda y Antón, Gaspar de. "Las maderas de América en la arquitectura naval del siglo XVIII." *Revista de Historia Naval* 10, no. 38 (1992a): 7–32.

Aranda y Antón, Gaspar de. *Los bosques flotantes: Historia de un roble del siglo XVIII*. Madrid: ICONA, 1992b.

Aranda y Antón, Gaspar de. "La construcción naval en Francia durante los siglos XVII y XVIII." *Revista de Historia Naval* 13, no. 49 (1995): 61–76.

Aranda y Antón, Gaspar de. *La Administración Forestal y los Montes de Ultramar durante el Siglo XIX*. Madrid: ICONA, 1995.

Arnold, Ellen F. *Negotiating the Landscape: Environment and Monastic Identity in the Medieval Ardennes*. Philadelphia: University of Pennsylvania Press, 2013.

Astill, Grenville, and Annie Grant, eds. *The Countryside of Medieval England*. Oxford: Basil Blackwell, 1988.

Badal, E. "La recolección de piñas durante la prehistoria en la Cueva de Nerja (Málaga)." In *De neandertales a cromañones. El inicio del poblamientohumano en las tierras valencianas*, edited by V. Villaverde, 101–104. Valencia: Universidad de Valencia, 2001.

Bamford, Paul W. *Forests and French Sea Power, 1660–1789*. Toronto: University of Toronto Press, 1956.

Bankoff, Greg. "'Tree as the Enemy of Man': Changing Attitudes of the Forests of the Philippines, 1565–1898." *Philippine Studies* 52, no. 3 (Sep. 2004): 320–344.

Bankoff, Greg. "One Island too Many: Reappraising the Extent of Deforestation in the Philippines Prior to 1946." *Journal of Historical Geography* 33, no. 2 (Apr. 2007): 314–334.

Bankoff, Greg. "'Deep Forestry': Shapers of the Philippine Forests." *Environmental History* 18, no. 3 (July 2013): 523–556.

Bankoff, Greg, and Peter Boomgaard, eds. *A History of Natural Resources in Asia: The Wealth of Nature*. Basingstoke, UK: Palgrave Macmillan, 2007.

Barney, Stephen A., W.J. Lewis, J.A. Beach, Oliver Berghoff, and Muriel Hill, eds. *The Etymologies of Isidore of Seville*. Cambridge: Cambridge University Press, 2006.

Barreiro Mallón, Baudilio. "Masa arbórea y su producto en Asturias durante la Edad Moderna." In *El medio rural español: Cultura, paisaje y naturaleza*, by J. Cavero, et al. Salamanca: Universidad de Salamanca, 1992.

Barreiro Mallón, Baudilio. "Montes comunales y vida campesina en las regiones cantábricas." *Studia historica. Historia moderna*, no. 16 (1997): 17–56.

Barrera, Antonio. "Empire and Knowledge: Reporting from the New World." *Colonial Latin American Review* 15, no. 1 (Jun. 2006): 39–54.

Barrera-Osorio, Antonio. *Experiencing Nature: The Spanish American Empire and the Early Scientific Revolution*. Austin, TX: University of Texas Press, 2006.

Barton, Simon. *A History of Spain*. New York: Palgrave Macmillan, 2004.

Bauer Manderscheid, Erich. *Los montes de España en la historia*. 3rd ed. Madrid: Fundación Conde de Valle de Salazar, 2003.

Bechmann, Roland. *Trees and Man: The Forest in the Middle Ages*, translated by Katharyn Dunham. New York: Paragon House, 1990.

Bernal Rodríguez, Antonio Miguel. "La tierra comunal en Andalucía durante la Edad Moderna." *Studia historica. Historia moderna*, no. 16 (1997): 106–111.

Béthencourt Massieu, Antonio. "El real astillero de Coatzacoalcos (1720–1735)." *Anuario de Estudios Americanos* 15 (1958): 371–428.

Bordeje y Morencos, Fernando. "El poder marítimo en la concepción política del Marqués de la Ensenada." In *La Marina de la Ilustración: Ciclo de Conferencias, Nov. 1988*, 5–22. Madrid: Instituto de Historia y Cultura Naval, 1989.

Borges, Jorge Luis, and Adolfo Bioy Casares. "On Exactitude of Science." In *A Universal History of Infamy*, by Jorge Luis Borges, translated by Norman Thomas di Giovanni, 131. London: Penguin Books, 1975.

Birrell, Jean R. "The Medieval English Forest." *Journal of Forest History* 24, no. 2 (Apr., 1980): 78–85.

Blair, Emma Helen, and James Alexander Robertson, eds. *The Philippine Islands, 1493–1898: Explorations by Early Navigators, Descriptions of the Islands and Their Peoples, Their History and Records of the Catholic Missions, as Related in Contemporaneous Books and Manuscripts, Showing the Political, Economic, Commercial, and Religious Conditions of Those Islands from Their Earliest Relations with European Nations to the Beginning of the Nineteenth Century*. 55 vols. Cleveland, OH: A.H. Clark Co., 1903–1909.

Boomgaard, Peter. "Exploitation and Management of the Surinam Forests, 1600–1975." In *Changing Tropical Forests: Historical Perspectives on Today's Challenges in Central and South America*, edited by Harold K. Steen and Richard P. Tucker, 252–264. Durham, NC: Duke University Press, 1992.

Boomgaard, Peter. "The VOC Trade in Forest Products in the Seventeenth Century." In *Nature and the Orient: The Environmental History of South and Southeast Asia*, edited by Richard Grove, Vinita Damodaran, and Satpal Sangwan, 375–395. Delhi: Oxford University Press, 1998.

Braudel, Fernand. *The Mediterranean and the Mediterranean World in the Age of Philip II*, translated by Siân Reynolds. 2 vols. New York: Harper Colophon Books, 1972.

Braudel, Fernand. *The Structures of Everyday Life*. Vol. 1 of *Civilization and Capitalism 15th–18th Century*, translated by Siân Reynolds. New York: Harper and Row, 1981.

Bravo Lozano, Jesús. *Montes para Madrid. El abastecimiento de carbón vegetal a la villa y corte entre los siglos XVII y XVIII*. Madrid: Caja de Madrid, D.L., 1993.

Buis, Jaap. *Historia Forestis: Nederlandse bosgeschiedenis*. 2 vols. Utrecht: Landbouwhogeschool-Wageningen, 1985.

Buisseret, David, ed. *Monarchs, Ministers and Maps: The Emergence of Cartography as a Tool of Government in Early Modern Europe*. Chicago: University of Chicago Press, 1992.

Burns, Kathryn. *Into the Archive: Writing and Power in Colonial Peru*. Durham: Duke University Press, 2010.

Butzer, Karl W. "Paleolithic Adaptations and Settlement in Cantabrian Spain." *Advances in World Archaeology* 5 (1986): 201–252.

Butzer, Karl W. "From Columbus to Acosta: Science, Geography and the New World." *Annals of the Association of American Geographers* 82, no. 3 (September 1992): 543–565.

Butzer, Karl W. "Ecology in the Long View: Settlement Histories, Agrosystemic Strategies and Ecological Performance." *Journal of Field Ecology* 23, no. 2 (Summer 1996): 141–150.

Butzer, Karl W., Juan F. Mateu, Elisabeth K. Butzer, and Pavel Kraus. "Irrigation Agrosystems in Eastern Spain: Roman or Islamic Origins?" *Annals of the Association of American Geographers* 75, no. 4 (December 1985), 479–509.

Caffrey, Patrick. "The Forests of Northeast China, 1600–1953: Environment, Politics, and Society." PhD diss., Georgetown University, 2002.

Cano, Tomé. *Arte para fabricar y aparejar naos (1611)*, edited by Enrique Marco Dorta. La Laguna, Canary Islands: Instituto de Estudios Canarios, 1964.

Cappa, Ricardo. *Estudios críticos acerca de la dominación española en América. Tomo X–XII: Industria Naval.* Madrid: Gregorio del Amo, 1894.

Carlé, María del Carmen. "El bosque en la Edad Media (Asturias-Leon-Castilla)." *Cuadernos de historia de España* (Buenos Aires) 59–60 (1976): 297–374.

Carrión, J.S., C. Finlayson, S. Fernández, E. Allué, J.A. López-Sáez, P. López-García, G. Gil-Romera, and P. González-Samperiz. "A Coastal Reservoir of Biodiversity for Upper Pleistocene Human Populations: Palaeoecological Investigations in Gorham's Cave (Gibraltar) in the Context of the Iberian Peninsula." *Quaternary Science Reviews* 27, no. 23–24 (2008): 2118–2135.

Carrión, José S., ed. *Paleoflora y paleovegetación de la peninsula Ibérica e Islas Baleares.* Murcia: Ministerio de Economía y Competitividad-Universidad de Murcia, 2012.

Carrión, José S., Santiago Fernández, Penélope González-Sampériz, Graciela Gil-Romera, Ernestina Badal, Yolanda Carrión-Marco, Lourdes López-Merino, José A. López-Sáez, Elena Fierro, and Francesc Burjachs. "Expected Trends and Surprises in the Lateglacial and Holocene Vegetation History of the Iberian Peninsula and Balearic Islands." *Review of Paleobotany and Palynology* 162 (2010): 458–475.

Casado Soto, José Luis. *Los barcos españolas del siglo XVI y la Gran Armada de 1588.* Madrid: Editorial San Martín, 1988.

Casado Soto, José Luis. "Historia del bosque." In *El bosque en Cantabria*, edited by Carlos Aedo, 35–80. Santander: Universidad de Cantabria, Asamblea Regional de Cantabria, D.L., 1990.

Casado Soto, José Luis. *Barcos y astilleros: La construcción naval en Cantabria.* Santander: Puerto de Santander, 1993.

Casado Soto, José Luis. "Barcos Para la Guerra: soporte de la monarquía hispánica." *Cuadernos de Historia Moderna* 5 (2006): 15–53.

Cipolla, Carlo M. *Guns Sails, and Empires: Technological Innovation and the Early Phases of European Expansion 1400–1700*. London: Minerva Press, 1965.

Clayton, Lawrence. *Caulkers and Carpenters in a New World: The Shipyards of Colonial Guayaquil*. Athens, OH: Ohio University, Center for International Studies, 1980.

Clemente Ramos, Julián, coord. *El medio natural en la España medieval: Actas del I congresso sobre ecohistoria e historia medieval* [Celebrado en Cáceres entre el 29 noviembre y el 1 diciembre, 2000]. Cáceres: Universidad de Extremadura, Servicio de Publicaciones, 2001.

Cobo de Guzmán y Lechuga, Jesús. *Estudio sobre las Ordenanzas de Montes del año 1748 y del expediente sobre el régimen y administración de los montes de Segura de la Sierra y de su provincia marítima de 1811*. Madrid: Caja de Jaén, Obra Socio Cultural, D.L., 1994.

Collins, Roger. *Visigothic Spain, 409–711*. Oxford: Blackwell, 2004.

Colón, Cristóbol. *Cartas y documentos completos*. Madrid: Alianza Universidad, 1984.

Constable, Olivia Remie. *Trade and Traders in Muslim Spain: The Commercial Realignment of the Iberian Peninsula, 900–1500*. Cambridge: Cambridge University Press, 1994.

Cook, Weston F. *The Hundred Years War for Morocco: Gunpowder and the Military Revolution in the Early Modern Muslim World*. Boulder, CO: Westview Press, 1994.

Cornell, Vincent. "Socioeconomic dimensions of reconquista and jihad in Morocco: Portuguese Dukkala and the Sa'id Sus, 1450–1557." *International Journal of Middle East Studies* 22 (1990): 379–418.

Corvol, Andrée. *L'homme aux bois: histoire des relations de l'homme et de la forêt, XVIIe–XXe siècle*. Paris: Fayard, 1987.

Corvol, Andrée, Pierre Chaunu, and Roland Mousnier. *L'homme et l'arbe sous l'Ancien Régime*. Paris: Economics, 1984.

Cowans, Jon, ed. *Early Modern Spain: A Documentary History*. Philadelphia: University of Pennsylvania Press, 2003.

Crespo Rico, Miguel Ángel, José Ramón Cruz Mundet, José Manuel Gomez Lago, and José Ángel Lema Pueyo, *Colección Documental del Archivo Municipal de Mondragón. V.I (1260–1400)*. San Sebastián: Eusko Ikaskuntza, 1992.

Cristina Penalba, M. "The History of the Holocene Vegetation in Northern Spain from Pollen Analysis." *Journal of Ecology* 82, no. 4 (December 1994): 815–832.

Cronon, William. *Changes in the Land: Indians, Colonists, and the Ecology of New England*. New York: Hill and Wang, 1983.

Crosby, Alfred W. *Ecological Imperialism: The Biological Expansion of Europe, 900–1900*. Cambridge: Cambridge University Press, 1986.

Crosby, Alfred W. *The Measure of Reality: Quantification and Western Society, 1250–1600*. Cambridge: Cambridge University Press, 1997.

Cruz Aguilar, Emilio de la. "La provincia marítima de Segura de la Sierra." *Boletín del Instituto de Estudios Giennenses* 107 (1981): 51–82.

Dean, Warren. *With Broadax and Firebrand: The Destruction of the Brazilian Atlantic Forest*. Berkeley: University of California Press, 1995.

Díaz González, Francisco Javier. *La Real Junta de Obras y Bosques en la época de los Austrias*. Madrid: Dykinson, 2002.

Díaz Ordóñez, Manuel. *Amarrados al Negocio: Reformismo Borbónico y suministro de Jarcia para la Armada Real (1675–1751)*. Madrid: Ministerio de Defensa, 2009.

Dickinson, William R. "Changing Times: The Holocene Legacy." *Environmental History* 5, no. 4 (2000): 483–502.

Diego Liaño, C., and J.C. García Cordón. "La corona y los pueblos en la explotación de los montes de Cantabria: Deforestación y gestión del bosque en la segunda mitad del siglo XVIII." *Cuadernos de la Sociedad Española de Ciencias Forestales*, no. 16 (2003), 215–220.

Disney, A.R. *The Portuguese Empire*. Vol. 2 of *A History of Portugal and the Portuguese Empire: From Beginnings to 1807*. Cambridge: Cambridge University Press, 2009.

Domergue, Claude. *Les mines de la Péninsule Ibérique dans l'antiquité romaine*. Rome: École Française de Rome, 1990.

Dursun, Selçuk. "Forest and the State: History of Forestry and Forest Administration in the Ottoman Empire." PhD diss., Sabanci University, 2007.

Edwards, John. *The Spain of the Catholic Monarchs, 1474–1520*. Oxford: Blackwell, 2000.

Eliasson, Per, and Sven G. Nilsson, "'You Should Hate Young Oaks and Young Noblemen': The Environmental History of Oaks in Eighteenth- and Nineteenth-Century Sweden." *Environmental History* 7, no. 4 (Oct. 2002): 659–674.

Elliott, J.H. *The Count-Duke Olivares: The Statesman in an Age of Decline*. New Haven: Yale University Press, 1986.

Elliott, J.H. *Spain and Its World, 1500–1700*. New Haven: Yale University Press, 1989.

Elvin, Mark. *The Retreat of the Elephants: An Environmental History of China*. New Haven: Yale University Press, 2004.

Elvin, Mark, and Liu Tsui-jung, eds. *Sediments of Time: Environment and Society in Chinese History*. Cambridge: Cambridge University Press, 1998.

Expediente sobre el régimen y administración de los montes de Segura de la Sierra y de su provincia. Madrid: Imprenta de Miguel de Burgos, 1825, originally 1811.

Fernández-Armesto, Felipe. *Ferdinand and Isabella*. New York: Dorset Press, 1975.

Fernández-Armesto, Felipe. *The Spanish Armada: The Experience of War in 1588*. Oxford: Oxford University Press, 1988.

Fernández Castro, María Cruz. *Iberia in Prehistory*. Oxford: Blackwell, 1995.

Fernández de Navarrete, Martín. *Biblioteca Marítima Española*. Madrid, 1851.

Fernández Duro, Cesáreo. *La armada invencible*. Madrid: Sucs. De Rivadeneyra, 1884–1885.

Fernández-González, Francisco. "The Spanish Regulations for Shipbuilding (Ordenanzas) of the Seventeenth Century." *International Journal of Naval History* 8, no. 3 (December 2010): 1–30.

Font Tullot, Inocencio. *Climatología de España y Portugal.* Madrid: Instituto Nacional de Meteorología, 1983.

Fritzbøger, Bo. "Environmental Agency: Forest Histories from Denmark c.1500–2000." In *Transcending Boundaries: Environmental Histories from the Øresund Region*, vol. 9 of *Skrifter med historiska perspektiv*, edited by Fredrik Björk, Per Eliasson, and Bo Poulsen, 74–87. Malmö, Sweden: Malmö University, 2009.

Fuente, Alejandro de la. *Havana and the Atlantic in the Sixteenth Century.* Chapel Hill: University of North Carolina Press, 2008.

Funes Monzote, Reinaldo. *From Rainforest to Cane Field in Cuba: An Environmental History since 1492*, translated by Alex Martin. Chapel Hill: University of North Carolina Press, 2008.

Garayo, J.M. "Deforestación del territorio: El hayedo de los montes de la Parzonería General de Encía (siglos XVIII–XX)." *Agricultura y Sociedad*, no. 62 (1992): 73–109.

García de Castro, Francisco Javier. *La marina de Guerra de Castilla en la Edad Media (1248–1474).* Valladolid: Universidad de Valladolid, 2014.

García Hurtado, Manuel-Reyes, ed. *La Armada española en el siglo XVIII: Ciencia, hombres y barcos.* Madrid: Silex Ediciones, 2012.

Gil, Luis, Pablo Fuentes-Utrilla, Álvaro Soto, M. Teresa Cervera, and Carmen Collada. "English Elm is a 2,000-year-old Roman Clone." *Nature* 431 (October 2004): 1053.

Gil Sánchez, Luis. "La voz 'montes' y la transformación histórica del espacio arbolado," *Cuadernos de la Sociedad Española de Ciencias Forestales* 16 (2003): 19–29.

Gil Sánchez, Luis, and M. Torre Antón, eds. *Atlas forestal de Castilla y León.* 2 vols. Valladolid: Junta de Castilla y León. Consejería de Medio Ambiente, 2007.

Gil Sánchez, Luis. *Pinares y rodenales: La diversidad que no se ve.* Madrid: Real Academia de Ingeniería, 2008.

Gilbert, John M. *Hunting and Hunting Reserves in Medieval Scotland.* Edinburgh: J. Donald, 1979.

Glete, Jan. *War and the State in Early Modern Europe: Spain, the Dutch Republic, and Sweden as Fiscal-Military States, 1500–1660.* London: Routledge, 2002.

Glick, Thomas F. *Islamic and Christian Spain in the Early Middle Ages.* 2nd ed. Leiden: Brill, 2005.

Gómez Rivero, Ricardo. "La superintendencia de construcción naval y fomento forestal en Guipúzcoa." *Anuario de historia del derecho español* 56 (1986), 591–636.

Gómez Urdáñez, José Luis. *El Proyecto Reformista de Ensenada.* Lleida: Milenio, 1996.

González Hernández, José Miguel. "Nature Conservation in Spain." In *Thirteenth Annual Cabrillo Festival Historical Seminar*, 1–15. San Diego, CA: Cabrillo Historical Association, 1984.

González Muñoz, María del Carmen. "Vigo y su comarca en los siglos XVI y XVII." In *Vigo en su historia*, edited by Fernando Acuña Castroviejo, Alvaro Conqueiro, and Xosé María Alvarez Blázquez, 151–276. Vigo: Caja de Ahorros de Municipal de Vigo, 1980.

Goodman, David. *Spanish Naval Power, 1589–1665: Reconstruction and Defeat.* Cambridge: Cambridge University Press, 1997.

Gordi i Serrat, Josep, and Jaume Vilaginés Segura. "La evolución del paisaje forestal de la Llanura del Vallés (Barcelona) hasta el Siglo XII." *Cuadernos de la Sociedad Española de Ciencias Forestales* no. 16 (2003): 167–172.

Graham, Hamish. "For the Needs of the Royal Navy: State Interventions in the Communal Woodlands of the Landes during the Eighteenth Century." *Proceedings of the Western Society for French History* 35 (January 2007): 135–148.

Greene, Jack P., and Philip D. Morgan, eds. *Atlantic History: A Critical Appraisal.* Oxford: Oxford University Press, 2009.

Groome, Helen J. *Historia de la Politica Forestal en el Estado Español.* Madrid: Agencia de Medio Ambiente, 1990.

Grove, Richard. *Green Imperialism: Colonial Expansion, Tropical Island Edens, and the Origins of Environmentalism, 1600–1860.* Cambridge: Cambridge University Press, 1995.

Grove, Richard, Vinita Damodaran, and Satpal Sangwan, eds. *Nature and the Orient: The Environmental History of South and Southeast Asia.* Delhi: Oxford University Press, 1998.

Guilmartin, John F. *Gunpowder and Galleys: Changing Technology and Mediterranean Warfare at Sea in the Sixteenth Century.* Cambridge: Cambridge University Press, 1974.

Harbron, John D. *Trafalgar and the Spanish Navy.* Annapolis: Naval Institute Press, 1988.

Harding, Richard. *The Emergence of Britain's Global Naval Supremacy: The War of 1739–1748.* UK: Suffolk, 2010.

Harrison, Richard. *Spain at the Dawn of History: Iberians, Phoenicians, and Greeks.* New York: Thames and Hudson, 1988.

Hernando Ortego, Francisco Javier. "La lucha por el Monte de El Pardo. Rey, municipio y uso del espacio en el Madrid del Antiguo Régimen." *Cuadernos de Investigación Histórica* 12 (1989): 169–196.

Herrera García, Antonio. "Labradores, ganadores, y aprovechamientos comunales: Algunos de su conflictividad en las tierras sevillanas durante el Antiguo Régimen." *Agricultura y Sociedad* no. 17 (1980): 255–291.

Hopfner, Hellmuth. "La evolución de los bosques de Castilla la Vieja en tiempos históricos: contribución a la investigación del primitivo paisaje de la España central." Translated by M. de Terán. *Estudios Geogáficos* 15, no. 56 (August 1954): 415–430.

Hoskins, W.G. *Making of the English Landscape.* London: Hodder and Stoughton, 1955.

Hunter, Richard, and Andrew Sluyter. "How Incipient Colonies Create Territory: The Textual Surveys of New Spain, 1520s–1620s." *Journal of Historical Geography* 37 no. 3 (July 2011): 288–299.

Imber, Colin. "The Navy of Süleyman the Magnificent." *Archivum Ottomanicum* VI (1980): 211–282.

Iriarte Chiapusso, María José, and Álvaro Arrizabalaga Valbuena. "El bosque en el País Vasco prehistórico." *Cuadernos de la Sociedad Española de Ciencias Forestales*, no. 16 (2003): 85–90.

Iriarte-Goñi, Iñaki, and María Isabel Ayuda. "Wood and Industrialization: Evidence and Hypotheses from the Case of Spain, 1860–1935." *Ecological Economics* 65, no. 1 (March 2008): 177–186.

James, N.D.G. *A History of English Forestry*. Oxford: Blackwell, 1981.

Jaritz, Gerhard, and Verena Winiwarter. "On the Perception of Nature in a Renaissance Society." In *Nature and Society in Historical Context*, edited by Mikulas Teich, Roy Porter and Bo Gustafsson, 91–111. Cambridge: Cambridge University Press, 1997.

Jordan Reyes, Miguel. "La deforestación de la isla de Cuba durante la dominación española, 1492–1898." PhD diss., Universidad Politécnica de Madrid, 2006.

Kamen, Henry. *The War of Succession in Spain 1700–1715*. London and Bloomington, 1969.

Kamen, Henry. *Spain in the Later Seventeenth Century, 1665–1700*. New York: Longman, 1980.

Kamen, Henry. *Philip of Spain*. New Haven: Yale University Press, 1997.

Kamen, Henry. *Philip V of Spain: The King Who Reigned Twice*. New Haven: Yale University Press, 2001.

Kamen, Henry. *Empire: How Spain Became a World Power, 1492–1763*. New York: Perennial, 2002.

Langton, John, and Graham Jones, eds. *Forests and Chases of England and Wales c.1500–c.1850: Towards a Survey and Analysis*. Oxford: St. John's College, 2005.

Larkin, John A. "Philippine History Reconsidered: A Socioeconomic Perspective." *American Historical Review* 87, no. 3 (June 1982): 595–628.

Laso, María-Pilar, and Erich Bauer. "La propiedad forestal en España." *Revista de Estudios Agrosociales*, no. 49 (1964), 7–53.

Luzón Nogué, José María. "Instrumentos mineros de la España Antigua." In *La minería Hispana e Iberoamericana. Contribución a su investigación histórica*, vol. 1, VI Congreso Internacional de Minería, 10 vols., 221–258. León: Cátedra de San Isidoro, 1970.

Lynch, John. *Spain under the Habsburgs, Volume 2: Spain and America, 1598–1700*. New York: New York University Press, 1984.

Lynch, John. *Bourbon Spain, 1700–1808*. Oxford: Basil Blackwell, 1989.

Lynch, John. *Spain 1516–1598: From Nation State to World Empire*. Oxford: Blackwell, 1991.

MacKay, Angus. *Spain in the Middle Ages: From Frontier to Empire, 1000–1500*. New York: St. Martin's Press, 1977.

Maffei Hutter, Lucy. "A madeira do Brasil na construção e reparos de embarcações." *Revista de Instituto de Estudios Brasileiros* 26 (1986), 47–64.

Manuel Valdés, C.M., and A. Rojo y Alboreca. "Valsaín Forest in the XVIII Century: An Example of Forest Management in the Pre-industrial Era." *Investigación Agraria, Sistemas y recursos forestales, fuera de serie* no. 3 (1994): 217–229.

Marks, Robert. *Tigers, Rice, Silk, and Silt: Environment and Economy in Late Imperial South China*. Cambridge: Cambridge University Press, 1997.

Marks, Robert. *China: Its Environment and History*. Lanham, MD: Rowman and Littlefield, 2012.

Martin, Colin, and Geoffrey Parker. *The Spanish Armada*. New York: Norton, 1988.

Martínez Ruiz, Enrique. "Los montes en la cultura islámica. El bosque del Al-Andalus desde el siglo VIII al XIII." In *Los Montes y su Historia: Una Perspectiva Política, Económica y Social*, edited by Francisco Marín Pageo, Juan Domingo Santos, and Ana Calzado Carretero, 63–88. Huelva: Universidad de Huelva, 1999.

Martínez Ruiz, Enrique, dir. *Diccionario de historia moderna de España, vol. II: La administración*. Madrid: Ediciones Istmo, 2007.

Martins, Xavier M. "Some Aspects of Portuguese Shipbuilding in India during 16th–17th Centuries." *Indica* 41, no. 2 (September 2004): 141–150.

Maruri Villanueva, Ramón. "Ensenada y el Real Astillero de Guarnizo." *Brocar: Cuadernos de investigación histórica* 25 (2001): 123–136.

McCarthy, William J. "The Yards at Cavite: Shipbuilding in the Early Colonial Philippines." *International Journal of Maritime History* 7, no. 2 (December 1995): 149–162.

McNeill, J.R. *Atlantic Empires of France and Spain: Louisbourg and Havana, 1700–1763*. Chapel Hill: The University of North Carolina Press, 1985.

McNeill, J.R. *The Mountains of the Mediterranean World: An Environmental History*. Cambridge: Cambridge University Press, 1992.

McNeill, J.R. "Woods and Warfare in World History." *Environmental History* 9, no. 3 (July 2004): 388–410.

Menéndez Pidal, Ramón. *Historia de España, tomo XXIX: La época de los Borbones – La nueva monarquía y su posición en Europa (1700–1759)*. Madrid: Espasa-Calpe, 1985.

Merino Navarro, José. *La armada española en el siglo XVIII*. Madrid: Fundación Universitaria Española, 1981.

Michel, Mario, and Luis Gil. *La transformación histórica del paisaje forestal en la Comunidad Autónoma de Euskadi*. Vitoria: Servicio de Publicaciones del Gobierno Vasco, Colección LUR, no. 18, 2013.

Miller, Shawn W. *Fruitless Trees: Portuguese Conservation and Brazil's Colonial Timber*. Stanford: Stanford University Press, 2000.

Miller, Shawn W. *An Environmental History of Latin America*. New York: Cambridge University Press, 2007.

Mintz, Sidney W. *Sweetness and Power: The Place of Sugar in Modern History*. New York: Viking Press, 1985.

Mira Caballos, Esteban. "El Sistema Naval Español en el Siglo XVI: Las Armadas del Imperio." *Revista de Historia Naval*, no. 74 (2001): 39–54.

Molina, Tirso de. *La Prudencia en la Mujer* (1633). In *Obras dramáticas completas de Tirso de Molina (En tres tomos)*, vol. 3, edited by Blanca de Los Ríos, 3 vols., 893–951. Madrid: Aguilar, 1958.

Moore, Jason W. "'Amsterdam Is Standing on Norway' Part I: The Alchemy of Capital, Empire, and Nature in the Diaspora of Silver, 1545–1648." *Journal of Agrarian Change* 10, no. 1 (January 2010a): 33–68.

Moore, Jason W. "'Amsterdam Is Standing on Norway' Part II: The Global North Atlantic in the Ecological Revolution of the Long Seventeenth Century." *Journal of Agrarian Change* 10, no. 2 (April 2010b): 188–227.

Morga, Antonio de. *Sucesos de las Islas Filipinas*, translated and edited by J.S. Cummins. London: The Hakluyt Society and Cambridge University Press, 1971.

Morton, F.W.O. "The Royal Timber in Late Colonial Bahia." *Hispanic American Historical Review* 58, no. 1 (February 1978): 41–61.

Mukerji, Chandra. "The Great Forest Survey of 1669–1671: The Use of Archives for Political Reform." *Social Studies of Science* 37 no. 2 (April 2007): 227–253.

Mukerji, Chandra. *Impossible Engineering: Technology and Territoriality on the Canal du Midi*. Princeton: Princeton University Press, 2009.

Muñoz Goyanes, Guillermo. *Crónica sobre bosques y montes de la Península Hispánica*. Madrid: Fundación Conde del Valle de Salazar, 1983.

Nader, Helen. *Liberty in Absolutist Spain: The Habsburg Sale of Towns, 1516–1700*. Baltimore: The Johns Hopkins University Press, 1990.

Neville, Ann. *Mountains of Silver and Rivers of Gold: The Phoenicians in Iberia*. Oxford: Oxbow Books, 2007.

Odriozola Oyarbide, María Lourdes. *Construcción Naval en el País Vasco, siglos XVI–XIX: Evolución y Análisis Comparativo*. San Sebastián: Diputación Foral de Gipuzkoa, Departamento de Economía y Turismo, 2002.

Ortega Pereyra, Ovidio. *El Real Arsenal de La Habana: la construcción naval en La Habana bajo la dominación colonial española*. La Habana, Cuba: Editorial Letras Cubanas, 1998.

Ortega Santos, Antonio. "Commons and Rural Communities: An Environmental History of Conflicts in [a] Mediterranean Ecosystem." Presented at *Common Ground, Converging Gazes: Integrating the Social and the Environmental in History*, International Conference held at l'École des Hautes Études en Sciences Sociales, Paris, September 11–13, 2008.

Ozanam, Didier. "La política exterior de España en tiempo de Felipe V y de Fernando VI." In *Historia de España, tomo XXIX: La época de los Borbones – La nueva monar-*

quía y su posición en Europa (1700–1759), translated by José Luis López Muñoz, edited by Ramón Menéndez Pidal, 441–699. Madrid: Espasa-Calpe, 1985.

Özveren, Y. Eyüp. "Shipbuilding, 1590–1790." *Review (Fernand Braudel Center)* 23, no. 1, Commodity Chains in the World-Economy, 1590–1790 (2000): 15–86.

Padrón, Ricardo. *The Spacious Word: Cartography, Literature, and Empire in Early Modern Spain*. Chicago: University of Chicago Press, 2004.

Parker, Geoffrey. "Maps and Ministers: The Spanish Habsburgs." In *Monarchs, Ministers and Maps: The Emergence of Cartography as a Tool of Government in Early Modern Europe*, edited by David Buisseret, 124–152. Chicago: University of Chicago Press, 1992.

Parker, Geoffrey. *Philip II*. 3rd ed. Chicago: Open Court, 1995.

Parker, Geoffrey. *The Military Revolution: Military Innovation and the Rise of the West, 1500–1800*. 2nd ed. Cambridge: Cambridge University Press, 1996.

Parker, Geoffrey. *The Grand Strategy of Felipe II*. New Haven: Yale University Press, 1998.

Parry, J.H. *The Spanish Seaborne Empire*. New York: Knopf, 1966. Reprint, Berkeley: University of California Press, 1990.

Pèlachs Mañosa, Albert, and Joan Manuel Soriano López, "Las fuentes paleobotánicas y la historia forestal: el ejemplo de los valles de la Coma de Burg y Vallferrera (Pallars Sobirà-Lleida)." *Cuadernos de la Sociedad Española de Ciencias Forestales* no. 16 (2003), 155–160.

Peluso, Nancy Lee. "The History of State Forest Management in Colonial Java." *Forest and Conservation History* 35, no. 2 (April 1991): 65–75.

Pereda Hernández, Miguel-Juan. "Conservación y repoblación de arbolado en Almansa a mediados del siglo XVI." *Congreso de historia de Castilla-La Mancha* 1, no. 7, "Conflictos sociales y evolución económica en la Edad Moderna" (1985): 89–96.

Perlin, John. *A Forest Journey: The Story of Wood and Civilization*. New York: W.W. Norton & Company, 1989. Reprint, Woodstock, VT: The Countryman Press, 2005.

Petrejus, E.W. "The Dutch Flute." In *The Great Age of Sail*, edited by Joseph Jobé, translated by Michael Kelly, 81–131. Lausanne: Edita, 1967.

Phillips, Carla Rahn. *Six Galleons for the King of Spain: Imperial Defense in the Early Seventeenth Century*. Baltimore: Johns Hopkins University Press, 1986.

Phillips, Carla Rahn. "'The Life Blood of the Navy': Recruiting Sailors in Eighteenth Century Spain." *The Mariner's Mirror* 87, no. 4 (November 2001): 420–445.

Phillips, Carla Rahn. "Naval Recruitment and Basque Resistance in Early Modern Times." *ITSAS Memoria: Revista de estudios marítimos del País Vasco* no. 5 (2006): 173–185.

Phillips, Carla Rahn. *The Treasure of the San José: Death at Sea in the War of the Spanish Succession*. Baltimore: Johns Hopkins University Press, 2007.

Phillips, Carla Rahn, and William D. Phillips Jr., *Spain's Golden Fleece: Wool Production and the Wool Trade from the Middle Ages to the Nineteenth Century*. Baltimore: Johns Hopkins University Press, 1997.

Phillips, William D. Jr. "Spain's Northern Shipping Industry in the Sixteenth Century." *The Journal of European Economic History* 17 no. 2 (Fall 1988): 267–301.

Phillips, William D. Jr., and Carla Rahn Phillips. *The Worlds of Christopher Columbus*. Cambridge: Cambridge University Press, 1992.

Phillips, William D. Jr., and Carla Rahn Phillips. *A Concise History of Spain*. Cambridge: Cambridge University Press, 2010.

Pinchot, Gifford. *The Fight for Conservation*. New York: Doubleday, Page & Co., 1910.

Piñera y Rivas, Álvaro de la, and Jacqueline Thial Boisière. "La construcción naval en España durante el siglo XVIII." *Revista de Historia Naval* 79 (2002): 17–33.

Poole, Stafford. *Juan de Ovando: Governing the Spanish Empire in the Reign of Felipe II*. Norman, OK: University of Oklahoma Press, 2004.

Postnikov, Aleksey V. "The Russian Navy as Chartmaker in the Eighteenth Century." *Imago Mundi* 52 (2000): 79–95.

Pulido Bueno, Ildefonso. *José Patiño: El inicio del gobierno político-económico ilustrado en España*. Huelva: I.P. Bueno, 1998.

Quintero González, José. *El arsenal de La Carraca, 1717–1736*. Madrid: Ministerio de Defensa, 2000.

Quintero González, José. *Jarcias y lonas: El renacimiento de la armada en la bahía de Cádiz*. Cádiz: Grupo Publicaciones del Sur, 2003.

Quintero González, José. "La Carraca y La Habana: Aproximación al intercambio de pertrechos entre arsenales." *Revista de Historia Naval* no. 93 (2006): 57–66.

Rábade Blanco, José María, ed. *Atlas Forestal de España*. Madrid: Grupo Tragsa, 2002.

Rackham, Oliver. *Trees and Woodland in the British Landscape*. London: J.M. Dent, 1976.

Rackham, Oliver. *The History of the Countryside*. London: J.M. Dent, 1986.

Radell, David R., and James J. Parsons. "Realejo: A Forgotten Colonial Port and Shipbuilding Center in Nicaragua." *Hispanic American Historical Review* 51, no. 2 (May 1971): 295–312.

Radkau, Joachim. *Nature and Power: A Global History of the Environment*, translated by Thomas Dunlap. Cambridge: Cambridge University Press, 2008.

Rey Castelao, Ofelia. *Montes y política forestal en la Galicia del Antiguo Régimen*. Santiago de Compostela: Universidade de Santiago de Compostela, 1995.

Richards, John F. *The Unending Frontier: An Environmental History of the Early Modern World*. Berkeley: University of California Press, 2003.

Rodríguez González, Agustín Ramón. *Trafalgar y el conflicto naval anglo-español del siglo XVIII*. San Sebastián de los Reyes (Madrid): Actas, 2005.

Rodríguez Villa, J.M. "Historia forestal y desarrollo forestal sostenible." *Cuadernos de la Sociedad Española de Ciencias Forestales* no. 16 (2003): 297–301.

Rubio Serrano, José Luis. *Arquitectura de las naos y galeones de las flotas de Indias, vol. I (1492–1590), vol. II (1590–1690)*. Málaga: Ediciones Seyer, 1991.

Russell-Wood, A.J.R. *The Portuguese Empire, 1415–1808: A World on the Move*. Baltimore: The Johns Hopkins University Press, 1998.

Sack, Robert David. *Human Territoriality: Its Theory and History*. Cambridge: Cambridge University Press, 1986.

Sánchez Baena, Juan José, Celia Chaín Navarro, and Lorena Martínez Solís, eds. *Estudios de historia naval: Actitudes y medios en la Real Armada del siglo XVIII*. Madrid: Ministerio de Defensa, 2011.

Serrano Álvarez, José Manuel. "Juan de Acosta y la construcción naval en La Habana (1717–1740)." *Revista de Historia Naval* 93 (2006): 7–31.

Scammell, Geoffrey. "British Merchant Shipbuilding, c. 1500–1750." *International Journal of Maritime History* XI, no. 1 (June 1999): 27–52.

Schiebinger, Londa. *Plants and Empire: Colonial Bioprospecting in the Atlantic World*. Cambridge, MA: Harvard University Press, 2007.

Scott, S.P. *The Visigothic Code (Forum Judicum)*. Boston: Boston Book Company, 1910.

Sierra Vigil, J.M. "Bases de partida para estudios sobre los montes españoles en la Edad Media; problemas a resolver, periodización cronológica." *Cuadernos de la Sociedad Española de Ciencias Forestales* no. 16 (2003): 249–254.

Smidt, Andrea J. "Enlightened Absolutism and New Frontiers for Political Authority: Building towards a State Religion in Eighteenth-Century Spain." In *The Limits of Empire: European Imperial Formations in Early Modern World History, Essays in Honor of Geoffrey Parker*, edited by Tonio Andrade and William Reger, 33–57. Farnham, UK: Ashgate, 2012.

Smith, Catherine Delano. *Western Mediterranean Europe: A Historical Geography of Italy, Spain, and Southern France since the Neolithic*. London: Academic Press, 1979.

Smith, Richard L. *Ahmad al-Mansur: Islamic Visionary*. New York: Pearson Longman, 2006.

Smout, T.C. *People and Woods in Scotland: A History*. Edinburgh: Edinburgh University Press, 2003.

Soriano Martí, J. "La documentación Medieval y la sostenibilidad de los aprovechamientos forestales mediterráneos." *Cuadernos de la Sociedad Española de Ciencias Forestales* no. 16 (2003): 73–78.

Stradling, R.A. *Europe and the Decline of Spain*. London: George Allen & Unwin, 1981.

Studniki-Gizbert, Daviken, and David Schecter. "The Environmental Dynamics of a Colonial Fuel-rush: Silver Mining and Deforestation in New Spain, 1522–1810." *Environmental History* 15, no. 1 (2010): 94–119.

Tarazona Grasa, Carlos. *La Guardería Forestal en España: Conservación de nuestros bosques a través de los tiempos*. Barcelona: Lunwerg, 2002.

Terán, Manuel de, Lluís Solé i Sabarís, Orlando Ribeiro, José Manuel Casas Torres, and Salvador Llobet. *Geografía de España y Portugal*. 2 vols. Barcelona: Montaner y Simón, 1952.

Thomas, Keith. *Man and the Natural World: A History of the Modern Sensibility*. New York: Pantheon Books, 1983.

Thompson, I.A.A. *War and Government in Habsburg Spain, 1560–1620*. London: Athlone Press, 1976.

Totman, Conrad. *The Green Archipelago: Forestry in Pre-industrial Japan*. Berkeley: University of California Press, 1989.

Tracy, James D., ed. *The Rise of Merchant Empires: Long-distance Trade in the Early Modern World, 1350–1750*. Cambridge: Cambridge University Press, 1990.

Tracy, James D., ed. *The Political Economy of Merchant Empires: State Power and World Trade, 1350–1750*. Cambridge: Cambridge University Press, 1991.

Urteaga, Luís. *La tierra esquilmada: Las ideas sobre la conservación de la naturaleza en la cultura española del siglo XVIII*. Madrid: CSIC, 1987.

Valbuena-Carabaña, María, Unai López de Heredia, Pablo Fuentes-Utrilla, Inés González-Doncel, and Luis Gil. "Historical and recent changes in the Spanish forests: A socio-economic process." *Review of Paleobotany and Palynology* 162, no. 3 (October 2010): 492–506.

Valdez-Bubnov, Iván. "War, Trade, and Technology: The Politics of Spanish Shipbuilding Legislation, 1607–1728." *International Journal of Maritime History* 21, no. 2 (December 2009): 75–102.

Vassberg, David. *Land and Society in Golden Age Castile*. Cambridge: Cambridge University Press, 1984.

Vega, Lope de. *Virtud, pobreza y mujer*. In *Comedias Escogidas de Frey Lope Félix de Vega y Carpio*, vol. 4, edited by Juan Eugenio Hartzenbusch, Juan Pérez de Montalván, Antonio Gil y Zárate, and Adolfo de Castro, 213–232. Madrid: Hernando y Compañía, 1901.

Vega, Lope de. "Soneto XLVI: A la Jornada de Inglaterra a bordo del 'San Juan,'" *Rimas (1602)*, edited by Ramón García González. Alicante: Biblioteca Virtual Miguel de Cervantes, 2003. http://www.cervantesvirtual.com/servlet/SirveObras/13527274323929275754491/p0000001.htm?marca=A%20la%20jornada%20de%20inglaterra#238 (accessed July 2, 2014).

Vicens Vives, Jaime, and Jorge Nadal. *An Economic History of Spain*. Translated by Frances M. López-Morillas. Princeton: Princeton University Press, 1969.

Vicente Maroto, María Isabel. "Agustín de Ojeda y la construcción de navíos a finales del siglo XVI." In *La ciencia y el mar*, edited by María Isabel Vicente Maroto and Mariano Esteban Piñeiro, 311–344. Valladolid: Sever-Cuesta, 2006.

Vicente Maroto, María Isabel. "The Art of Shipbuilding in Spain's Golden Century." *History of Technology* 30 (2010):79–94.

Vilà Valentí, Joan. *La Península Ibérica*. Barcelona: Ariel, 1968.

Warde, Paul. *Ecology, Economy, and State Formation in Early Modern Germany*. Cambridge: Cambridge University Press, 2006a.

Warde, Paul. "Fear of Wood Shortage and the Reality of the Woodland in Europe, c.1450–1850." *History Workshop Journal* 62, no. 1 (Autumn 2006b): 28–57.

Warde, Paul. "The Environmental History of Pre-industrial Agriculture in Europe." In *Nature's End: History and the Environment*, edited by Sverker Sörlin and Paul Warde, 70–92. Basingstoke, UK: Palgrave Macmillan, 2009.

Warde, Paul. "The Invention of Sustainability." *Modern Intellectual History* 8, no. 1 (2011): 153–170.

Watson, Andrew M. *Agricultural Innovation in the Early Islamic World: The Diffusion of Crops and Farming Techniques, 700–1100*. Cambridge: Cambridge University Press, 1983.

White, Sam. *The Climate of Rebellion in the Early Modern Ottoman Empire*. Cambridge: Cambridge University Press, 2011.

Williams, Michael. *Deforesting the Earth: From Prehistory to Global Crisis, An Abridgment*. Chicago: University of Chicago Press, 2006.

Woodward, David, ed. *Cartography in the European Renaissance*. 2 vols. Vol. 3 of *The History of Cartography*. Chicago: University of Chicago Press, 2007.

Yahya, Dahiru. *Morocco in the Sixteenth Century: Problems and Patterns in African Foreign Policy*. Atlantic Highlands, NJ: Humanities Press, 1981.

Young, Charles R. *The Royal Forests of Medieval England*. Philadelphia: University of Pennsylvania Press, 1979.

Young, David Bruce. "Forests, Mines, and Fuel: The Question of Wood and Coal in Eighteenth-Century France." *Proceedings of the Western Society for French History* 3 (October 1975): 328–336.

Index

Acapulco 100, 103, 114, 153, 154
Acorns 14n26, 22, 32, 36, 44, 59, 143
Acosta, José de 88
Acosta, Juan de 196
Actionable intelligence 71, 76, 83, 167
Adriatic Sea 118
Agriculture
  expansion of 21, 32, 48, 53, 69, 72, 83, 184, 242
  in Islamic Spain 52
  of Moriscos 69
Aguas River 188, 191
Albacete 72, 227
Albania 8, 27
Albay province, Philippines 112, 155
Alberoni, Giulio, Cardinal 174–175
Alçatte, Antonio de 79–82, 136–137, 227, 244
Alçatte, Carlos de 136
Alfonso I 48
Alfonso X 16n29, 51, 60
Alfonso XI 60, 69
Algeciras 49
Algiers 25, 123
Alicante 10, 49, 52, 72, 227
Almería 49
Almirantazgo (Admiralty) 43, 165, 176–178, 180, 190, 201, 203, 222
Almirante, Prince Felipe 177–180, 183–184, 187–188, 203
America
  early European descriptions of 87–91
Andalusia 13, 46, 48, 54, 59, 61, 83, 167, 202, 219
Aragón 10, 44, 50n16, 55–56, 78, 126, 157, 162, 167, 189–192, 201
Aragón River 188–189, 191
Aranjuez 70, 215, 219
Arévalo, Philippines 112
Arica, port of Potosí 99
Armada, Great, of 1588 16, 39, 41, 74, 77–79, 95, 127, 131–132, 162, 243
Armada de Barlovento 109, 145
Armada de la Guardia del Estrecho 123
Armada del Mar del Sur 100, 104, 149, 153
Aróstegui, Martín de 127, 133
Asteasu 120, 136
Asturias 18, 31, 47, 54, 62, 66–67, 120, 128, 130, 137, 142, 167, 169–170, 183, 214, 217, 220, 229–230
Asunción del Paraguay 99, 198
Autrán, Ciprián 176n34, 182–184, 198, 201
Ávila 72, 223n48
Azores 77, 131, 131n34

Baetic mountains 10, 13
Bahia, Brazil 117, 124
Baldíos 53
Balkans 118
Baltic, region 8, 27, 65, 124–126, 137, 147
Bances León, Lorenzo de 151–152
Barcelona 49–50, 52, 65–66, 68, 76–82, 130n30, 136–137, 140, 157–158, 162, 174, 181–182, 188–190, 192, 201, 228
Barros, Cristóbal de 41, 66–67, 76–77, 83, 121, 127–132, 134–135, 137, 159, 209–210, 215, 242–244
Basque Country 10, 14n26, 16, 18, 18n31, 47, 54, 62–64, 76, 94–95, 127, 140, 148, 167, 175, 180
Benefits, of plantations 142–143, 145, 171, 173, 206, 211, 213
Bermúdez de Castro, José 170, 179, 179n50, 222, 224
Bicol 112, 114
Bilbao 50, 78, 159, 230
Black Sea 27, 118
Bordic, Diego 183–184
Borneo 110, 112
Brazil 117, 124, 139–140, 149
Bulubulu, forests of 105, 198
Bureaucracy 6–7, 41, 63, 71, 75–76, 121, 126, 138, 158–159, 236, 241, 243
Burgos 44, 57, 183

Cádiz 35, 124, 147, 154, 176, 182–186, 188, 198, 201, 214, 218–220, 230–232, 234–235, 237
Callao 99, 102, 104, 198
Camarines coast, Philippines 112, 155

Campeche 147, 193
Cano, Tomé 94
Cantabria 1, 5, 10, 13, 13n, 18, 32, 42, 42n, 46–47, 51, 54, 62–63, 74, 95, 106, 133, 146, 149, 167–168, 178–179, 212, 220, 230
Carriedo, valley of 1, 7, 13, 13n
Carlos I
long-term naval commitments 41, 55, 62, 65
and shipbuilding 46, 63, 64n80, 75
and geographic information 70–71, 90
Carlos II 7, 42, 156–158, 162–163
Carlos III 39, 43, 177n37, 236–238, 245
Carrera de Indias 41, 46–47, 63, 76–77, 85, 95, 108, 131, 133, 146, 148
Cartagena 13, 52, 82, 176, 180–182, 184, 186, 214, 218, 221, 226–229, 232, 235, 237, 244
Cartagena de Indias 147–149, 154
Cartas pueblas (town charters) 53, 57–60
Carvajal, José de 205, 235
Casa de la Contratación (House of Trade) 73, 87, 90–91, 108, 154, 174
Castelló de la Plana 58–59, 79
Castile 1, 10, 14–15, 18n31, 44, 51, 55–56, 58, 61, 68, 69–70, 108, 116, 126, 140, 146
Castro Urdiales 42n, 50–51
Catalonia 10, 18, 27, 42, 51–52, 60, 62, 65–66, 68, 75, 77, 79–83, 120, 130, 130n30, 136–137, 140, 157–158, 167, 171, 180, 187–192, 201–202, 228–229
Catastro de Ensenada 13, 232
Cavite 9, 99, 109, 112, 114, 153–156, 198–199
Cebu 109
Centralization 4–6, 24–25, 39, 45, 46, 60, 137, 166, 201, 225, 236
Charcoal 16n29, 22–23, 25, 27, 33, 35, 62, 70, 129
Charles V, Holy Roman Emperor *See* Carlos I
Chile 99, 104, 106, 153
Chinese, in Philippines 110, 112, 114–115, 154
Cinca River 189–190, 194
Coatzacoalcos 100, 176, 195–196, 201
Coconuts 98, 106
Cofrentes, Valley of 181–182, 201
Colbert, Jean-Baptiste, French forest ordinance by 27, 144
Colindres 51
Columbus, Christopher 87–89
Common good 25, 44–45, 62, 160
Commons 14, 16, 35, 53, 58, 59, 61, 67, 83, 96–97, 135, 172, 207
Conservation 2, 2n4, 19, 21, 24, 38, 44, 68, 76, 83, 85, 96, 120–122, 128, 130, 132, 142, 145, 158, 160, 163, 169–171, 173, 204, 206–207, 231, 236–237, 241, 243
Coppicing *See* pruning
Córdoba 13, 33, 59, 62
Corregidores 44–45, 61, 64–67, 75, 78, 106, 151, 154, 216, 220, 223
corte de madera 114, 155–156
Cortes, of Castile 24, 50n16, 65
Coto de Doñana 231, 234
Criollos (American-born Europeans) 98, 121, 146, 148, 152
Cuatro Villas 42, 42n, 50–51, 66, 120, 129–131, 137, 141–142, 168–170, 178
Cuba 70, 87, 107–109, 139, 146–150, 176, 193, 197–198, 205, 217, 235, 243
Cuenca 13, 49, 72
Cuero y Arze, Juan de 1–2, 5, 9

Dávila, Pedrarias 102
De Pando, Catalina 1–2, 5–7, 9, 13
Deforestation 29, 34n84, 35, 50, 56, 61, 64–66, 68–70, 122, 137–138, 152, 160, 163, 169, 176, 187, 207, 216, 241
Dehesa 53, 90
Departments, naval
establishment of 176, 176n32
Djerba 65
Draft labor *See corte de madera; See* forced labor
Duero River 13, 48, 242

Ebro River 10, 48–49, 52, 81, 137, 162, 165, 188–189, 192, 194, 242
El Escorial 70, 74, 108
El Pardo 69–70, 215
England 23, 26–27, 50, 70, 76–78, 107, 109, 116, 121, 123, 125, 139–141, 157–158, 163, 175, 177, 205, 235, 238, 244
Enrique III 61
Enrique IV 69
Ensenada, Zenón de Somodevilla, Marqués de la (secretary of Almirantazgo)

and the "widow's oak" 1–2, 5
government reforms 203–205, 232–236
and the 1748 ordinances 3, 43, 197, 204, 206, 213, 220, 226
and the *Almirantazgo* 43, 165, 177–178, 180, 188–189, 191
Española, La 87, 97, 107
Estados de montes 222, 224–225, 229–230, 234
Evelyn, John 27

Fábrica de Tabacos (Royal Tobacco Factory) 183–185
Farnese, Isabel, Queen of Spain 174, 177n37, 215
Felipe II
and forest bureaucracy 6, 66, 75–76, 83, 242
and forest depletion 19, 66–8, 77
and mapping 40, 73–74
and shipbuilding 46, 63, 68, 76–78, 84, 113, 131
and naval strategy 65–66, 122–123
and forest plantations 65, 77
and royal palaces 70
and geographic information 70, 72, 90, 232
and forest reconnaissance 78–79
and Portuguese throne 76, 131, 131n34
Felipe III 123–124, 127, 131–132, 137, 145
Felipe IV 124, 126–127, 137, 139–141, 146, 157
Felipe V 163, 165n1, 187, 193, 203, 215–216, 237
Fernández de Oviedo, Gonzalo 90
Fernando of Aragón 44, 46, 54
Fernando III 51
Fernando VI 39, 84, 186, 203–205, 236–238
Ferrol 51, 78, 161, 176, 178–180, 182, 214, 218, 220, 229–230, 232, 235, 237, 244
Fines 7, 9, 42, 59, 62, 66, 84–85, 106, 120, 128–130, 134–135, 144, 153, 159–161, 170–173, 197, 208, 211, 213, 216, 221, 223
Fire 25, 31–33, 36, 48, 59–61, 143–144, 156, 161, 179–181, 188, 191, 198, 210–211, 215, 243
Firewood 20, 23, 34n84, 44, 56–57, 59, 61–62, 67, 69, 72, 97, 108, 129, 143, 213
Forced labor 42, 85, 105, 114, 118, 121, 154, 243
See also corte de madera
Forests
Abuses 24, 169–171, 178–179, 212, 220
Inspections 5, 42, 59, 75–77, 79–83, 92, 121, 128–130, 132, 135–137, 158–159, 161, 165–166, 169, 171–172, 178–179, 181–195, 199, 203, 205, 207–209, 216, 222–234, 236–239, 243–244
and property 13, 16, 27, 29, 35, 53–54, 58, 66–67, 79, 97, 135, 137, 145, 166, 172, 190, 207–208, 226, 230
France 6, 14, 27, 38, 50, 67, 70, 107, 117, 121, 123, 125, 139–140, 156–158, 175, 180, 205, 235, 237, 244
Fuero Juzgo (*Liber iudiciorum*) 35–36, 59
Fueros 14–15, 18, 53, 57, 60, 62, 126, 134–135, 188, 208
Functionality, forest 4, 39–41

Galicia 10, 18, 42, 51, 62, 66–68, 75, 78, 120, 130, 132–133, 159–162, 170, 175, 179, 207–208, 230
Gállego River 188–191, 201
Gaztañeta, Antonio 17, 168–171, 173–175, 193
Gener, Francisco 184–185
Grazing 22, 25, 32–34, 53, 58
Great Britain 7, 125, 175, 178, 238
Grimaldo, José de 169–170
Guadalquivir River 13, 34, 48–49, 165, 167, 183–187, 214, 231, 242
Guadiana River 13, 34, 48, 242
Guards, forest 25, 59, 167n5, 172, 180, 211, 213, 219–220, 233
Guarnizo 51, 129, 168, 174, 178–180, 196, 235
Guayaquil 87, 99, 103–106, 118, 148–149, 151–154, 198, 205
Guernica oak 14–16, 18–19, 241
Guetaria 51, 140
Guipúzcoa 8, 18n31, 42, 51, 63–64, 66–67, 75, 84, 120, 127–128, 130–131, 133–137, 140, 154, 158, 168, 170, 183, 212, 217
Gutiérrez de Rubalcava, José 231–232
Gutiérrez Ruvalcava, Alexo 181, 184

Havana 40, 87, 96, 99, 106–109, 118, 145–151, 154, 176, 193, 195–198, 205, 235
Hemp 52, 94, 106, 116, 165, 187–192, 194, 227–228
Herding *See* livestock
Holland *See* Netherlands
Huelva 33, 49, 230, 234
Hunting 22, 61, 69–70, 72, 83, 88, 128

Ibiza 34, 49, 52, 226, 229
Idiáquez, Alonso de 136
Idiáquez, Domingo de 120, 134–136, 139, 158
Iloilo 112, 114
Iron
  foundries 2, 16n27, 50–51, 142n70, 179, 213
  supplies of 52, 86, 99, 102, 109, 114, 128, 147, 198
Isabel of Castile 44, 46, 55
Isidore of Seville 23, 36–37
Isla Hermosa (Taiwan) 155
Isunza, Pedro de 77, 81–82

Jagua, Cuba 149–150, 193, 198, 201
Jamaica 148
Jeréz de la Frontera 219, 231–232, 234
Jover, Blas 219–225
Juan, Jorge 235
Juana, Queen of Castile 55, 97, 206
Júcar River 10, 13, 48–49, 181–182, 201, 226–227, 242

La Carraca 182–186, 198, 214, 231, 244
Las Casas, Bartolomé de 87, 103
Legislation, forest
  to curb destructive behavior 25, 35–36, 44, 55–62, 68, 107–108
  for shipbuilding 28–29, 45, 63–65, 109, 141–147, 152–153, 159–162, 197–198, 204, 206–219
  implementation of 222–234
Lérida 52, 240
Liber iudiciorum See Fuero Juzgo
Libro de Montería 60
Licenses 44–45, 58, 76n, 79, 143, 147, 153, 160, 197, 208, 210–211, 216, 219–221
Lima 104, 106, 151
Lisbon 13, 117, 131
Livestock 9, 18, 22, 24–25, 32–36, 44, 48, 53–54, 56, 59, 61–62, 68–69, 79–80, 97, 107, 129, 143–144, 146, 159, 171, 184, 198, 209–210, 220
Lizárraga, Domingo de 120, 136
López de Legazpi, Miguel 89, 109–110, 112, 115
Louis XIV 26, 42, 156–158, 163
Luzon 109–110, 112, 115, 154, 156, 199

Madrid 1, 13n, 24, 45, 69–70, 75, 132, 177, 184, 188–189, 191, 198, 215–217, 223–224
Mahogany 8, 102, 106–108, 146–147, 198
Málaga 32, 34, 49, 52, 78, 183, 232, 234
Mallorca 73, 225–226, 229
Maltés, José 181–182, 193–194, 201
Manila 9, 87, 99–100, 102–103, 109–110, 112, 114, 116, 153, 156, 199
Manrique de Lara, Nicolás 168, 170–174, 179
Mapping 1, 4, 5, 5n, 40–41, 71, 73–75, 79–83, 87, 150, 166, 192–196, 199–200, 233, 240
Maracaibo 147–148, 193
Mariana of Austria 157, 160–161, 207–208, 244
Marinduque 112, 155
Masbate 112, 155
Mediterranean Sea 8, 10, 13, 18, 25, 30–35, 37, 46, 48–52, 65, 76, 80–81, 118–119, 123, 127, 136, 158, 162–163, 174, 193, 242
Mesta 48, 53, 69
Mexico 97–100, 114, 154, 193, 196, 198
Mindanao 109–110, 154
Mindoro 112, 115, 155
Mining 20, 33, 35, 50, 52, 63, 96, 98, 112, 149, 195
Molina, Tirso de 15
Montes See forests
Morella 80–81, 227–229
Morga, Antonio de 113, 152
Moriscos 69, 123
Motrico 168
Moya, Marquesado de 181–182, 193
Municipal laws See *fueros*
Murcia 10, 13, 72–73, 167, 221, 226–227, 229, 237

Naples 63, 65, 175, 177, 237
Navarre 54, 62, 68, 137, 162, 167, 170, 179, 189, 191–192

Navarro, Juan José, Marqués de la Victoria 93, 192–194, 193n93, 233, 237
Netherlands 27, 38, 70, 74, 94, 116–117, 121, 123–126, 139–140, 145, 147, 149, 153, 157–158, 175
New Spain 96, 104, 106, 110, 114, 145, 154
Nicaragua 100, 102–103
Nicolalde, Francisco José de 158–159
North Sea 123–124, 126–127, 137
Norway 109
Nueva Segovia 156

Oakum 94, 98, 106
Ojeda, Agustín de 130–133, 136, 159n122, 244
Olivares, Gaspar de Guzmán, Count-Duke of 124, 126, 138–140, 146
Ordonnance des eaux et fôrets See Colbert
Ordinances, forest *See* legislation, for shipbuilding
Otón 112, 114
Ottoman Empire 25, 46, 65, 118
Ovando, Juan de 71, 90

Panama 99–100, 102, 104, 106, 147, 149
Panay 110, 112, 156
Pangasinán 112, 155–156, 199n114
Paredes, Juan de 160
Pasajes 51, 131, 133, 140, 175, 180
Patiño, José 43, 174–178, 182–184, 195, 201, 208, 213
Pedro I 50
Peñíscola 81, 228–229
Pensacola 195
Pérez de Bustamante, Toribio 25, 29, 141–145, 168, 170, 173, 207, 209–212, 244
Philippines 9, 41, 87, 89, 98–99, 103, 109–115, 124, 153–156, 198–200
Pineda, Sebastián de 115, 154–156
Pinzón, Juan 185–187, 190
Piracy 25, 85, 103, 105–107, 116, 123, 145, 149
Pitch 6, 8–9, 35, 49, 94, 102–103, 106, 109, 147, 165, 187–192, 194, 201
Plantations, forests
management of 44, 56, 64, 129, 135, 143–144, 159–161, 171–174, 208–213, 243
Portugal 13, 30, 67, 77, 109, 117, 128–129, 140, 242
Portugalete 51, 133
Pruning 70, 209–212, 221–223, 225, 227, 243
Coppicing 23, 36, 62
Puerto Rico 148

Questionnaires 40, 41, 71–75, 87, 90, 232
Quotas, planting 7, 9, 67, 84–85, 120, 128, 130, 133–134, 136, 159, 170, 172–173, 186, 205, 226, 229

Real Junta de Obras y Bosques 70
Realejo 87, 100, 102–105, 148, 154, 198
Reconnaissance missions *See* forest inspections
Reconquest 47–49, 51, 53–54, 97
Regidores 56, 61, 68, 134, 136, 144, 172, 220
Relaciones geográficas de Indias See questionnaires
Relaciones topográficas de España See questionnaires
Resin, pine *See* pitch
Resistance 9, 42, 67, 120, 126–127, 129, 135–136, 142, 156, 167n5, 169, 187, 199–200, 204, 217, 219–220, 243
Riga 109, 126
Rigging 52, 94, 98, 103, 109, 115–116, 147, 167n6, 188–190, 198, 201
Rio de Janeiro 117
Río de la Plata 99
Ríos Coronel, Hernando de 114, 155
Riva Herrera y González de Acevedo, Fernando de la 137
Riva Herrera, Hernando de la 45, 83, 128–132, 134, 137, 210
Royal Palaces, construction 69–70, 83, 108, 198, 215

Salaries, of superintendents 60, 134, 136, 158, 208, 220
Samar 9, 114, 156
San Lorenzo de El Escorial *See* El Escorial
San Sebastián 50, 78, 131, 135–137, 135n49, 159
San Vicente de la Barquera 42n, 50–51
Santander 42n, 50–51, 77–78, 128–129, 159, 168, 175, 230, 235
Santo Domingo 131
Real Audiencia de 145, 154

Santoña 51, 175, 180
Sardinia 52, 174, 188
Segura River 13, 48, 163, 167, 183–185, 221, 226–228, 242
Segura, Sierra de 49, 163, 167, 183–188, 190–192, 201, 214, 217, 228, 231–232, 234, 238, 239n, 244
Selaya (Cantabria) 1, 5, 7, 9, 13
Setera, Jorge 79–83
Seville 13, 23, 49, 51–52, 61–62, 73, 83, 89, 128, 154, 183, 185–186, 203n2, 220, 231–232, 234
Sheepherding *See* livestock
Shipbuilding
  Types of wood sought for 7–8, 104–105, 108–109, 115
  Amount of timber used in 29, 94–96
  Design 66–67, 132, 198–199
  Methods 91–94, 97–99
Shipworm (*Teredo navalis*) 98, 102–103, 105, 115–116, 156
Sicily 48, 63, 157, 174–175, 177, 237
Sierra de Guadarrama
Slavery 9, 36, 100, 102–103, 106–108, 147, 152
Superintendents, of forests and plantations
  responsibilities 7, 9, 42, 45, 67, 120–121, 126–130, 134, 158–159
Sustainability 3n5, 20, 58

Tajo 13, 48, 242
Tar See pitch
Tehuantepec 100, 195
Teredo navalis See shipworm
Territorialization, internal 4–5, 4n, 16, 19, 38–41, 43, 47, 64, 81, 83, 85–86, 130, 135, 138, 161–163, 166–167, 172, 187, 201–202, 204–205, 213–215, 217, 237–245
Timber
  'Green' 92, 153, 156
  Building and construction 18, 20, 23, 44, 57, 67, 72, 108, 113, 128, 143, 153, 159, 171, 183, 197–198, 214
  Compass 23, 92, 115
  Cost of 96
  Transportation of 9–10, 13–14, 16n27, 52, 79, 81–82, 86, 93, 113n80, 114, 147, 152, 155, 171, 182–183, 185–187, 192, 198, 213n26, 214, 222, 227–228, 231, 241–243
Tinajero de Escalera, Bernardo 193
Tordesillas, Treaty of 109
Tortosa 49, 52, 80–81, 137, 172, 180–181, 188–190, 192, 228
Turia River 181, 194, 242

Ulloa, Antonio de 235
Urdaneta, Andrés de 110
Urquiola, Antonio de 131, 133–135, 139

Valcanera 79–80, 82
Valdés y Castro, Juan de 165–167, 181, 184–194, 201, 237, 244
Valencia 10, 13, 50n16, 52, 58, 69, 75–76, 79–82, 136, 162, 167, 173, 181–182, 201–202, 221, 226, 229
Valivana 79–80, 82
Vallecilla y Ochoa del Casal, Martín de 133, 244
Valsaín 70
Varas y Valdés, Francisco de 176n34, 184–185, 187–191, 220
Vega, Lope de 13n, 16
Venice 19, 81, 118
Veracruz 100, 147, 176, 195–196
Vinaroz 80–82, 182, 227–228
Visayas 109–110, 112, 115
Vizcaya 8, 14–15, 18, 18n31, 42, 46n5, 50–51, 63–64, 66–67, 75–77, 83–84, 106, 108, 120, 127–133, 135–136, 154, 169–170, 179, 183, 193, 212

War of Succession, Spanish 42, 159, 168, 173–174, 177, 177n39, 188, 193, 244
Wood *See* firewood; *See* timber
Woodlands *See* forests

Zamora 13, 61
Zaragoza 191

www.ingramcontent.com/pod-product-compliance
Lightning Source LLC
Chambersburg PA
CBHW070908230926
51613CB00013B/150